Structural Patterns and Proportions
in Vergil's *Aeneid*

# Structural Patterns and Proportions in Vergil's Aeneid

**A STUDY IN MATHEMATICAL COMPOSITION**

by George E. Duckworth

The University of Michigan Press

Ann Arbor

Copyright © by The University of Michigan 1962

All rights reserved

Library of Congress Catalog Card No. 62-12162

Published in the United States of America by
The University of Michigan Press and simultaneously
in Toronto, Canada, by Ambassador Books Limited
Manufactured in the United States of America

Paperback ISBN: 978-0-472-75111-2

To Dorothy

# PREFACE

THE *AENEID* OF VERGIL is the great national epic of ancient Rome; it portrays the journey of Aeneas and the Trojans from Troy to Italy and their trials and victories after they reached their "promised land"; it gives the archaeology and topography of early Latium and Rome, and by means of prophecy and foreshadowing it presents the outstanding events of Roman history and the achievements of Augustus in Vergil's own day. The miracle of the poem is Vergil's ability to treat three different topics simultaneously—the legendary narrative of Aeneas, themes and personages of Roman history, and the praise of Augustus who has brought a new era of peace to the Roman world. The epic rises far above the patriotic and historical level in the poet's dramatic treatment of character and event and in his introduction of loftier themes of philosophy and religion; it is an epic not only of Rome but of human life as well. Vergil's superb poetic power, as seen in imagery, sound effects (such as alliteration and assonance), and complex metrical patterns, contributes to the greatness and splendor of the poem. The *Aeneid* is one of the most consciously planned and carefully constructed epics of world literature.

In a work of such magnitude, with so many threads inextricably and harmoniously interwoven, it should occasion no surprise that one or more basic designs of symmetry and variety, of parallelism and contrast, underlie the composition of the poem, both as a whole and in its separate parts. All readers of the *Aeneid* are conscious of these qualities to a degree, but the extent to which structural pattern and architectonic design dominate the epic has not been fully realized. Moreover, these structural features are not merely an adornment for their own sake but are devised to emphasize and make more significant the meaning of the poem; structure and content go hand in hand, and Vergil's conscious artistry in combining the two is an additional proof of his supreme achievement as an epic poet.

This book is devoted primarily to the structural aspects of the *Aeneid*, both large and small. The work was neither planned nor anticipated, but results in large part from some amazing discoveries made recently when I was engaged upon a larger and more comprehensive book, *Vergil, the Poet of Augustan Rome*. In this work, which will not be completed for several years, I had planned a chapter on the structure of the *Aeneid*. As I was investigating the composition of the individual books, I discovered by accident the basic symmetry of the *Aeneid*, and it is a mathematical symmetry. *Vergil composed the Aeneid on the basis of mathematical proportion; each book reveals, in small units as well as in the main divisions, the famous numerical ratio known variously as the Golden Section, the Divine Proportion, or the Golden Mean ratio.*

The existence of this particular mathematical proportion everywhere in the *Aeneid* may seem to some readers, if not fantastic, at least highly controversial, and yet it must be accepted as a fact. As my preliminary findings led to numerous additional discoveries, the subject became too significant in its ramifications to lie buried in typescript for a number of years and too extensive to find a place in the chapter originally planned on the structure of the *Aeneid*. I therefore decided to present the new material as fully as possible in a separate work, with the analysis of the structure of the *Aeneid* as a whole (Chapter 1) and the investigation of the structural patterns of the individual books (Chapter 2) as necessary preliminaries to the description of the mathematical proportions and their value for our understanding of the poem. The presence of the same Golden Mean ratio in both the *Eclogues* and the *Georgics* is discussed in Chapter 3; Le Grelle in 1949 had explained the structure of *Georgics* I by means of the Golden Section, and many additional examples in the *Eclogues* and in *Georgics* II-IV prove that numerical symmetry is characteristic of Vergil's poetry in general.

The discovery in the *Aeneid* of hundreds of Golden Mean ratios, analyzed by patterns in Chapter 4, has many important results, as described in Chapter 5: it not only throws new light upon Vergil's method of composition and the surprising correspondence between the passages containing the ratios and the divisions and subdivisions of each book of the epic, but it helps to solve certain significant problems about the text of the poem and the interpretation of specific passages; furthermore, it touches upon questions of even broader interest, such as the authenticity of the *Appendix Vergiliana* and the use of this same mathematical ratio by other Latin poets of the first century B.C.

I have used as the basis for my investigations the Oxford Classical Text of Vergil by F. A. Hirtzel, but for many problems, such as interpolations, spurious passages, suggested revisions or transpositions, and matters of paragraphing, I have had recourse to other modern editions which are listed in the bibliography. The bibliography also gives more detailed information concerning the many books and articles to which I refer in an abbreviated form in the footnotes. To avoid possible confusion, I might add here a note on the use of the letters *a, b,* etc., when joined to lines of the text: (1) in a few passages considered non-Vergilian, the letters refer to whole lines, e.g., *Aen.* I, 1a-1d are the four verses preceding *arma virumque cano;* see below, Chapter 5, "Spurious Passages"; (2) elsewhere *a* and *b* denote parts of lines, e.g., *Georg.* I, 463a refers to the first part of 463 (*sol tibi signa dabit*), 463b to the remainder of the verse.

I wish on this occasion to express my gratitude to the John Simon Guggenheim Memorial Foundation for the fellowship which enabled me to devote the academic year 1957-58 to the study of Vergil and the writing of the first part of the book on *Vergil, the Poet of Augustan Rome.* It was in April, 1958, that the existence of the Golden Section in the *Aeneid* first came to my attention, and the present volume is the result of my investigations, both classical and mathematical, since that time. A brief summary of my findings, with the title "Mathematical Symmetry in Vergil's *Aeneid*," has already appeared in *TAPhA* 91 (1960), pp. 184-220.

I am indebted to Professor William Feller of the Department of Mathematics, Princeton University, for helpful suggestions on the subject of probabilities and proportions, but I accept full responsibility for all the mathematics in both the text and tables below. I am likewise most grateful to Professor Robert Duff Murray, Jr., of the Department of Classics, Princeton University, for his careful reading of this book and his interest and encouragement at all stages of its composition; also to Professor Robert J. Getty, of the Department of Classics, the University of North Carolina, and many other friends who have called to my attention interesting bibliographical items on the importance of the Golden Section.

The publication of the book has been made possible by a grant from the Princeton University Research Fund and a grant from the American Council of Learned Societies as a result of a contribution from the United States Steel Foundation. For these grants I am deeply appreciative. I wish to thank also Mr. Edwin Watkins and Mr. John Dimoff of the University of Michigan Press for kind and helpful co-operation in the publishing of this volume.

George E. Duckworth

Princeton
June, 1962

# CONTENTS

Chapter 1. The Architecture of the *Aeneid* ........................................... 1
    The Alternating Rhythm ........................................................... 1
    The Parallelism of the Halves ..................................................... 2
    The *Aeneid* as a Trilogy .......................................................... 11
    "In medio mihi Caesar erit" ....................................................... 14
        Notes to Chapter 1 ............................................................ 16

Chapter 2. Structural Patterns in the Books ........................................... 20
    Alternation and Contrast .......................................................... 20
    The Framework or Recessed Panel Pattern ........................................ 21
    Tripartite Structure ............................................................... 25
        Notes to Chapter 2 ............................................................ 34

Chapter 3. The Golden Section in the *Eclogues* and the *Georgics* .................. 36
    Le Grelle and *Georgics* I ........................................................ 36
    The Golden Mean Ratio ........................................................... 37
    Proportions in the *Eclogues* ..................................................... 39
    Proportions in the *Georgics* ..................................................... 41
        Notes to Chapter 3 ............................................................ 43

Chapter 4. The Proportions in the *Aeneid* ............................................. 45
    The Nature of Mathematical Composition ......................................... 45
    Short Passages: Bipartite Pattern .................................................. 48
    Short Passages: Tripartite (Nonframework) Pattern ................................ 51
    Short Passages: Tripartite Framework Pattern ..................................... 52
    Short Passages: Four or Five Parts ................................................ 54
    The Main Divisions of the Books .................................................. 58
    The Main Divisions in Proportion .................................................. 59
    Supplementary List of Ratios ...................................................... 59
    Summary of the Proportions ....................................................... 60
    The *Aeneid* as a Whole .......................................................... 63
        Notes to Chapter 4 ............................................................ 65

Chapter 5. The Value of the Proportions ............................................... 68
    Vergil's Method of Composition ................................................... 68
    Vergil and Pythagoreanism ........................................................ 73
    The Problem of the Half-Lines .................................................... 77
    Interpolations ..................................................................... 81
    Spurious Passages ................................................................. 83
    Transpositions .................................................................... 86
    Paragraphing ..................................................................... 87
    The Problem of the Revision ...................................................... 90
    The Authenticity of the Minor Poems ............................................. 93
        Notes to Chapter 5 ............................................................ 97

| | | |
|---|---|---|
| Conclusion | | 103 |
| | Notes to Conclusion | 104 |

Appendices .................................................................. 105

| | | |
|---|---|---|
| | Appendix A. Catullus LXIV | 107 |
| | Appendix B. Lucretius | 107 |
| | Appendix C. The Aristaeus Story (*Georg.* IV, 281-558) | 109 |
| | Appendix D. Horace | 109 |
| | Appendix E. Maphaeus Vegius | 110 |
| | Appendix F. Metrical Patterns and Golden Mean Ratios | 111 |
| | Appendix G. Addendum on *Aen.* II, 76 | 117 |

Tables ...................................................................... 119

| | | |
|---|---|---|
| I. | Short Passages: Bipartite Pattern | 121 |
| II. | Short Passages: Tripartite (Nonframework) Pattern | 132 |
| III. | Short Passages: Tripartite Framework Pattern | 139 |
| IV. | Short Passages: Four or More Parts, Usually Interlocked | 152 |
| V. | Proportions in the Main Divisions | 166 |
| VI. | The Main Divisions in Proportion | 169 |
| VII. | Supplementary List of Ratios (for Tables I-IV) | 170 |
| VIII. | The *Aeneid* as a Whole | 173 |
| IX. | Chart-Index of the *Aeneid* | 175 |
| X. | Ratios More Accurate with Half-Lines as Fractions | 200 |
| XI. | Ratios Less Accurate with Half-Lines as Fractions | 202 |
| XII. | The Interpolations | 203 |
| XIII. | A Guide to Paragraphing | 205 |
| XIV. | Proportions in the *Appendix Vergiliana* | 209 |
| XV. | Comparison of *Appendix* Ratios with Those in Vergil, Catullus LXIV, and Lucretius I | 216 |
| XVI. | Chart-Index of the Longer Poems in the *Appendix* | 217 |
| XVII. | Proportions in Catullus LXIV | 226 |
| XVIII. | Chart-Index of Catullus LXIV | 228 |
| XIX. | The Main Divisions of the *De Rerum Natura* in Proportion | 230 |
| XX. | Proportions in the *De Rerum Natura*, Book I | 232 |
| XXI. | Chart-Index of Book I of *De Rerum Natura* | 235 |
| XXII. | Proportions in *Georg.* IV, 281-558 | 238 |
| XXIII. | Chart-Index of *Georg.* IV, 281-558 | 239 |
| XXIV. | Proportions in the *Satires* and *Epistles* of Horace | 242 |
| XXV. | Fibonacci Series in the *Ars Poetica* of Horace | 244 |
| XXVI. | Proportions in the "Thirteenth Book" of Maphaeus Vegius | 245 |
| XXVII. | Chart-Index of the "Thirteenth Book" | 247 |

Abbreviations ............................................................... 251

Bibliography ................................................................ 253

Index ....................................................................... 259

# Chapter 1
# THE ARCHITECTURE OF THE *AENEID*

VERGIL describes in *Georg*. III, 13 ff., the *templum* which he planned to erect in honor of Octavian; this is a "temple of song," whether we accept the usual and more probable interpretation that it signifies a historical epic which Vergil intended to write about the achievements of Octavian, the idea of the *Aeneid* not yet having been developed, or whether we refer it, with Büchner, to Vergil's poetry in general which he offers to Octavian as his hero and a god.[1] Vergil speaks of his poetry here in terms of an architectural structure, as he did likewise, according to the Donatus-Suetonius Life, when he said that he propped up certain passages of the *Aeneid* with temporary supports (*tibicines*) until the *solidae columnae* arrived.[2] The *Aeneid* is a long and complex epic, composed with an unflagging devotion to symmetry and balance, similarity and contrast, proportion and structural harmony; these features appear not only in the relation of shorter passages to the individual books but also in the manner in which each book, a unit in itself, fits into its place as part of a greater unity and contributes meaning to the epic as a whole. For the analysis of such a work, the term "architecture," already employed by several writers, seems singularly appropriate.[3]

The Donatus-Suetonius Life (23) also informs us that Vergil first made a prose outline of the *Aeneid* divided into twelve books and then composed whatever part appealed to him, taking nothing in order. Even if this important information had not come down from antiquity we should have assumed that this was the poet's procedure. Many sections of the various books seem to have been composed as structural units, but the artistic manner in which each fits into its context and the significant relations of the books to each other and to the whole would be most difficult to achieve without the use of the preliminary outline mentioned by Suetonius. Since Vergil arranged his epic material in this prose outline before he began to write, the architecture of the *Aeneid* undoubtedly occupied his attention from the very first; it is hardly surprising, therefore, that he composed an epic which reveals not merely one structural pattern but three which overlap and enrich each other.

In analyzing the *Aeneid* as a whole I shall discuss in order Vergil's method of alternating the books, the division of the poem into two halves with the resultant parallelism of the books in each half, and the *Aeneid* as a trilogy—divided into three parts of four books each, with the first and third sections framing the central portion which contains the material of greatest historical and patriotic interest. Vergil died before he had given to the work its final revision, but one point seems clear: whatever early changes the poet may have made in the order of the books,[4] the structural analysis which follows proves conclusively that the present arrangement was the final one; any revisions planned by Vergil could not possibly have affected the elaborate and harmonious architectonic patterns of the *Aeneid* as we have it today.

## THE ALTERNATING RHYTHM

Vergil in his earlier works had already displayed a striking interest in variety and contrast. More than one type of alternation has been pointed out for the ten *Eclogues;* those with odd numbers have Italian, local scenery, while those with even numbers have scenery beyond Italy, scenery that is more ideal;[5] also, the odd-numbered poems are dialogues, those with even numbers are monologues (or, in the case of VIII, two monologues); in other words we find here a clear-cut alternation between dramatic poems and nondramatic poems or songs.[6]

The four books of the *Georgics* fall into two halves, I-II and III-IV, and the principle of alternation is perhaps seen most clearly in the conclusions of the books; I and III end on a gloomy note—

in I the portents after Caesar's death, the dangers of civil war, and the need of Octavian to save the Roman people, and in III pestilence and death; the conclusions of II and IV present a happier picture—praise of country life in II, and in IV the regeneration of the bees (including a tribute to Rome in II, 534 ff., and to Octavian in IV, 560 ff.). Since the praise of country life at the end of II is also a picture of a new Golden Age, an age of peace,[7] the emphasis at the end of each book is as follows: I, War; II, Peace; III, Death; IV, Resurrection—an eloquent testimony of the manner in which Vergil has elevated the *Georgics* from the level of a poetical treatise on farming to an epic with universal significance.

The principle of alternation appears even more clearly in the *Aeneid*. The most important books and those with the greatest tragic impact are the books with even numbers:

 II. The fall of Troy
 IV. The tragedy of Dido
 VI. The trip to the Underworld
VIII. The visit to early Rome
  X. The great battle, with the deaths of Pallas, Lausus, and Mezentius
 XII. The final conflict and the death of Turnus

These are the famous books which stand out in the reader's memory, and the odd-numbered books, important and essential as they are, have a lighter nature and serve to relieve the tension. Books V and XI, each separating books of great dramatic power, are almost interludes; V begins with funeral games for Anchises and XI with a truce and burial of the dead, but in each the tension increases; in V we have the attempted burning of the ships and in XI the cavalry battle and the death of Camilla. These two books are also, curiously enough, among the longest of the poem; in the first half V is exceeded only by VI, and likewise XI is the longest in the second half (and in the poem as a whole) with the exception of XII.[8]

The plan of the *Aeneid* may be designated thus:

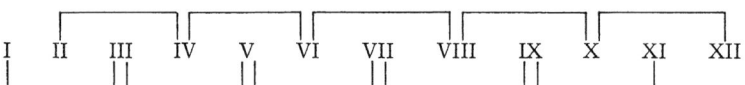

Vergil, by means of this alternating rhythm, has made his more serious and tragic books stand out in bold relief. The alternation between the even-numbered and the odd-numbered books has been considered the basic division of the *Aeneid* by some scholars,[9] and has been viewed in various ways: Conway says that Vergil's achievement "was to combine in alternation the methods and motives of epic poetry with those of Greek tragedy."[10] Stadler looks upon the even-numbered books as books of depth, dealing chiefly with the hero, his mission, and Fate, the other books being books of breadth, more concerned with other characters and events.[11] Both views seem correct and they are complementary; the fact that the books with even numbers are those of greatest tragedy and deepest significance is of the utmost importance for our understanding of Vergil's procedure. The alternating rhythm is the simplest of the three structural patterns which combine to create the architecture of the epic.

## THE PARALLELISM OF THE HALVES[12]

The most obvious division of the *Aeneid* is that into two halves—the journey of Aeneas from Troy to Latium (I-VI), announced by Vergil at the beginning of the poem (I, 1-4):

> arma virumque cano, Troiae qui primus ab oris
> Italiam fato profugus Lavinaque venit
> litora—multum ille et terris iactatus et alto
> vi superum, saevae memorem Iunonis ob iram,

and Aeneas' adventures and victories after his arrival (VII-XII), to which the poet likewise alludes in the prooemium (5-7):

> multa quoque et bello passus, dum conderet urbem
> inferretque deos Latio—genus unde Latinum
> Albanique patres atque altae moenia Romae.

Here he gives also the political and religious themes of the poem and foreshadows his later descriptions of the Alban kings and Roman heroes.

The twofold structure of the epic is implied again by Vergil's reference in VII, 44 f., to the remainder of the poem as a *maior rerum ordo*, a *maius opus*. Many thus look upon the first half as a Roman "Odyssey" of wanderings and the second half as a Roman "Iliad" of battles. Mackail condemns this view of the *Aeneid* as superficial and deplorable in its lack of appreciation, and says, "with all its debt to both the Homeric poems, it is an organic unity and a masterpiece of creative art."[13] The tendency in most secondary schools to read only *Aeneid* I-VI, or parts therefrom, and the publication of numerous school editions containing only the first six books have, of course, accentuated the twofold division and made it more difficult for the average reader to realize that Vergil has bound together the two halves by numerous similarities and contrasts. In spite of the many differences in subject matter, each book in the first half of the poem is closely related to the corresponding book in the second half.

This structural device had likewise appeared earlier in both the *Eclogues* and the *Georgics*. In addition to the alternations mentioned above, certain *Eclogues* in the second half of the collection correspond to poems in the first half, and in reverse order; a brief outline will make this clear:

| | |
|---|---|
| I and IX: | country life and the confiscations of territory |
| II and VIII: | the passion of love |
| III and VII: | music; responsive singing matches |
| IV and VI: | loftier themes of religion and philosophy; the world to come (IV), the world that was (VI) |

These eight *Eclogues* thus form a frame about V, the songs concerning the dead and deified Daphnis.[14] As V honors the shepherd who became a god, so X, a later addition to the collection, honors Cornelius Gallus, the friend who appears as a shepherd.[15] The first nine pastoral poems are thus viewed as a harmonious whole, with X added not only to honor Vergil's friend Gallus, but also to provide a poem to balance V and, as Richardson suggests, to heighten the interest at the end of the collection.[16] The arrangement is as follows:

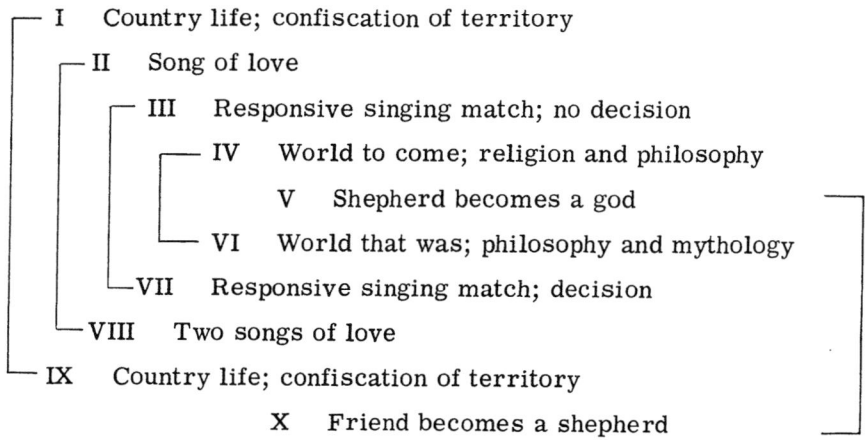

The first nine *Eclogues* have also been arranged in triads, with X the final poem blending the shepherds and realism of Triads One (I, II, III) and Three (VII, VIII, IX) with the gods and fantasy of Triad Two (IV, V, VI).[17] This analysis accepts also the close relationship between I and IX, II and VIII, etc., and Triad Two, containing "the grander, more cosmic themes," has V as its central poem—a place of honor similar to that in the framework pattern. The triadic structure of the *Eclogues* may be illustrated by the following diagram:

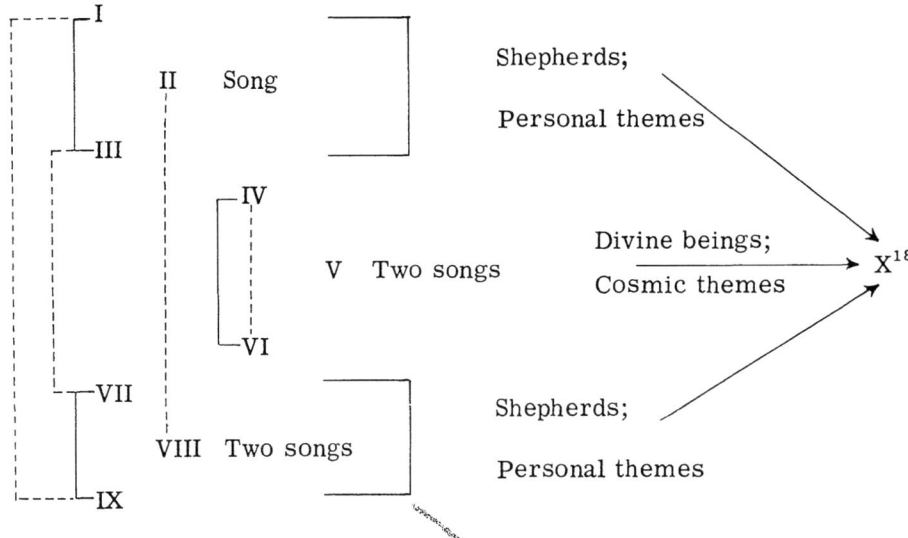

Vergil's arrangement of the *Eclogues* thus combines alternation, framework (or recessed panel) pattern, and triadic structure in an intricate design and foreshadows his use of these same devices in the books of the *Aeneid*.

The composition of the *Georgics* is equally complex and reveals a striking unity of structure and content. Vergil has arranged the four books in units of two books each, I-II on inanimate nature (fields, trees and vines), and III-IV on living creatures (herds and flocks, and bees). Although the technical matter in each half differs widely, there exist in the descriptive passages many similarities and contrasts between I and III and between II and IV:

and at the same time there are certain links between I and IV and between II and III; the arrangement thus becomes

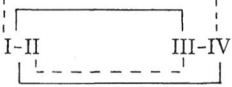

The most significant of the parallels and contrasts are the following:[19] Books I and III have extended prologues, in each of which Octavian plays a major role; in I, 24-42, he is described as a god; in III, 10-48, Vergil will erect in his honor a temple of song, an allusion to the poet's original intention to praise Octavian in a historical epic. Books II and IV have short prefaces of eight and seven verses respectively, that in II invoking Bacchus, that in IV invoking Apollo; the shorter prefaces are appropriate here, as each book forms the second half of a larger unit. The passage on astronomy, or "geography of the heavens," in I, 231-258, is balanced by that on the geography of Libya and Scythia in III, 339-383, with the references to Scythia and Libya in I, 240 f., serving as a connecting link. In II, 136-176, we have the glowing tribute to the majesty of Italy, and in IV,

116-146, the simple charms of Italy. The alternation between the pessimistic conclusions of I and III and the brighter endings of II and IV, producing the impressive sequence: War, Peace, Death, Rebirth, has already been mentioned.

Drew is inaccurate when he says that there are no correspondences between I and IV or II and III.[20] Just as *labor* is stressed in I, 125-146, so in IV we have an emphasis on the *labor durus* of the beekeeper (cf. 114) and the *labor* of the bees themselves (cf. 184: *labor omnibus unus*); also, in IV, 125-146, an exact numerical equivalent to I, 125-146, the labors of the old gardener of Tarentum are described; this passage is far more significant for the poem as a whole than is often realized, as the *senex* symbolizes the industrious but happy life of the poor man who "lived the life of kings" (132: *regum aequabat opes animis*).[21] Aristaeus, to whose story the second half of IV is devoted, is invoked in I, 14 f.—an interesting link between the beginning and the end of the poem. Perret sees a symbolic relation between the conclusions of I and IV; as Octavian, the *iuvenis* of I, 500, is to restore the Roman people, so Aristaeus, the *iuvenis* of IV, 445 (= Octavian), regenerates his swarm of bees (= Romans); each acts under divine guidance and to each divine honors are promised (I, 503 f.; IV, 325).[22] Another possible link between I and IV, this time of a numerical nature, is the fact that Maecenas is addressed in line 2 in both books, whereas the mention of his name in II and III occurs in line 41 and brings together these two books in the same mechanical but unobtrusive fashion.

The parallelism of the individual pastorals and the many balancing passages in the *Georgics* prepare us for the many correspondences between the two halves of the *Aeneid*. Conway, who first stressed the fact that the books of the second half correspond to those of the first half, lists important similarities and contrasts in each pair of books.[23]

| | |
|---|---|
| I and VII: | arrival in a strange land; friendship offered |
| II and VIII: | each the story of a city—one destroyed by Greeks, the other to be founded with the help of Greeks |
| III and IX: | Aeneas inactive and action centers around Anchises (III); Aeneas absent and action centers around Ascanius (IX) |
| IV and X: | Aeneas in action—inner conflict between love and duty (IV); outer conflict with the enemy (X) |
| V and XI: | each begins with funeral ceremonies and ends with death—Palinurus (V) and Camilla (XI) |
| VI and XII: | Aeneas receives his commission in VI, executes it in XII |

These parallel features are too numerous to be due merely to accident.

Conway looks upon VI as the "crowning Book, which Vergil has placed in the centre, to unite all that stand before it and all that stand after"; it is "the keystone of the whole poem" and "contributes a sense of unity to the epic."[24] Aeneas' visit to the underworld is thus very different in function as well as in content from Odysseus' interview of the shades in the *Odyssey*, an incidental episode which lacks the philosophical, religious, and national significance which Vergil has given to *Aeneid* VI.

A new analysis of the architecture of the *Aeneid* was proposed by Perret in 1952.[25] Just as VII-XII depict the story of Aeneas in Italy, so I-V give the story of Aeneas at Carthage, and the first half of the poem derives its unity from this fact. Perret looks upon VI much as does Conway, calling it "le sommet de l'*Énéide*," and "la synthèse du poème."[26] He isolates VI and purposely excludes it from his structural analysis;[27] also, he makes no attempt to relate the books of the second half to those of the first half. His views deserve a careful examination before we return to a consideration of the parallelism of the books as postulated by Conway.

According to Perret, the architecture of the *Aeneid* is as follows:

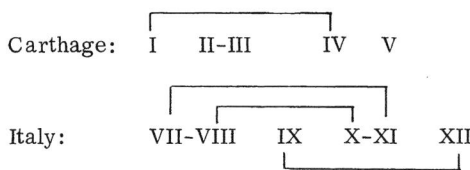

In the first half, the story of Dido in I and IV is interrupted by Aeneas' narrative. Perret believes that V rather than IV serves as the conclusion to the story of Carthage. This he bases on similarities and contrasts between I and V: intervention of Juno (Aeolus in I, Iris in V); catastrophe (storm in I, fire in V); Neptune rebukes the winds in I, Ascanius the Trojan women in V; Aeneas comforts his men in I, is comforted by them in V; Venus appeals to Jupiter in I, to Neptune in V. These parallels are striking but hardly prove that the latter part of V should be viewed as "une reprise" of I, or that V is the conclusion of the Carthaginian episode. We shall find even more numerous and impressive parallels between I and VII.

Perret divides the second half of the *Aeneid* into two groups of three books each, with the second section beginning with X, the assembly of the gods and Aeneas' return to combat. VII and VIII are linked together, being books of negotiations and embassies, and as VII ends with a picture of pre-Trojan Italy, so VIII ends with a picture of Roman Italy. One must suggest, however, that the conclusion of VIII is balanced far better by the description of Roman heroes in VI. Perret likewise links X and XI as books of combat which stress the glorious death of young heroes, Pallas in X and Camilla in XI. But is it possible to disregard XII in this connection? Certainly, X and XII should be viewed as the important books of combat, and at the conclusion of each a major opponent falls at the hands of Aeneas, Mezentius in X and Turnus in XII.

Perret relates VII to XI, VIII to X, and IX to XII, an arrangement not unlike that of the *Eclogues*, except that IX, framed by the two preceding and the two following books, does not here have the importance which scholars attribute to *Eclogue* V. Perret's main points are these:

VII and XI: both concern the Latins; King Latinus weak in both; embassy from Trojans to Latins in VII, from Latins to Trojans in XI; description of Camilla at end of VII, her death at end of XI

VIII and X: both concern Aeneas' allies, Arcadians and Etruscans; departure of Aeneas (VIII) and his return (X), each accompanied by a prodigy; Hercules of the *ara maxima* (VIII) mourns the approaching death of Pallas (X)

IX and XII: both are reserved for Turnus and Trojan valor; Turnus appears also in X and XI, but our full picture of the hero comes from IX (in the Trojan camp) and XII (combat with Aeneas);[28] likewise, IX depicts the valor of the Trojans (to which the episode of Nisus and Euryalus contributes) and XII that of Trojan Aeneas

Perret's analysis of the *Aeneid* gives us valuable insights into the interrelations of the various books, especially in the second half of the poem, and reveals again Vergil's love for alternation and contrast. The possible existence of overlapping and interlocking designs should not be ignored, and the ability of different scholars to see quite dissimilar schemes merely gives added testimony of the structural richness of the epic.[29] I do not deny the validity of many of Perret's comments on the last six books, but his grouping provides a supplementary and more complex pattern imposed upon the fundamental design which links VII-XII to I-VI. He seems to have missed the basic architecture of the *Aeneid*; the unsuccessful attempt to attach V to I-IV, the isolation of VI, and the failure to relate VII-XII closely to I-VI all weaken his position. He admits that the architecture of the poem must be "très harmonieuse et très simple,"[30] but his arrangelacks both harmony and simplicity.

Vergil's own description of VII-XII as a *maius opus* implies that I-VI are an enriched and amplified prelude to his main theme, and Perret's careful and interesting analysis of VII-XII has presented additional proof of the attention which Vergil gave to the structure and content of the second half of the poem. Brilliant characterizations are numerous—Mezentius, Lausus, Pallas, Nisus, Euryalus, Camilla, and, above all, Turnus. The outstanding episodes—Aeneas' visit to the site of Rome, the tragic deaths of Nisus and Euryalus, the slaying of Pallas with its fateful result for Turnus, the defeat of the wounded Mezentius as he attempts to avenge his son's death—are all firmly embedded in the main structure and are essential parts of it. Two of the greatest books of the *Aeneid* are undoubtedly IV and VI, but the latter, as we have already seen, is great not only for its content but because of its central position in the structure of the whole. But what of X and XII? These two books must rank high in any consideration of the poem. Book X pictures the tragic deaths of Pallas, Lausus, and Mezentius and provides an effective counterpart to Dido's suicide in IV—the tragedy of war balancing the tragedy of love. Mackail compares XII with II, IV, and VI and says that the final book "reaches an even higher point of artistic achievement and marks the utmost of what poetry can do, in its dramatic value, its masterly construction, and its faultless diction and rhythm."[31] Even those who do not rate XII so highly must admit that it provides an effective and dramatic conclusion to the poem and serves as an adequate balance for VI.

The correspondence at the conclusion of the two books is especially striking.[32] Anchises in VI, 851 ff., gives the duty of the Roman, ending with the famous line (853): *parcere subiectis et debellare superbos,* "spare the conquered and overpower the arrogant"—an exhortation to *clementia* and *iustitia*. At the very end of XII, Aeneas yields to Turnus' appeal for mercy and is about to show clemency when he sees the swordbelt of Pallas; Turnus' slaying of Pallas in X was characterized by *superbia* (cf. X, 445: *iussa superba;* 514 f.: *superbum nova caede*), and justice demands that Turnus die. Aeneas as a symbol of the ideal Roman thus fulfills the words of his father in VI, 853.

In the first half of the poem, I and IV are separated by II and III (Aeneas' narrative); in like manner VII and X are separated by VIII and IX (Aeneas' absence). V is an interlude between the tragedy of IV and the lofty and serious themes of VI, and similarly XI provides a lessening of tension between the tragic fighting in X and the final conflict in XII. The grouping of the books is therefore as follows:

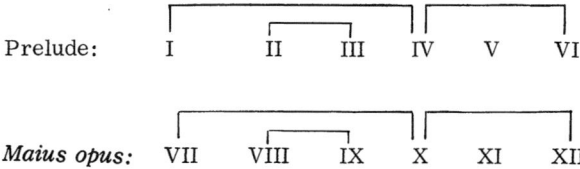

Prelude:       I     II    III    IV    V    VI

*Maius opus:*  VII   VIII  IX     X     XI   XII

Whereas this scheme for the first half resembles that of Perret in part, the second half is very different and, what is more important, it is an exact counterpart of the first half. This supports Conway's view of the correspondence of the books in each half.

Vergil has composed his epic in two large parallel panels, with an alternating rise and fall of tension and with each book of the second panel balancing the corresponding book of the first. The twelve books of the *Aeneid* may be presented in the following diagrammatic form:[33]

    I  Juno and storm

        II  DESTRUCTION OF TROY

            III  Interlude (of wandering)

                IV  TRAGEDY OF LOVE

                    V  Games (lessening of tension)

                        VI  FUTURE REVEALED

VII Juno and war

VIII BIRTH OF ROME

IX Interlude (at Trojan camp)

X TRAGEDY OF WAR

XI Truce (lessening of tension)

XII FUTURE ASSURED

I agree that VI is the keystone, the focal point of the poem as a whole, but it should not be isolated between two contrasted panels; it forms the climax of the first panel, as XII concludes the second. The diagram presented above reveals the manner in which the books of the second panel balance those of the first, but to gain an adequate impression of the numerous parallels and contrasts which exist within each pair of corresponding books, a more detailed analysis is necessary. The parallel columns which follow give the many similarities and contrasts (including those pointed out by Conway) and illustrate the dramatic rise and fall of the action:

| I | VII |
|---|---|
| Juno and storm[34] | Juno and war |
| Invocation of Muse | Invocation of Muse |
| Trojans *laeti* (35) | Trojans and Aeneas *laeti* (36, 130, 147, 288) |
| Juno laments lack of power | |
| Juno arouses storm on sea with aid of Aeolus and forces of nature | |
| Arrival in strange land | Arrival in strange land |
| Banquet on shore | Banquet on shore |
| Prophecy of Roman greatness (Jupiter to Venus) | Prophecy of Roman greatness (Faunus to Latinus) |
| Venus in disguise meets Aeneas—reveals identity | |
| Trojans already known | Trojans already known |
| Pictures of Trojan past | Statues of Latin past |
| Dido receives Trojans | Latinus receives Trojans |
| Ilioneus speaks for Aeneas | Ilioneus speaks for Aeneas[35] |
| Friendship and gifts offered | Friendship and gifts offered |
| | Juno laments lack of power[36] |
| | Juno arouses war on land with aid of Allecto and human forces |
| | Allecto in disguise visits Turnus—reveals identity |
| Venus prevails over Juno | Juno prevails over Venus |
| Closing of Gates of War (in Jupiter's prophecy) | Opening of Gates of War |
| Effect of Cupid on Dido (685 ff.) | Effect of serpent on Amata (349 ff.) |
| Movement of book—misery to happiness | Movement of book—happiness to misery |

## II
### DESTRUCTION OF TROY

Story of Carthage interrupted
Laocoon killed by two serpents
Greeks destroy
Trojans suffer from Greeks
Helplessness of aged Priam
Luxury of Priam's palace
Destruction of Troy
Aeneas center of stage
Anchises prominent—father
    of Aeneas
Venus as goddess appears to
    Aeneas
Gods against Troy
Ascanius—fire about head,
    comet
At end, Aeneas carries on
    shoulders his father
    (symbolic of past)

## VIII
### BIRTH OF ROME

Story of Trojan camp interrupted
Two serpents killed by Hercules
Greeks help to found
Trojans profit from Greeks
Helpfulness of aged Evander
Simplicity of Evander's home
Picture of later Rome
Aeneas center of stage
Evander prominent—father
    of Pallas
Venus as goddess appears to
    Aeneas
Gods for Rome (at Actium)
Augustus—fire about head,
    comet
At end, Aeneas carries on
    shoulder the shield
    (picture of future)

## III

Interlude (of wandering)

Aeneas has minor role
Anchises important
Apollo helps Aeneas
Anchises misinterprets
    oracle
Helenus and Andromache—
    joyful episode, but
    Andromache grieves for son
Astyanax-Ascanius equation
    (489 ff.)
Trojans escape from danger—
    Scylla, Charybdis, Cyclops

## IX

Interlude (at Trojan camp)

Aeneas absent
Ascanius important
Apollo advises Ascanius
Turnus misinterprets meta-
    morphosis of ships
Nisus and Euryalus—
    tragic episode, with grief
    of Euryalus' mother
Euryalus-Ascanius equation
    (297 ff.)
Trojans escape from danger—
    Turnus in camp

## IV
### TRAGEDY OF LOVE—DIDO

Venus and Juno (agreement)
Jupiter intervenes
Inner conflict of Aeneas
Affection yields to duty
Aeneas sheds tears, but
    fate prevails (449)[37]
Dido wants *parvulus Aeneas* if
    she cannot have Aeneas
*culpa* of Dido—results in death

## X
### TRAGEDY OF WAR—PALLAS, LAUSUS, MEZENTIUS

Venus and Juno (conflict)
Jupiter refuses to intervene
Outer conflict of Aeneas
Pity yields to justice
Hercules sheds tears, but
    fate prevails (464 ff.)
Venus wants Ascanius if she
    cannot save Aeneas
*culpa* of Turnus—leads to death
    in XII

Turning point—Aeneas' decision
   and effect on Dido
At end, suicide of Dido—
   cannot live without Aeneas

Turning point—death of Pallas
   and effect on Aeneas
At end, death of wounded Mezentius—
   cannot live without Lausus

## V

Games (lessening of tension)

Funeral games
Aeneas quiets disputes
Entellus reproached—
   agrees to fight
Friendship of Nisus and Euryalus
Nisus seeks to gain prize
   for friend
Cavalry display (*ludus Troiae*)

Increase of tension—
   burning of ships
Juno sends down Iris
At end, death of Palinurus
Palinurus killed by minor
   deity (Somnus)

## XI

Truce (lessening of tension)

Burial of dead
Latinus unable to avert dissension
Turnus reproached—
   agrees to fight
Hatred of Drances and Turnus
Turnus wants to keep prize
   from enemy
Increase of tension—renewal of
   fighting
Cavalry battle

Diana sends down Opis
At end, death of Camilla
Arruns killed by minor
   deity (Opis)

## VI

FUTURE REVEALED

Dramatic progression—retarda-
   tions and suspense, climaxed
   by revelation of Rome's destiny
Meeting with Dido—concludes
   Dido story (IV)
Anchises reveals later Roman
   history
Aeneas receives his commission
Anchises recommends clemency
   and justice (853)
At end, death of Marcellus
   consecrates new order

## XII

FUTURE ASSURED

Dramatic treatment of combat—
   retardations and suspense,
   climaxed by victory of Aeneas
Avenging of Pallas—concludes
   Pallas story (X)
Reconciliation of Jupiter and Juno
   creates later Roman people
Aeneas fulfills his commission
Aeneas symbolizes clemency and
   justice (940 ff.)
At end, death of Turnus seals
   doom of old order

    The existence of so many similarities and contrasts in books of such varied subject matter reveals a high degree of conscious art. We find here the symmetry and contrast and alternation of tension which are so peculiarly characteristic of Vergil.[38] Furthermore, not only do we have (again to use Conway's phrase) an alternation of lighter and more serious books, but a second type of alternation may now be seen: in the numerous contrasts and similarities which exist in the corresponding books of each half, the similarities appear to predominate in I and VII, III and IX, V and XI, i.e., in the odd-numbered books, whereas in II and VIII, IV and X, VI and XII the contrasts seem more numerous. The *Aeneid* contains a far more subtle fusion of the two principles of alternation and correspondence than Conway himself realized.

## THE AENEID AS A TRILOGY[39]

We saw above that the *Eclogues,* in addition to the alternation in the type of the poems and their correspondence in a framework pattern around V, may be divided into triads with Triads One (I-III) and Three (VII-IX) enclosing the central triad (IV-VI) on more serious themes. So, too, in the *Aeneid,* Vergil combines with the alternation of the books and their division into two corresponding panels a third and most important architectonic device—a tripartite division of the epic into three groups of four books each; in this way he counteracts the somewhat artificial division of the work into two halves. To many readers this arrangement may appear less obvious, but it is even more significant as a part of Vergil's over-all plan for the *Aeneid;* it points up and stresses the meaning of the epic as a poem not merely of Aeneas and the Trojans but of Rome and Augustus.

The tripartite division of the *Aeneid* is not new; it has been stressed by Stadler and Pöschl, who speak of the change from dark to light and back to dark,[40] and, more recently, by Büchner, who analyzes the three parts as follows: I-IV, Aeneas at Carthage; V-VIII, arrival in Latium and preparation for battle; IX-XII, the conflict itself.[41] Camps speaks of "an underlying division of the poem into three main portions, the episodes of Aeneas' story which have Dido and Turnus for secondary heroes standing one on either side of a large central section wherein the wider significance of the story is expounded."[42] Camps's statement provides an excellent summary of Vergil's division of the poem into three sections, but he, as well as Stadler, Pöschl, and Büchner, has failed to point out the full implications of the tripartite arrangement for Vergil's use of epic and tragic material and for the significance of the poem as a whole.

The interpretation of the three sections by Stadler and Pöschl as dark, light, and dark is open to criticism[43] but points the way to Vergil's plan for his epic. The *Aeneid* is the story of Aeneas and also the story of the destiny of Rome under Augustus. The latter provides the central core of the work (V-VIII) and concludes with the victories and triumphs of Augustus at the end of VIII.[44] The first section is the tragedy of Dido, and the third is the tragedy of Turnus. Books I and IV are the Dido books and they enclose II and III, which give the story of Aeneas at Troy and his wanderings. In like manner, IX and XII are predominantly the Turnus books; Turnus is active in X and XI, and the slaying of Pallas in X is decisive in bringing about his own death, but these two books are primarily devoted to other characters, X to Pallas, Lausus, and Mezentius, XI to Drances and Camilla.

Pöschl looks upon I, 8-296 (the storm and the Jupiter-Venus scene) as symbolic of the entire *Aeneid* and especially of the first or "Odyssean" half.[45] When we view the epic as composed of thirds, this same opening passage anticipates and symbolizes the main theme in each of the three parts. The mention of Carthage (I, 13-22) looks ahead to the tragedy of Dido and to the later hostility of the two nations; cf. I, 13: *Karthago Italiam contra* and IV, 628: *litora litoribus contraria.* The theme of Rome and Carthage thus stands at both the beginning and the end of the first third. The prophecy of Jupiter (I, 257-296) anticipates V-VIII, the central core of the poem, and especially the conclusions of VI and VIII; Jupiter alludes to Rome and Roman history, Romulus, the destiny of Rome (cf. 279: *imperium sine fine*), and Augustus as the creator of a new Golden Age of peace.[46] Jupiter describes (291 ff.) this new age under Augustus and the closing of the *Belli portae*. In the reference to the enchainment of *Furor impius* (294-296) we can see the victory over both *impietas* and *furor*—and Mezentius, the *contemptor divum*, defeated by Aeneas at the end of X, symbolizes *impietas*, while Turnus, over whom Aeneas is victorious at the end of XII, stands for *violentia* and *furor*. Also, as Juno and her hostility toward Aeneas are stressed in I, 8-33, so in XII, 818-840, we have the reconciliation of Jupiter and Juno which assures the future greatness of the Romans, descended from the mingled races of the Trojans and the Latins.

Much has been written about Vergil's use of tragic drama, and Rand has aptly said: "Tragedy is an essential part of Vergil's poem—he was forever joining together what critics would keep asunder."[47] The tripartite division of the *Aeneid* enables us to see more clearly the manner in which Vergil has framed his central message by the two tragedies of Dido and Turnus; both are portrayed most sympathetically and both meet death, not merely because they stand in the way of

Aeneas and his mission and are, therefore, the victims of Divine Will, but also because each does the wrong thing and pays the penalty for his action.[48] The two tragedies are not lacking in historical significance, however; just as Dido's death symbolizes the overthrow of Carthage by Rome (cf. the simile of the burning city in IV, 669 ff.), so the defeat of Turnus suggests the later union of Romans and Latins and Roman supremacy in Italy.

The threefold division of the *Aeneid* also throws new light upon Vergil's use of epic material, especially that drawn from Homer. In the first and third sections of the poem many Homeric episodes, descriptions, and similes are incorporated into the action, but they are in general short passages (e.g., the story of Polyphemus in III and the breaking of the truce in XII). The epic material in these sections is adapted to the tragic nature of the context and is used primarily for the delineation of character, e.g., Aeneas' rage after the slaying of Pallas in X, not unlike that of Achilles after the death of Patroclus,[49] and in XII the effect of the breaking of the truce of both Turnus and Aeneas; Turnus (324 f.) avoids his opponent and rejoices in the opportunity to kill other warriors, but Aeneas (464 ff.) considers himself bound by the terms of the truce and seeks Turnus alone.

In the central section (V-VIII) Vergil makes a very different use of Homeric material; in each of the books he adapts a lengthy passage:

|  |  |
|---|---|
| V. | Funeral games (from *Iliad* XXIII) |
| VI. | Underworld scene (from *Odyssey* XI) |
| VII. | Catalogue (from *Iliad* II) |
| VIII. | Shield (from *Iliad* XVIII) |

Here the long episodes are reworked and transformed for the glorification of Rome and its history, the portrayal of ancient Italy, and the greatness of the new era under Augustus who has triumphed over his enemies and introduced a new Golden Age of peace. The funeral games in Sicily in V are no exception but contain much of interest for the Romans of Vergil's day.[50] This adaptation of long Homeric episodes for historical, patriotic, and nationalistic purposes is something unique and appears only in the central section of the poem. One other long episode, the night expedition in IX, is based upon Homer (the *Doloneia* in *Iliad* X), but this may be considered the exception to prove the rule: the story of Nisus and Euryalus, appearing in the opening book of the third part of the poem (the tragedy of Turnus), is itself a miniature tragedy, in which two characters meet disaster and death as the result of their own actions;[51] this sets the tone for the final section of the *Aeneid*, when Turnus at the end of XII likewise pays the penalty for his own wrong act—the insolent treatment of the body of Pallas.

We are therefore entitled to look upon the *Aeneid* as a trilogy, with each third of the poem divided into four parts, or acts. The two tragedies of Dido and Turnus provide the framework for the central four books, which stress the patriotic and nationalistic themes, and at the very center of this section we find the speech of Anchises with its emphasis on Roman heroes, the achievement of Augustus, and the task of the Roman (VI, 760-853). *Aeneid* VI, called "the keystone of the whole poem," has been considered important in the twofold division of the epic; when we view the poem as a trilogy, we see that Vergil has stressed the importance of VI and especially Anchises' speech by its central position in the second section. In like manner, Eugene O'Neill in his trilogy *Mourning Becomes Electra* emphasized the sea background of the Mannon family and the symbolic motive of the sea as a means of escape and release by placing "the one ship scene at the center of the second play."[52]

The alternating rhythm of the *Aeneid* and the more serious nature of the even-numbered books take on added significance when we look at the poem as a trilogy. Conway pointed out that each of the books with even numbers has a culminating point—II and IV in calamity, VI and VIII in revelation, X and XII in triumph.[53] This grouping of the serious books has little importance in a twofold arrangement but becomes more meaningful in a tripartite division; the important books in each section of the trilogy are linked together by parallels and contrasts, just as are the corresponding books of each half in the twofold scheme.

Books II and IV are books of tragedy, the one of Priam and Troy, the other of Dido and Carthage, and each marks a tragic turning point in the life of Aeneas; II ends with his departure from Troy and the loss of Creusa, IV with his departure from Carthage and the suicide of Dido; in each instance he leaves unwillingly and under divine instruction. Numerous striking parallels appear in the delineation of Priam and Dido.[54] Books VI and VIII are books of revelation concerning the later history of Rome and are complementary; in VI Anchises describes famous men of legend and history, from the Alban kings to Augustus, founder of the new Golden Age, and in VIII Vergil portrays first the site of Rome[55] and then presents on the shield scenes closely connected with Rome the city, from Romulus and Remus to the triumph of Octavian after his defeat of Antony and Cleopatra.[56] Books X and XII are books of victory: at the end of X Aeneas defeats Mezentius who, as *contemptor divum*, stands for impiety; at the end of XII Aeneas triumphs over Turnus who, characterized by *violentia*, represents *vis consili expers*; the victory of Aeneas is the victory of *vis temperata* which, as Horace says, is favored by the gods.[57] The divine element is prominent in both X and XII: X begins with the speeches of Jupiter, Venus, and Juno, and the final reconciliation of Jupiter and Juno takes place at the end of XII, just before the defeat and death of Turnus. Finally, X and XII, like II and IV, are also books of tragedy and death, with XII bringing to a close the tragic story of Turnus, just as IV concludes the tragic tale of Dido.

The trilogy of the *Aeneid* may be outlined as follows, with the significant books of each part linked:

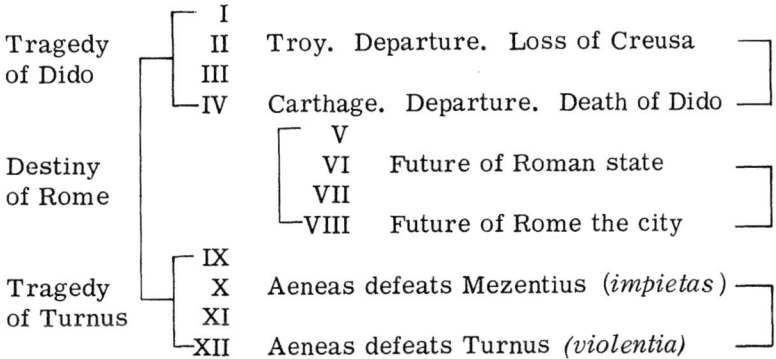

This analysis of the tripartite nature of the poem reveals anew the subtle technique of Vergil as a creative artist and the careful attention which he paid to the architecture of the *Aeneid*. The alternating rhythm of the books is a basic feature of the structure, and the even-numbered books are as important in their relations to each other in the tripartite arrangement (II and IV, VI and VIII, X and XII) as in the division into halves (II and VIII, IV and X, VI and XII). The tripartite division is not a substitute for the twofold arrangement but is superimposed upon it; Vergil seems to have thought of his poem as composed both of two halves and three thirds, and the latter arrangement was perhaps a deliberate attempt to avoid too sharp a break into an "Odyssey" of wanderings and an "Iliad" of battles.

The analysis of the *Aeneid* as a trilogy also reveals more clearly the manner in which Vergil used both epic and tragic material. His treatment of Homeric episodes in the central portion of the poem differs strikingly from that in the more tragic sections. The great impact which Greek tragedy made upon him is seen in the way in which his central message of Rome's history and mission is framed by the two tragedies of Dido and Turnus, and the tripartite division may itself be influenced by the structure of the Greek dramatic trilogy.[58] Finally, the threefold division of the *Aeneid* shows most effectively how Vergil has underscored his major themes and has presented in his epic not merely the story of Aeneas but also that of Rome and Augustus.

## "IN MEDIO MIHI CAESAR ERIT"

In *Georg.* III, 13 ff., Vergil describes the temple to honor Octavian, usually believed to refer to a historical epic in which he intended to praise the achievements of Octavian; cf. 16: *in medio mihi Caesar erit templumque tenebit*, "Caesar shall be in the center and shall occupy the shrine." Vergil did not compose the historical poem which he then had in mind, but he achieved the same result in his mythological epic about Aeneas and the Trojans which, as we have seen, is also a poem about Rome and Augustus. When we view the *Aeneid* as a trilogy, it becomes apparent that his desire to have Caesar *in medio* has been fulfilled. Augustus is at the very heart and center of the poem (VI, 788-807) and he appears at the conclusion of the central portion on the shield, where both his victory at Actium and his triumphs at Rome are described, with references also to three hundred shrines throughout the city and to the newly built temple of Apollo on the Palatine (VIII, 675-728). It should be noted also that on the shield itself Caesar occupies the central position; cf. *in medio* (VIII, 675).

In 27 B.C. Augustus received from the Senate and the Roman people a golden shield, on which he was honored for four virtues—*virtus, clementia, iustitia,* and *pietas*.[59] The description of his deeds in VIII stresses both his *virtus* (the victory at Actium) and his *pietas* (the restored temples and the new temple to Apollo).[60] The other two virtues of the shield appear at the end of XII, where, as I pointed out above, Aeneas, who symbolizes Augustus in many other respects (cf. the *virtus* and *pietas* displayed throughout the poem and, more specifically, his promise in VI, 69-74, to erect a temple to Apollo on the Palatine), shows both *clementia* and *iustitia*—he considers sparing Turnus' life and then kills him when he sees Pallas' belt; this illustrates VI, 853: *parcere subiectis et debellare superbos* (as the two previous verses, 851 f., refer to *virtus* and *pietas*, another link between the conclusion of VI and that of VIII).

The fact that Augustus appears *in medio* also in Horace's Roman Odes (III, 1-6) and in the friezes of the Ara Pacis is not unrelated to the structure of the *Aeneid*—its divisions both into halves and into thirds.

Horace's six Roman Odes divide into halves: three Roman and general, and three more specifically Augustan, concerning Augustus (4) and his military (5), religious, and social policies (6); they correspond as do the books of the *Aeneid* in each half:

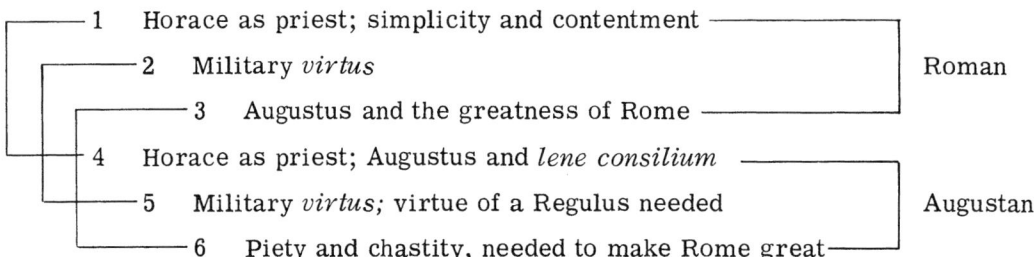

The final message to the Romans in the Sixth Ode is religious and social, complementing the famous passage in *Aen.* VI, 851-853, wherein Vergil presents a military and imperial ideal. Also, Anchises' description of Roman heroes in *Aen.* VI, 760-853, falls in two parts, each with three sections—a total of six, thus corresponding loosely to the six Roman Odes; the first or Julian half (760-807) describes the Alban kings and Romulus and culminates in the description of Augustus restoring the Golden Age of Saturn and his comparison to Hercules and Bacchus;[61] the Roman half (808-853) begins with Numa and the other kings, lists the many heroes of the Republic, and ends with the famous passage on the duties of the Roman. In VI, 760-853, Vergil seems indebted to Horace's Roman Odes in both structure and content.[62]

The Roman Odes divide also into thirds, the first two and the last two poems framing the two central odes—the longest, most Pindaric, and most significant poems in the collection; these two central odes deal with *iustitia* and *lene consilium* (= *clementia* ?) and stress Rome and Augustus:

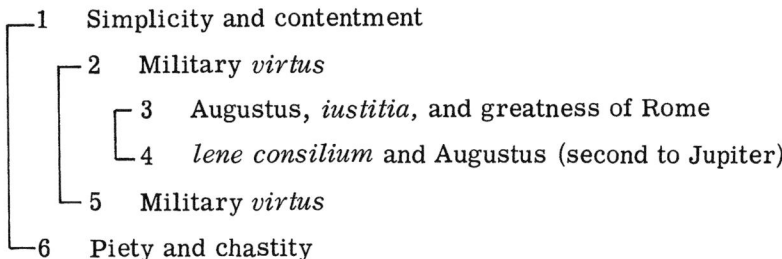

Here we have a tripartite division, with Rome and Augustus occupying the same central position as they do in *Aeneid* V-VIII. Since Vergil undoubtedly planned his books to divide into both halves and thirds at the beginning of his work on the *Aeneid*, Horace must have known of his friend's structural plans before he himself composed the Roman Odes. In this case, the intricate structure of the six odes may well derive from that of the *Aeneid*.

The Ara Pacis Augustae, now restored near the Mausoleum of Augustus in what was once the Campus Martius, achieved in art what Vergil and Horace had accomplished in poetry, and again balance and symmetry played an important part. The six friezes are separated into two groups of three each by the two entrances, and we have a Roman half and a Julian and Augustan half; in each half the historical frieze is framed by legend and symbol:

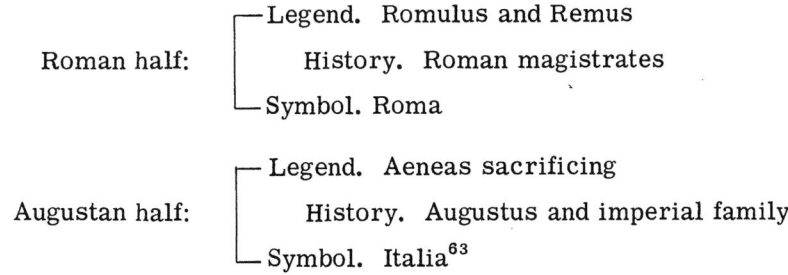

Or we may see in the friezes of the famous monument a tripartite division:

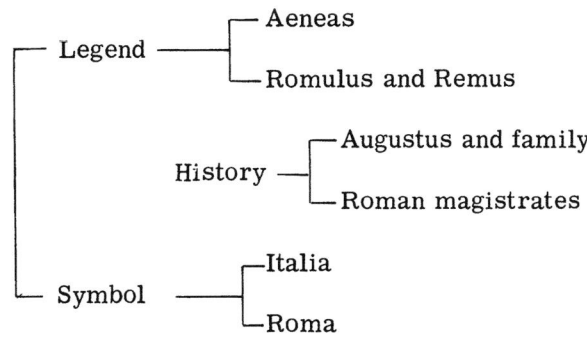

Here, as in *Odes* III, 3 and 4, and in the central third of the *Aeneid*, Augustus and Rome are *in medio*. The architects and sculptors of the Ara Pacis, as they planned the monument, must certainly have had in mind the poetic structure of both the *Aeneid* and the Roman Odes.[64]

# Notes To Chapter 1

1. *P. Vergilius Maro,* cols. 270 f.

2. *Vita Donati,* 24. The arabic numbers refer to the sections of the Life in standard editions, e.g., Hardie, *Vitae Vergilianae Antiquae.*

3. Cf. Conway, *Harvard Lectures,* p. 129; Perret, *Virgile,* p. 111; Duckworth, *AJPh* 75 (1954), p. 1. Mackail, *The Aeneid,* p. xliii, referring to *Aeneid* VII-XII as the main subject of the epic, compares the second half of the poem to "a basilica, approached through a triple-bayed narthex (Books I, III, V), with two splendid and elaborately adorned flanking halls (Books II and IV), and a great central dome (Book VI). These among them fill an equivalent space to the basilica itself, and are wrought with it into a single architectural composition." For Maury's analysis of the *Eclogues* as a temple, with *Eclogue* V the shrine where Caesar was honored as the dead and deified Daphnis, see below, note 15.

4. See Crump, *The Growth of the Aeneid,* pp. 30 ff., 118, for the view that *Aeneid* III was originally the first book, with the events described in the third person, and that this was followed by the happenings in Sicily, the present Book V, as the second book; cf. also Pease, *Publi Vergili Maronis Aeneidos Liber Quartus,* pp. 56-59 and bibliography in note 458; Büchner, *P. Vergilius Maro,* cols. 405 f.

5. Conway, *Harvard Lectures,* pp. 16 f., 139; on this see Rand, *The Magical Art of Virgil,* pp. 89 f., 160 ff.

6. This has been noticed by many writers; cf., e.g., Cartault, *Étude sur les Bucoliques,* pp. 53 f.; Klotz, *RhM* 64 (1909), pp. 325 ff.; Saint-Denis, *IL* 6 (1954), pp. 186 ff. For an additional type of balanced alternation, that between interest in subject and interest in form, see Richardson, *Poetical Theory,* pp. 120 f.

7. Cf. 460: *iustissima tellus,* i.e., Justice who lived on earth in the original Golden Age has now returned, and life in Italy is again the life of *aureus Saturnus* (538); see Rand, *The Magical Art of Virgil,* pp. 264 ff.

8. For a possible reason for the excessive length of V and XI, see below, Chapter 4, note 40.

9. Cf. Conway, *Harvard Lectures,* p. 141: "the contrast of the grave and the less grave; of a sense of tension and a sense of leisure. . . . This is the real division of the *Aeneid.*" See Stadler, *Vergils Aeneis,* pp. 17-43.

10. *PBA* 17 (1931), p. 25; this seems more satisfactory than Conway's earlier statement *(Harvard Lectures,* p. 141) that "the books with odd numbers show what we may call the lighter or Odyssean type; the books with the even numbers reflect the graver colour of the *Iliad.*" But the more tragic nature of the even-numbered books should not lead us to assume, as does Knight, *Vergilius* 6 (1940), p. 20, that "there are thus six tragedies in the *Aeneid.*" Cf., however, Knight, *Roman Vergil,* pp. 135-138.

11. *Vergils Aeneis,* pp. 18-20, 40-43; Stadler terms the even-numbered books systolic, the others diastolic. Cf. also Büchner, *P. Vergilius Maro,* col. 419.

12. This section is a revision and expansion of my article in *AJPh* 75 (1954), pp. 1-15.

13. *The Aeneid,* p. xlv.

14. This correspondence was first noticed by Krause, *Quibus temporibus . . . Vergilius eclogas scripserit,* pp. 6 f.; his analysis was rejected by Cartault, *Étude sur les Bucoliques,* p. 53, note 2. A division of the ten pastorals into two halves is favored by Port, *Philologus* 81 (1925-26), pp. 287 f., and by Becker, *Hermes* 83 (1955), pp. 317 ff.; cf. Büchner, *P. Vergilius Maro,* cols. 236 f. Skutsch, *RhM* 99 (1956), pp. 193-201, accepts the correspondence of I-III and IX-VII but denies the relationship of IV and VI, since he views VI as merely a collection of Alexandrian themes.

15. See Maury, *Lettres d'Humanité* 3 (1944), pp. 71-147. Maury develops the correspondence between *Eclogues* I-IV and IX-VI into what he terms a "bucolic chapel," with four poems on each side, like columns, leading the way to V, the central and most important poem, the shrine where Caesar is honored in the guise of the deified Daphnis; X, when added, places Gallus, the suffering mortal, at the entrance to the chapel. See below, Chapter 5, "Vergil and Pythagoreanism."

16. *Poetical Theory,* p. 121.

17. See Hahn, *TAPhA* 75 (1944), pp. 239-241.

18. Cf. Hahn, *TAPhA* 75 (1944), p. 240: "This monologue [of Gallus] on the theme of unrequited love, jealousy, and despair at once recalls Corydon of the central *Eclogue* in Triad 1 and Damon of the central *Eclogue* in Triad 3, while at the same time the dying Gallus [i.e., 'dying of love,' cf. X, 10] lamented by nature and the gods recalls the dead Daphnis of the central *Eclogue* in Triad 2."

19. See also Drew, *AJPh* 50 (1929), pp. 242-254; Richardson, *Poetical Theory,* pp. 132 ff.; Perret, *Virgile,* pp. 69 ff.; Duckworth, *AJPh* 80 (1959), pp. 229 ff. Some parallels are forced; Drew seems wrong in viewing the conclusion of III as dealing with contemporary history (to balance the ending of I); cf. Richardson, pp. 147 f.

20. *AJPh* 50 (1929), p. 254.

21. On the significance of this passage for the *Georgics* as a whole, see Burck, *Navicula Chiloniensis,* pp. 156-172; cf. also Grimal, *Les jardins romains,* pp. 412 ff., who believes that the old man of Tarentum symbolizes Pythagoras; see Saint-Denis, *Virgile, Géorgiques,* p. 114 (on IV, 130).

22. *Virgile,* pp. 84 f. If we reject this symbolism, the conclusions of I and IV are still linked by references to the later divinity of Octavian (I, 503 f., IV, 562).

23. *Harvard Lectures,* pp. 139 f.

24. *Harvard Lectures,* p. 143. Cf. also Prescott, *The Development of Virgil's Art,* pp. 360 ff.; Letters, *Virgil,* p. 122.

25. *Virgile,* pp. 111-120.

26. *Virgile,* pp. 113 f.

27. Cf. Mendell, *YClS* 12 (1951), p. 226, for a similar treatment of VI: "The sixth book ... is itself a great focal point between two contrasted panels, Books I-IV and Books VIII-XII, with the quieter books of suspense, V and VII, as an inner frame."

28. Vergil's portrayal of Turnus in X and XI, however, is essential to our understanding of his character; see Duckworth, *Vergilius* 4 (1940), pp. 8 ff.; *CJ* 51 (1955-56), pp. 361 f.

29. To illustrate from a shorter passage, cf. the different patterns found in the Latin catalogue in VII; see below, pp. 20 f. and note 4.

30. *Virgile,* p. 117.

31. *CJ* 26 (1930-31), p. 17.

32. Cf. Terzaghi, *Virgilio ed Enea,* p. 167 and note 183.

33. The headings in capitals both here and in the parallel columns below indicate the more significant books (those with even numbers).

34. Cf. Pöschl, *Die Dichtkunst Virgils,* pp. 46 ff., who considers the opening scenes of I and VII symbolic of each half.

35. For verbal parallels in the two speeches of Ilioneus, see Hirst, *CQ* 10 (1916), pp. 91 f.

36. For the similarity of wording in Juno's two soliloquies (I, 37 ff., and VII, 293 ff.), see Fraenkel, *JRS* 35 (1945), p. 3.

37. The *lacrimae inanes* of IV, 449, are undoubtedly those of Aeneas; this is indicated not only by the *lacrimae inanes* of Hercules in X, 465, but also by the simile of the *quercus* in IV, 441-447: the tree is shaken by the blasts of the wind and sheds its leaves, but it is firmly rooted and immovable; so Aeneas is shaken by the appeals of Anna and sheds tears, but his resolve remains fixed. See Rand, *The Magical Art of Virgil,* pp. 361 f.; Pease, *Publi Vergili Maronis Aeneidos Liber Quartus,* pp. 367 f.; and especially Pöschl, *Die Dichtkunst Virgils,* pp. 76 ff.

38. These features are, of course, not limited to Vergil but appear in all great works of literature; cf. Whitman, *Homer and the Heroic Tradition,* pp. 254 f.: "Probably all aspects of formal symmetry depend ultimately upon these two categories of similarity and opposition, as Plato seemed to know when in the *Timaeus* he finished off his cosmology with the two spheres of Sameness and Difference, which revolve in opposite directions. . . . For Homer the framework of identity and antithesis is fundamental."

39. This section is a revision and expansion of my article in *TAPhA* 88 (1957), pp. 1-10.

40. Stadler, *Vergils Aeneis,* pp. 50-61; Pöschl, *Die Dichtkunst Virgils,* pp. 279 f.; cf. p. 280: "Dunkel—Licht—Dunkel: dies also ist der Rhythmus, der das Epos in seiner Gesamtheit beherrscht." Mackail had earlier pointed out the tripartite division in *The Aeneid,* p. 298: "The traditional and superficial view of the Aeneid as falling into two halves may be supplemented and rectified by another view of it as a structure made up of three acts."

41. Büchner, *P. Vergilius Maro,* col. 418; see also col. 420.

42. Camps, *CQ* N.S. 4 (1954), p. 215. Camps, however, does not have the usual tripartite arrangement (I-IV, V-VIII, IX-XII) but divides as follows: I-IV; V-VI and VIII-IX; and VII, 286-817, plus X-XII; he looks upon VII, 25-285, as the center of the *Aeneid,* where the references to Aeneas as the man of destiny invite us to look both backward and forward (cf. p. 214), and accordingly he excludes this passage from his analysis. But lines 760-853 of *Aeneid* VI seem more appropriate as the center of the poem; see above. In *CQ* N.S. 9 (1959), pp. 53 ff., Camps discusses at greater length Books V-IX as the central portion of the poem; the importance of Turnus in IX, however, indicates that this book belongs with the final section in which Turnus plays a major role.

43. All is not brightness in the center section of the poem; cf. Tracy, *Phoenix* 4 (1950), pp. 4 f., on the darkness and gloom of VI. Are the activity of Allecto and the outbreak of war in VII to be looked upon as light rather than darkness? And in the third part of the poem, is no light to be seen in Aeneas' victories in X and XII?

44. This may explain Drew's curious statement in *The Allegory of the Aeneid,* p. 60: "The rest of the epic is, in a certain sense, anti-climax." Cf. also Caiati, *Vita di Virgilio,* p. 155, who

looks upon the last four books as uninteresting and the least successful portion of the poem.

45. *Die Dichtkunst Virgils,* pp. 23 ff., 41 ff.

46. Formerly I looked upon I, 257-296, as the first of the three great Roman passages in the epic. It is this, of course, but it would perhaps be equally accurate to view it as a preliminary to the two longer Roman passages in the central third of the poem (VI, 756-887, and VIII, 626-728).

47. *CJ* 26 (1930-31), p. 46; for references on Vergil and tragedy, see Duckworth, *TAPhA* 87 (1956), p. 295, note 51; cf. also Knight, *Vergilius* 6 (1940), pp. 20-24; *Roman Vergil,* pp. 133-135.

48. Vergil refers specifically to the *culpa* of Dido (IV, 19 and 172); on Turnus, cf. Duckworth, *Vergilius* 4 (1940), pp. 5-17; *CJ* 51 (1955-56), pp. 361-363 and (for bibliography) p. 364, note 27.

49. Cf. Achilles' anger when he goes forth to avenge the death of Patroclus in *Iliad* XXI: in 27 ff. he takes twelve captives alive and then slays Lycaon, rejecting his plea of guest-friendship; so Aeneas (X, 517 ff.) takes alive eight captives and rejects the entreaties of Magus, who offers money if Aeneas will spare his life. The slaying of Magus is therefore more justified than that of Lycaon.

50. The games in V, although suggested by those in *Iliad* XXIII, must be considered Roman and Augustan as well as Homeric; Augustus' interest in athletic contests, expecially boxing, is mentioned by Suetonius *(Aug.* 45), and Vergil's contemporaries would undoubtedly have related the contests of V to those which they themselves had seen; see Constans, *L'Énéide de Virgile,* pp. 170-174. The *ludus Troiae* in 545-603, which follows the four contests, strikes a distinctly Roman note; Mackail, *The Aeneid,* p. 166, calls it "one of those interludes or episodes which connect the epic closely with contemporary history, with the new Empire and the Imperial family."

51. See Duckworth, *CJ* 51 (1955-56), pp. 359 f. Actually, the tragic nature of the episode is to be explained in part by its indebtedness to the *Rhesus,* and many of the supposedly Homeric echoes may well be indirect; see Fenik, *The Influence of Euripides on Vergil's Aeneid,* pp. 54 ff. Vergil's use of long Homeric episodes may thus be limited to *Aeneid* V-VIII, the central portion of the epic, where the Roman and Augustan elements are most strongly emphazised.

52. Cf. Clark, *European Theories of the Drama,* p. 533. The exact center of the trilogy would be Act Three of "The Hunted"; actually, the ship scene is Act Four.

53. *PBA* 17 (1931), p. 25; see also Stadler, *Vergils Aeneis,* pp. 55 f.

54. On the parallelism of theme and imagery in *Aeneid* II and IV, see Fenik, *AJPh* 80 (1959), pp. 1-24.

55. Vergil describes the site of Rome as he imagines it to have been in the days of Evander, but he also alludes to places which would suggest to the Roman reader the city of his own day and the building program of Augustus; cf. Grimal, *REA* 50 (1948), pp. 348-351.

56. It is most appropriate that the visit to the site of Rome in VIII is followed by the scenes on the shield of events which later took place in the city; cf. Cartault, *L'Art de Virgile,* p. 634. First we have the empty stage, and then the stage is filled in (1) with happenings from early Roman history and (2) with Octavian's victory and triple triumph. Fowler's attempt in *Aeneas at the Site of Rome,* pp. 103 ff., to find unity in the scenes on the shield by viewing them as escapes of the Roman people from great dangers seems unnecessary; it is Rome the city which provides the unifying factor.

57. *Odes* III, 4, 65-68; cf. Duckworth, *CJ* 51 (1955-56), p. 361; *TAPhA* 87 (1956), p. 303.

58. This last is not a necessary assumption, however. The *Iliad* has three major parts or rhythms: I-VII, the Greeks are still victorious in spite of Achilles' withdrawal; VIII-XVII, the Trojans force the Greeks to the ships and Patroclus is slain; XVIII-XXIV, Achilles returns to battle, the Greeks drive back the Trojans, and Achilles slays Hector. The *Odyssey* falls into two halves, but each half has three parts; cf. Heinze, *Virgils epische Technik,* p. 458, note 3. On the elaborate correspondence of the various parts of the *Iliad* to each other by "ring composition" or "Geometric structure," see Whitman, *Homer and the Heroic Tradition,* pp. 249-284 and the chart at the end.

59. Cf. Gagé, *Res Gestae Divi Augusti,* p. 144 [34, 2 = VI, 16-22]. On these four Augustan virtues, see Markowski, *Eos* 37 (1936), pp. 109-128; Charlesworth, *PBA* 23 (1937), pp. 111-114.

60. Drew, *The Allegory of the Aeneid,* pp. 25 ff., believes that all the scenes on the shield in VIII suggest the virtues mentioned on the *clupeus aureus* presented to Augustus; see Duckworth, *TAPhA* 87 (1956), p. 307, note 86.

61. *Aen.* VI, 788 ff. Cf. Horace, *Odes* III, 3, 9-16, where Augustus on Olympus is associated with Pollux, Hercules, Bacchus, and Romulus.

62. For a detailed discussion (with bibliography) of the six Roman Odes and their relation to *Aen.* VI, 760-853, see Duckworth, *TAPhA* 87 (1956), pp. 299-308.

63. Or Terra Mater. Van Buren, *JRS* 3 (1913),

pp. 134-141, argues strongly for Italia. For more recent discussions of the problem, see Duckworth, *TAPhA* 87 (1956), p. 314, note 104.

64. The recent theory of Weinstock, *JRS* 50 (1960), pp. 44-58, that the monument "is certainly not the Ara Pacis Augustae" (p. 58), even if accepted, does not invalidate the relations of the altar to the poetry of Vergil and Horace. For an analysis of Book IV of the *Odes* and the similarity of Horace's themes to those of the Ara Pacis, see Benario, *TAPhA* 91 (1960), pp. 339-352.

# Chapter 2
# STRUCTURAL PATTERNS IN THE BOOKS

IN THE ARCHITECTURE of the *Aeneid* as a whole Vergil utilized three different structural principles—an alternation between books of a lighter and more serious nature, a balanced symmetry between the books in each half, with the correspondence achieved by both parallels and contrasts, and a tripartite pattern which enabled him to emphasize the more national and Augustan themes of the epic by placing them in the central portion between the tragedies of Dido and Turnus. When we examine the structure of the individual books, or parts of books, we find the same three principles at work. By means of alternation Vergil introduces effective contrasts; balanced passages framing a central focal point illustrate the type of structure known as the concentric or recessed panel pattern, and the use of tripartite divisions extends from short passages to the structure of entire books.

## ALTERNATION AND CONTRAST

In shorter passages within the individual books Vergil displays a striking interest in effective contrasts: scenes of the gods alternate with those of mortal activity, and peaceful scenes are contrasted with those on the battlefield; individual fighting varies with mass action, and brilliant aristeias of warriors on each side alternate not only with each other[1] but also with the brief scenes in which the fighters as a group re-establish an equilibrium.[2] In like manner, darkness and gloom contrast with brightness and joy, e.g., VI, 548-636 (Tartarus), and 637 ff. (Elysium). Sudden reversals are frequent, and a scene of happiness is often succeeded by immediate disaster; the Trojans are *laeti* in I, 35, just before Juno and Aeolus bring about the storm; the Trojans and Aeneas are again described as *laeti* in VII (36, 130, 147, 288) before Juno and Allecto stir up war. In IX, 366, Nisus and Euryalus leave the camp of the enemy in apparent safety and then meet disaster; in XII, 202 f., Latinus says:

> nulla dies pacem hanc Italis nec foedera rumpet,
> quo res cumque cadent;

the truce is broken and the fighting renewed almost immediately. The joy and delight of the Trojans as they take the wooden horse into Troy (II, 234 ff.) is followed by the return of the Greeks and the capture of the city. The horse which the Trojans think will bring them victory is in reality the agent of their destruction, and Aeneas' foreshadowing of their doom increases the tragic irony of the passage.[3]

Alternation and contrast are especially evident in those passages where Vergil gives a series of events or lists a number of persons. Book III consists of nine stops on the journey from Troy to Sicily and is an excellent illustration of the poet's use of tripartite structure, to be discussed below; it also displays the effective use of an alternating pattern, for in each third of the book we have two short and less important stops followed by one which is longer and more significant. The four contests in V reveal a similar alternation, the first (boat race) and third (boxing match) being longer and containing more details of characterization than the second (foot race) and the fourth (archery match). I do not include among the contests the *ludus Troiae* which is a separate spectacle not previously announced (cf. 64-70) and which forms the transition to the burning of the ships, in which Ascanius also takes an active part (cf. 667-674).

Vergil's use of alternation is most effective in VII, 641-817; this catalogue of Latin warriors begins with two very important leaders, Mezentius and Lausus; then follows a group of three less important ones (actually four, but the twins Catillus and Coras are treated together), then Messapus, a warrior of considerable importance, and next two groups of three unimportant leaders each, the second group more interesting than the first; the list closes with the significant leaders Turnus and Camilla.[4] The deaths of the four leaders described at the beginning and the end of the catalogue provide the conclusions of the final three books. In addition to the alternation between important and unimportant warriors, a geographical contrast may also be seen: Mezentius and Lausus come originally from southern Etruria but are now with Turnus in Latium; the next three come from Latium (Aventinus from Rome, Catillus and Coras from Tibur, and Caeculus from Praeneste); the six warriors who follow fall into three distinct groups: the first two (Messapus and Clausus) come from the more distant north, the next pair (Halaesus and Oebalus) from Campania in the south, and the final pair (Ufens and Umbro) from districts to the east of Latium; with Virbius we return to the Alban hills in Latium, and Turnus and Camilla come from western and southern Latium respectively. Vergil has thus woven together two different types of alternation in his presentation of the Latin warriors;[5] cf. the following diagram which combines the two patterns (the numbers which are underlined refer to the important warriors):

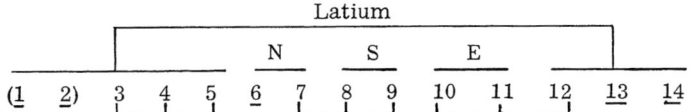

## THE FRAMEWORK OR RECESSED PANEL PATTERN

Vergil has composed many passages of the *Aeneid* in a very different manner, using a concentric or recessed panel pattern with balancing passages framing a central focal point. This type of symmetrical framework is frequently illustrated by the structure of Catullus LXIV[6] and appears both in the *Eclogues* and in the *Georgics* (especially in the Aristaeus and Orpheus stories in IV).[7] In certain of the *Eclogues*, Vergil has added a numerical symmetry to the concentric arrangement of corresponding passages around the central core; e.g., in *Eclogue* I the focal point is the generosity of the youth (40-45), the *deus* of 6 f.:

(5)    1-5 Introduction
(21)      6-26 Good fortune of Tityrus
(13)        27-39 Tityrus in Rome
(6)          40-45 Benefits from the youth
(13)        46-58 Tityrus at home
(20)      59-78 Plight of Meliboeus
(5)    79-83 Conclusion

Here we have an almost perfect numerical balance in the passages framing 40-45.

In *Eclogue* IV the central passage of 28 verses (18-45) is framed on each side by two passages of seven verses each and these by the preface and conclusion which total seven:

(3)    1-3 Preface: a higher theme
(7)      4-10 Golden Age and child
(7)        11-17 Pollio's consulship; child will rule
(28)          18-45 Three stages of child's growth
(7)        46-52 Parcae; honors; rejoicing
(7)      53-59 Vergil's desire to praise his deeds
(4)    60-63 Conclusion: child's smile and his destiny

Vergil's use of heptads here is striking, and the number 28 (the lines of the focal point) had in ancient mathematics a special perfection, being one of the very few numbers (like 6 and 496) which are the sum of their factors (i.e., $1 + 2 + 4 + 7 + 14 = 28$). This recessed panel pattern appears in other *Eclogues* and also, as was pointed out in Chapter 1, in the arrangement of the collection as a whole, with I and IX, II and VIII, etc. framing the central poem V, on the death and deification of Daphnis.

Many portions of the *Aeneid* are framed by balancing passages; a simple form of such construction occurs in II, where the Sinon episode (57-198) is preceded and followed by passages concerning Laocoon (40-56, 199-227), in VIII, where the story of Cacus (184-267) is framed by the rites in honor of Hercules (172-183, 268-305), and in VII, where the threefold activity of Allecto (341-539) is preceded by Juno's lament and the summoning of Allecto (286-340) and followed by the dismissal of Allecto and the result of her actions (540-600).

More elaborate concentric patterns around a focal point occur far more frequently than is usually realized. Mendell analyzes several such passages in I, IV, VII, and XI;[8] he considers this method of composition "an habitual practice" of Vergil in the *Aeneid,* one which reveals the continuing influence of Catullus and neoteric poetry. Murley, speaking of Catullus LXIV, states that the concentric pattern is typically Roman, being chiastic like a periodic sentence.[9] Actually, it is neither specifically Roman nor Hellenistic but should be viewed as a fundamental method of ancient composition; it goes back to Homer where it is termed "ring composition" or "Geometric structure." Whitman has now shown in full detail the extent to which circularity of design penetrates the *Iliad,* not merely in scenes and books and groups of books but in the poem as a whole.[10] The presence of the concentric patterns in Vergil's poetry should no longer be attributed merely to Catullus and neoteric poetry; they are also in part the result of his indebtedness to Homer, whose genius constructed a masterpiece of epic architecture, the incomparable unity of which has so often been misunderstood or ignored.

Mendell cites as an illustration of symmetrical framework in the *Aeneid* the meeting of Aeneas and Venus in I; the structure is as follows:

305-320 Introduction. Aeneas sets out. Venus appears

    321-324 Venus

        325-334 Aeneas

            335-370 Venus' long story

        371-386 Aeneas

    387-401 Venus

402-418 Conclusion. Venus disappears. Aeneas proceeds

The focal point of the episode is Venus' important speech about Dido and the founding of Carthage in 335-370.

Book XI is particularly interesting in that it is divided into three parts each with a concentric framework around a focal point: 1-224, the burial of the dead, with the embassy to Aeneas and Aeneas' peace appeal (100-138) as the central passage; 225-467, the council of the Latins, framing Latinus' speech (302-335); and 468-915, the cavalry battle, with the deeds of Camilla (648-724) as the focal point; Mendell's analysis of the final section is as follows:

468-519 Turnus prepares battle. Camilla volunteers

    520-531 Plan of battle

        532-596 Opis sent. Camilla's youth and prophecy of death

            597-647 General battle

                648-724 Deeds of Camilla

725-798 Battle joined by Tarchon's cavalry

799-867 Death of Camilla. Opis departs

868-895 Battle resumed

896-915 Turnus enters the battle to meet Aeneas

Mendell finds this pattern unusual, for, as he himself admits, the focal point is more often a significant speech or scene of emotional tension.

Mackail gives the three main sections of Book XI as 1-212, 213-497, and 498-867, and he views the conclusion (868-915) as "a coda of forty-eight lines describing the approach of the main force of the Trojan infantry, and the encampment of the armies at nightfall."[11] I prefer with Mendell to end the first two divisions at 224 and 467, but Mackail seems correct in ending the third section at 867. The story of XI proper concludes with Camilla's death, as X and XII end with the deaths of Mezentius and Turnus, and the remainder of the book is an epilogue preparing for and making a transition to XII. Also, with the conclusion of the third division at 867, a new and simpler recessed panel pattern now appears:

468-497 Preparations for battle

498-531 Camilla enters the conflict

532-596 Diana's speech about Camilla

597-647 Cavalry battle

648-867 Aristeia and death of Camilla

The famous speech of Diana to Opis about the life and imminent death of Camilla now becomes the focal point, and the unusual feature which Mendell himself noted in the third section of the book disappears.

In XI the corresponding parts framing the centers are of unequal lengths and Vergil here does not strive for numerical symmetry. In VI, 56-123 (a passage not cited by Mendell), we find an almost perfect balance of passages about the words of the Sibyl (83-97); her speech is important for the *Aeneid* as a whole—one of several passages in I-VI which arouse suspense concerning the events of VII-XII:[12]

(21)   56-76   Speech of Aeneas

(6)     77-82   Description of the Sibyl

(15)     83-97   Speech of the Sibyl

(5)     98-102   Description of the Sibyl

(21)   103-123   Speech of Aeneas

Here we have a numerical symmetry similar to that found in *Eclogues* I and IV, and the framing of the Sibyl's speech underlines its importance for the later action of the poem.

Mendell's analysis of Dido's suicide in IV, 450-705, is of interest; he says:

The story tends to become the tragedy of Dido rather than the testing of Aeneas. Virgil makes use of many devices to prevent this and not the least striking is the focal position of Aeneas in the great scene of Dido's suicide. The climax of action comes of course with the actual death, 630-692. But in the symmetrical form in which the whole is presented it is Aeneas rather than Dido who is the focal center, even though this center is dramatically a period of suspense.

450-473  Tone prologue
  474-521  Dido's plan
    522-553  Dido
      554-583  Aeneas
    584-629  Dido
  630-692  Consummation of Dido's plan
693-705  Juno epilogue[13]

Mendell's point about the significance of Aeneas as the focal center in 554-583 may be reinforced by the analysis of *Aeneid* IV in its entirety as a series of corresponding passages around a focal point in the very center of the book—Aeneas' famous speech in 333-361. However much the modern reader may sympathize with Dido, it is clear that Vergil looked upon the Carthaginian queen as a danger to be resisted; Aeneas' duty lay elsewhere, and the poet has emphasized the rightness of Aeneas' decision by placing his defense in the very center of the book as a whole. I shall omit many connecting links and list only the corresponding passages which provide the balanced framework;[14] the most important part is 276-415, where Aeneas' speech is framed with an approximate numerical symmetry:

1-8  Dido's love
  9-53  Speeches: Dido, Anna
    [54 ff.  Love and consummation]
    [173 ff.  Fama and results]
      265-275  Speech of Mercury
(28.8)   276-304  Narrative. Preparation for departure
(26)       305-330  Dido's speech
(2.4)        331-333a  Aeneas' emotions
(28.2)         333b-361  *Speech of Aeneas*
(3)          362-364  Dido's emotions
(23)       365-387  Dido's speech
(27.6)   388-415  Narrative. Preparation for departure
      [416 ff.  Dido's entreaties, desire for death, scene of magic, Dido's lament]
    560-570  Speech of Mercury
    [584 ff.  Curse of Dido. Preparations for suicide]
  651-685  Speeches: Dido, Anna
685-705  Dido's death

In the central portion the appearance of the perfect number 28 is striking, and it is worth recording that the two speeches of Dido (305-330, 365-387) total 49 verses (26 + 23); here again, as in *Eclogue* IV, we have multiples of seven in a prominent position. Book IV, as a tragic story of love and suicide, indebted not only to Greek tragedy but in part to the love affair of Jason and Medea in Apollonius' *Argonautica*,[15] reveals the most complete framework pattern about a central core to be found in the *Aeneid*.

## TRIPARTITE STRUCTURE

Tripartite structure is as old as European literature. The *Iliad* has three main divisions or movements: I-VII, the continuing victory of the Greeks; VIII-XVII, the Greeks driven back to the ships by the Trojans; and XVIII-XXIV, the return of Achilles to battle and the defeat and death of Hector. The *Odyssey* has a threefold interest: the story of Telemachus, the wanderings of Odysseus, and the victory over the suitors when both Odysseus and Telemachus return to Ithaca. The first half of the epic falls naturally into three parts: I-IV, the journey of Telemachus; V-VIII, Odysseus and the Phaeacians; IX-XII, Odysseus relates his wanderings. Odysseus' tale of his adventures comprises three episodes in each of three books, IX, X, and XII (XI is the book of the Underworld), and the third story in each book (IX, Polyphemus; X, Circe; XII, the cattle of Helios) is related at greater length. In magic "the god rejoices in an uneven number" *(numero deus impare gaudet, Ecl.* VIII, 75; cf. 73-77: *terna, triplici, ter, tribus, ternos),* and in poetry the number "three" is equally useful to the poet in his composition of passages and books.

I have discussed above the *Aeneid* as a trilogy; Vergil's use of tripartite divisions in short passages and in individual books is equally impressive. Sinon in II makes three speeches (77-104, 108-144, 154-194), each with its appropriate effect on his listeners; the activity of Allecto in VII is threefold: she incites Amata (341-405), Turnus (406-474), and the hounds of Ascanius (475-539), and the results of her actions are described in reverse order in 572-582 (shepherds, Turnus, Amata). In IX the tragic story of Nisus and Euryalus falls into three parts: the scene in the Trojan camp (176-313), the slaughter of the enemy (314-366), their departure and deaths (367-449).[16] In the council of the gods in X three divinities speak: Venus (18-62), Juno (63-95), and Jupiter, whose two speeches (6-15, 104-113) frame those of the goddesses; later in the same book three important warriors perish—Pallas, Lausus, and Mezentius. In the concluding portion of XII the fighting of Aeneas and Turnus falls into two stages (697-790, 843-952), separated by the important reconciliation of Jupiter and Juno (791-842).

*Aeneid* V is often divided into two main parts: 1-603, arrival and contests; 604-871, burning of the ships, departure, death of Palinurus. But the *ludus Troiae* (545-603) was a spectacle, not a contest, and it was not mentioned among the games announced by Aeneas (cf. 64-70). If we are to have a bipartite division of V, I should prefer 1-544 and 545-871, since Ascanius is prominent both in the spectacle and during the fire. An examination of V, however, reveals a simple type of recessed panel structure:

1-34 Arrival. Palinurus

   35-544 Contests

      545-603 *Ludus Troiae*

   604-778 Burning of the ships

779-871 Departure. Death of Palinurus

With this arrangement the Trojan display becomes the focal point, and this is particularly appropriate as it is the most Roman element in the book which begins the Roman and Augustan portion of the poem (V-VIII).[17] This concentric pattern suggests a tripartite division for V, as follows:

1-544    Arrival and contests

545-603 *Ludus Troiae*

604-871 Burning of ships, departure, death of Palinurus

The fact that the Trojan spectacle is short (59 verses) is not a valid reason against considering it a main division of the book; the second part of Book VIII (370-453, the night scene of Venus and Vulcan and the making of the armor) is almost as short (84 verses). But perhaps the strongest argument in favor of the tripartite arrangement of V is this: *every other book of the Aeneid falls naturally into three main divisions;* these are determined by the patterns of thought and are likewise of unequal length.[18]

In the other books the subdivisions are also tripartite in most instances. This is not true of Book V, as is seen when we add a more detailed outline of the book:

    I. 1-544    Contests

        1-34    Arrival. Palinurus

        35-544    Contests

            35-113  Preparations

            (1)  114-285  Boat race

            (2)  286-361  Foot race

            (3)  362-484  Boxing match

            (4)  485-544  Archery contest

    II. 545-603  *Ludus Troiae*

    III. 604-871  Burning of the ships. Departure

        604-663  Iris and the Trojan women

        664-778  Burning of the ships

        779-826  Venus and Neptune

        827-871  Death of Palinurus

In the case of Book XI, composed, as we saw above, of three recessed panels (1-224, 225-467, 468-867) and "a coda of forty-eight lines," each main section divides into three parts; this illustrates the close relationship between tripartite structure and the framework pattern; the focal point in each main division separates the corresponding passages which precede and follow it and thus becomes either the second of the three subdivisions or the central core of the second subdivision. The structure of XI is therefore as follows:

    I. 1-224    Truce and burial of the dead

        (1)  1-99    Mourning for Pallas

        (2)  100-138  Embassy of Latins

        (3)  139-224  Grief of Evander. Burial of dead

    II. 225-467  Council and speeches

        (1)  225-295  Speech of Venulus

        (2)  296-375  Latinus' speech and Drances' reply

        (3)  376-467  Turnus' speech. Renewed attack

    III. 468-867  Cavalry battle and death of Camilla

        (1)  468-497  Preparations for battle

        (2)  498-647  Camilla and the cavalry battle

            (a)  498-531  Camilla enters the conflict

            (b)  532-596  Diana's speech about Camilla

            (c)  597-647  Cavalry battle

        (3)  648-867  Aristeia and death of Camilla

Epilogue 868-915  Transition to Book XII

It is very fitting that an equally elaborate use of tripartite structure appears in Book III and its multiples, VI, IX, and XII; in each book the three main sections divide naturally into three subdivisions, and these in turn are sometimes composed of three still smaller portions. For the sake of brevity and clarity I shall continue to present my analyses in outline form.

Book III

I. 1-191     Aegean area
   (1) 1-68     Departure from Troy. Thrace
   (2) 69-120   Delos
   (3) 121-191  Crete

II. 192-505  Western Greece
   (1) 192-273  Strophades
   (2) 274-293  Actium
   (3) 294-505  Buthrotum (Helenus and Andromache)

III. 506-718 Magna Graecia and Sicily
   (1) 506-547  Journey to Italy
   (2) 548-587  Scylla and Charybdis
   (3) 588-718  Rescue of Achaemenides. Conclusion

The nine episodes of the journey from Troy to Sicily are divided geographically into groups of three, with the third in each group more important and narrated at greater length.[19]

Book VI

I. 1-235     At Cumae
   (1) 1-123    Cumae and the Sibyl
   (2) and (3) 124-155 Instructions concerning Misenus and the Golden Bough
      (2) 156-182   Death of Misenus
      (3)  183-211  The Golden Bough
      (2) 212-235   Burial of Misenus

II. 236-547  Journey to Underworld. Three encounters
      236-267  Preliminary sacrifices (cf. 153-155)
      268-336  First stage of journey
   (1) 337-383  Palinurus
      384-449  Charon, Cerberus, untimely deaths
   (2) 450-476  Dido
      477-493  Warriors
   (3) 494-534  Deiphobus
      535-547  At the entrance to Elysium

III. 548-901 The Underworld proper
   (1) 548-636 Tartarus
   (2) 637-702 Elysium
   (3) 703-755 Lethe[20]

Anchises' forecast of Roman history
   (1) 756-807 Julian line
      (a) 756-776 Alban kings
      (b) 777-787 Romulus
      (c) 788-807 Augustus
   (2) 808-853 Roman line
      (a) 808-818 Roman kings
      (b) 819-846 Roman heroes
      (c) 847-853 Task of the Roman[21]
   (3) 854-901 Marcellus. Conclusion

## Book IX

I. 1-175    DAY    Attack on Trojan camp
   (1) 1-76    Attack on camp
   (2) 77-122    Metamorphosis of the ships
   (3) 123-175    Turnus' speech. Preparations

II. 176-449    NIGHT    Nisus and Euryalus
   (1) 176-313    In the Trojan camp
   (2) 314-366    In the camp of the enemy
   (3) 367-449    Departure and death

III. 450-818    DAY    Battle at the Trojan camp
   (1) 450-589    Fighting
   (2) 590-671    Ascanius-Numanus episode
   (3) 672-818    Turnus inside the camp

## Book XII

I. 1-288    Breaking of the truce[22]
   (1) 1-80    Turnus resolves to fight
      (a) 1-9    Description of Turnus
      (b) 10-53    Turnus and Latinus
      (c) 54-80    Turnus and Amata
   (2) 81-215    Preparations for the truce
      (a) 81-133    Preparations
      (b) 134-160    Juno and Juturna
      (c) 161-215    Vows of Aeneas and Latinus
   (3) 216-288    Breaking of the truce

## STRUCTURAL PATTERNS IN THE BOOKS

II. 289-696 Turnus on the battlefield
    (1) 289-553 Turnus in battle. Parallel aristeias
    (2) 554-611 Assault on Laurentum. Death of Amata
    (3) 614-696 Turnus again resolves to face Aeneas

III. 697-952 Turnus and Aeneas
    (1) 697-790 First encounter
    (2) 791-842 Jupiter and Juno
    (3) 843-952 Death of Turnus

Each of the remaining six books also displays clearly a similar division into three main parts, usually with tripartite subdivisions.

### Book I

I. 1a-222 Prologue. Juno and the storm
    (1) 1a-80 Prologue. Juno and Aeolus
    (2) 81-156 The storm and Neptune
    (3) 157-222 Aeneas and the Trojans in Africa

II. 223-417 The Venus episodes
    (1) 223-296 Venus and Jupiter
    (2) 297-417 Venus and Aeneas

III. 418-756 The Trojans at Carthage
    (1) 418-519 Carthage and Dido
    (2) 520-656 The Trojans welcomed
    (3) 657-747 Cupid and the banquet
        748-756 Epilogue to introduce *Aeneid* II-III[23]

### Book II

I. 1-249 Sinon, Laocoon, and the wooden horse[24]
    (1) 1-39 Wooden horse on shore
    (2) 40-56 Laocoon
    (3) 57-194 Sinon
    (2) 195-227 Laocoon
    (1) 228-249 Wooden horse in city

II. 250-558 Fall of Troy
    (1) 250-369 Return of Greeks
    (2) 370-505 Capture of Troy
    (3) 506-558 Death of Priam

III. 559-804 Aeneas' departure

    (1) 559-633 Aeneas-Venus episode

    (2) 634-729 Aeneas at home

    (3) 730-804 Departure and loss of Creusa

Book IV

I. 1-172 Dido's love and its consummation

    (1) 1-89 Growth of Dido's love

    (2) 90-128 Juno-Venus scene

    (3) 129-172 Hunting scene and "coniugium"

II. 173-449 Aeneas' determination to leave

    (1) 173-278 Fama—Iarbas—Jupiter—Mercury

    (2) 279-415 Narrative

                  Speeches (Dido, Aeneas, Dido)

                  Narrative

    (3) 416-449 Attempted reconciliation fails

III. 450-705 Aeneas' departure and Dido's suicide

    (1) 450-552 Magic rites and Dido's lament

    (2) 553-583 Aeneas' departure

    (3) 584-705 Dido's curses and suicide

The symmetrical arrangement of the final division around the Aeneas scene in 553-583 as a focal point was discussed above;[25] in the second main division the speech of Aeneas in 333-361, framed by the two Dido speeches, is not only the center of the second subdivision but provides the focal point for the book as a whole. *Aeneid* IV, like XI, illustrates well the manner in which the tripartite divisions of the books may be combined in a symmetrical framework with the recessed panel structure. For this reason the main sections as given above seem preferable to those of Mackail, who divides as follows: 1-295, 296-503, 504-705.[26] Each of Mackail's three parts begins with *at regina;* these words in 296 and 504 do not mark the beginning of new divisions but rather provide a transition within the divisions from Aeneas (295) and Anna (503) to Dido. MacKendrick's recent analysis of IV as a tragedy in five acts plus a bipartite "central core" (173-218, 219-295) dealing with Rumor and Jupiter's command to Aeneas is likewise unconvincing.[27]

Book VII

I. 1-285 Arrival in Latium and welcome by Latinus

    (1) 1-106 Arrival. Latinus and omens

    (2) 107-169 Eating of tables. Embassy

    (3) 170-285 Reception of embassy. Three speeches

        (a) 192-211 Latinus

        (b) 212-248 Ilioneus

        (c) 249-285 Latinus

II. 286-460 Juno, Allecto, and outbreak of war
- (1) 286-322 Juno's lament
- (2) 323-539 The Allecto episode
  - 323-340 Juno summons Allecto
  - 341-539 Threefold activity of Allecto
    - (a) 341-405 Maddens Amata
    - (b) 406-474 Maddens Turnus
    - (c) 475-539 Maddens hounds of Ascanius
- (3) 540-640 Preparations for war

III. 641-817 Catalogue of Latin warriors

Three groups of three less important warriors each, enclosed by important leaders:

(a) Mezentius and Lausus, (b) Messapus, (c) Turnus and Camilla

## Book VIII

I. 1-369   Aeneas and Evander

1-101   Aeneas at his camp

102-369   DAY   Aeneas at Rome

- (1) 102-183 Welcome by Evander
- (2) 184-305 Festival of Hercules
- (3) 306-369 The site of Rome

II. 370-453   NIGHT   Venus and Vulcan. Making of armor

III. 454-731   DAY   Journey and the shield

- (1) 454-596 Departure from Pallanteum[28]
- (2) 597-625 Venus brings the armor
- (3) 626-728 Description of the shield

729-731 Conclusion

## Book X

I. 1-361   Return of Aeneas

- (1) 1-117   Council of the gods
  - (a) 1-17   Speech of Jupiter
  - (b)   18-62a Speech of Venus
  - (c)   62b-95 Speech of Juno
  - (a) 96-117 Speech of Jupiter
- (2) 118-255 Return of Aeneas. Catalogue of ships
- (3) 256-361 Landing and battle

II. 362-688  Death of Pallas

      (1) 362-478  Aristeia of Pallas

      (2) 479-605  Death of Pallas and effect on Aeneas

      (3) 606-688  Removal of Turnus from battle

   III. 689-908  Death of Lausus and Mezentius

      (1) 689-746  Aristeia of Lausus

      (2) 747-832  Death of Lausus

      (3) 833-908  Death of Mezentius

My purpose in presenting these analyses has been to show how the patterns of Vergil's thought and his development of the narrative produce so regularly a tripartite structure in both the larger and the smaller sections of each book. The main divisions are always of unequal length. Belling, interested in discovering a balanced symmetry in various books, attempts to reallocate the episodes in an endeavor to procure two equal halves; e.g., he divides Book II into three parts, as above (1-249, 250-558, 559-804), and then recombines the episodes, adding 250-401 to the first section and 402-558 to the third; he thus shows (by rejecting four verses) that the book falls into two halves of 400 verses each.[29] Such a procedure is both arbitrary and unconvincing.

We have already seen that the central section of the second main division of *Aeneid* IV, containing the speeches of Dido and Aeneas, has great significance for the book as a whole and for the entire poem. In the other books likewise Vergil uses his tripartite structure to emphasize certain important aspects of his combined Trojan and Roman themes. I list below the topic in the central position of the second main division of each book, or that in the second division as a whole, when it does not lend itself to a tripartite subdivision.

| | | |
|---|---|---|
| I. | (main division) | |
| | 223-417 | Venus and Jupiter (Rome and Augustus) |
| | | Venus and Aeneas (Carthage and Aeneas) |
| | | [Augustus-Aeneas relationship stressed?] |
| II. | 370-505 | Capture of Troy |
| III. | 274-293 | Actium (Roman theme) |
| IV. | 279-415 | Recessed panel, with Aeneas' defense (333-361) as focal point. Decisive for the poem as a whole |
| V. | (main division) | |
| | 545-603 | *Ludus Troiae* (Roman theme) |
| VI. | 450-476 | Meeting with Dido (conclusion of the Dido story) |
| VII. | 323-539 | Threefold activity of Allecto (important for the war in VII-XII and for the character of Turnus) |
| VIII. | (main division) | |
| | 370-453 | Venus, Vulcan, and making of armor |
| IX. | 314-366 | Nisus and Euryalus in camp of enemy (mistakes that caused their deaths) |
| X. | 479-605 | Death of Pallas. Not only the focal point of the book but decisive for the fate of Turnus; cf. XII, 940 ff. |

XI. 296-375            Latinus' speech and Drances' reply (cf. 355 f.: wedding of Aeneas and Lavinia necessary for peace)

XII. 554-611            Attack on Laurentum and death of Amata (important for effect on Turnus)

These key passages, several of them very significant for our understanding of the *Aeneid*, receive added emphasis from the central position in each book which results from Vergil's arrangement of his material into divisions and subdivisions of three.

In making the tripartite divisions of the books as outlined above, I had no thought of any additional type of symmetry or proportion. However, the fact that the main divisions of the books were of such uneven length suggested the possibility that these inequalities might be intentional. Since the Golden Section had already been found in both the *Eclogues* and the *Georgics*, it occurred to me that the same ratio might exist within the books of the *Aeneid*. It was then that I made the surprising discovery (surprising to me, at any rate, as I had no idea what would develop) that at least two of the three main divisions in each book, and sometimes all three, were in the approximate Golden Mean ratio. Additional investigation revealed that within each main division were smaller units (usually the tripartite subdivisions listed above), composed of others still smaller, all in the same approximate proportion. The almost incredible extent to which these proportions appear throughout the *Aeneid* will be presented in Chapter 4 and in Tables I-VII.

# Notes To Chapter 2

1. E.g., the balanced activity of Aeneas and Turnus in XII, 505-520, 529-547, is particularly effective with the rapid shift from the one warrior to the other.

2. Cf. X, 360 f., 431-433a, 755-757; XII, 548-553.

3. Cf. Duckworth, *Foreshadowing and Suspense*, pp. 63, 77.

4. I follow here the arrangement suggested by Hahn, *TAPhA* 63 (1932), pp. lxii f.; see also Brotherton, *TAPhA* 62 (1931), pp. 192-202, who sees in the catalogue twelve groups of forces with each group in the first six having a parallel in the last six and in reverse order (with the exception of Camilla who, for special reasons, concludes the list). Miss Hahn's brief criticisms of this structural parallelism seem valid, but do not preclude the presence of two different patterns in the same passage.

5. It should also be noted that, if we exclude from the list the important warriors (including Messapus), the remaining ten names are in alphabetical order; this was pointed out by Cook, *CR* 33 (1919), pp. 103 f.

6. Cf. Murley, *TAPhA* 68 (1937), pp. 305-317.

7. For an analysis of the structure of the *Eclogues* and the *Georgics*, see Richardson, *Poetical Theory*, pp. 100-163; for the pattern of the Aristaeus episode in *Georgics* IV, see also Saint-Denis, *Virgile, Géorgiques*, p. xxxix, and, for a slightly different analysis of the Orpheus and Eurydice story, cf. Norwood, *CJ* 36 (1940-41), pp. 354 f. (a concentric pattern around 481-503).

8. *YClS* 12 (1951), pp. 222 ff.; cf. also Lloyd, *AJPh* 78 (1957), pp. 397 f., on the structure of *Aen.* III, 568-683.

9. *TAPhA* 68 (1937), p. 308.

10. *Homer and the Heroic Tradition*, pp. 249 ff.; cf. p. 255: "books balance books and scenes balance scenes by similarity or antithesis, with the most amazing virtuosity."

11. *The Aeneid*, p. 417.

12. Cf. Duckworth, *Foreshadowing and Suspense*, pp. 112-115. The Sibyl's second speech (125-155) contains instructions limited to the action of Book VI.

13. *YClS* 12 (1951), p. 222.

14. For this analysis of *Aeneid* IV, I am indebted to Pease, *Publi Vergili Maronis Aeneidos Liber Quartus*, p. 30.

15. Dido is, of course, a very different and a far more tragic person than Apollonius' charming but immature maiden; it is most inaccurate to say, as does Page, *The Aeneid of Virgil*, p. xiii, note 3, that large portions of the *Aeneid* are "copied from the Argonautica." Actually, the resemblances to Apollonius' poem are in purely external and superficial details; cf. Fenik, *The Influence of Euripides on Vergil's Aeneid*, pp. 168 ff. For Vergil's indebtedness to Euripides' *Alcestis* and *Hippolytus*, see Fenik, pp. 32 ff., 151 ff.

16. It is also possible to view the Nisus and Euryalus story as a miniature tragedy in five acts: I, on the wall (176-223); II, at the Trojan council (224-313); III, in the camp of the enemy (314-366); IV, capture and death (367-449); V, lament of the mother of Euryalus (450-502). But the episode proper concludes with the apostrophe in 446-449, and the mother's lament seems rather an epilogue. Mendell, *YClS* 12 (1951), pp. 216 f., suggests that 459-476, as a framework passage to balance 159-175, should follow the grief and lament of Euryalus' mother, that the verses are "sketch notes" left by Vergil in the margin and wrongly inserted in their present position by the editors. This suggestion seems implausible; the mother's lament in 485 ff. implies knowledge of 465 f.

17. See Chapter 1, note 50.

18. In this respect they are very unlike *Georgics* III, which has a symmetrical division into two exact halves: 1-283 (on horses and cattle), and 284-566 (on sheep and goats), i.e., 283 + 283. *Georgics* IV divides into two almost equal halves: 1-280 (description and care of the bees), and 281-558 (the story of Aristaeus), i.e., 280 + 278. The final passage, 559-566, is the conclusion to the poem as a whole.

19. See Lloyd, *AJPh* 78 (1957), pp. 136 ff. Lloyd points out (p. 138, note 23) the similarity to the grouping of adventures into units of three in *Odyssey* IX, X, and XII. Büchner, *P. Vergilius Maro*, col. 336, with less plausibility views the journey to Italy and Scylla and Charybdis as one episode (506-569) and lists Drepanum (707-715) as the ninth and final episode. In the second main division most editors begin a new paragraph after 277, but 274 seems preferable as the beginning of the second subdivision, on Actium; cf. *Leucatae montis* (274), *Apollo* (275), *parvae urbi* (276) [=Actium]; see below, p. 89.

20. See Norwood, *CPh* 49 (1954), pp. 15-26, especially 19 f.; cf. also MacKay, *TAPhA* 86 (1955), pp. 180-189.

21. On the tripartite divisions in 760-853 and their relation to the Roman Odes of Horace, see Duckworth, *TAPhA* 87 (1956), pp. 304-308 and especially note 79.

22. Hirtzel begins a new paragraph after 286, but Fairclough, Mackail, and Durand seem correct in making the division after 288. See below, Table XIII: A Guide to Paragraphing.

23. Belling, *Studien über die Compositionskunst Vergils,* p. 173, looks upon 723 ff. as the introduction to *Aeneid* II-III; on the contrary, lines 723-747 provide the conclusion to the banquet.

24. The three parts of this first section form a recessed panel pattern (abcba) with the episode of Sinon as the focal point.

25. Mendell, *YClS* 12 (1951), p. 222, gives the focal point as 554-583, but I prefer, with Hirtzel and Mackail, to begin a new paragraph with 553; see below, Table XIII: A Guide to Paragraphing.

26. *The Aeneid,* p. 129

27. *CJ* 54 (1958-59), p. 199. MacKendrick's first three acts are the three parts of my first main division; his central core of Fama, Iarbas, Jupiter, and Mercury is the first section of my second main division; the remainder of the second division is his Act IV, and his Act V is my third main division. The tripartite arrangement seems preferable for three reasons: (1) the other books likewise fall naturally into three main divisions, usually with three subdivisions each; (2) his first three acts seem too short, and the book as a whole is thrown out of balance; and (3) according to MacKendrick, Aeneas' speech in 333-361 and the two speeches of Dido enclosing it form part of Act IV; I am convinced that Vergil designed these speeches for the very center of the book; cf. the recessed panel structure of 276-415 given above.

28. Most editors fail to indicate a new paragraph after 596, the end of the first subdivision. Lines 597 ff. describe the arrival near Caere, many miles from the site of Rome; see below, p. 89; cf. also Duckworth, *TAPhA* 91 (1960), p. 199, note 30.

29. Cf. Belling, *Studien über die Compositionskunst Vergils,* pp. 175-198, especially p. 195. In *Aeneid* I he finds two main divisions of 303 verses each, but these do not include 305-417 or 723-756 and likewise require the rejection of four verses (p. 174).

# Chapter 3

# THE GOLDEN SECTION IN THE ECLOGUES AND THE GEORGICS

BEFORE WE DISCUSS the Golden Mean ratios in the *Aeneid* and their value for the text of the poem and for Vergilian studies in general, we must turn to the earlier works where the same proportion appears. Chapter 3 describes the discovery of Vergil's use of the Golden Section in *Georgics* I, explains the nature of the ratio, and gives typical illustrations of the proportions which have been found both in the *Eclogues* and elsewhere in the *Georgics*.

## LE GRELLE AND GEORGICS I

Vergil's indebtedness to Hesiod is more evident in *Georgics* I than in the other three books of the poem. The first book contains both "Works" (43-203) and "Days" (259-463a), separated by a passage on astronomy (204-258). Many writers have seen a twofold division in the book, but the astronomical passage is so subtly blended both with what precedes and with what follows that they are uncertain whether to make the division between the two parts after 203, or 230, or 258.[1] There is clearly no division of the book into two equal halves, as occurs in both *Georgics* III and IV.[2] Some scholars are frankly puzzled by the structure of the book; it begins and ends with outstanding passages: 5b-42, invocation to the gods and praise of Octavian as a divinity, and 463b-514, the prodigies after Caesar's death and the desire of the people for Octavian as a savior, and these give the book a unified framework of historical and national significance, but the major portion, most of it technical, seems to lack systematic arrangement. Rand accepts the division into a "Works" (43-203) and a "Days" (204- "ending it is hard to say just where"), but suggests that Vergil is writing in "the disjointed manner of the gnomic sections in Hesiod";[3] Perret, who finds no arrangement or plan to the book but rather "une pulvérisation systématique de l'énoncé," likewise sees therein the naive style of Hesiod.[4]

The explanation is that Vergil, in composing *Georgics* I, has attempted a different kind of poetic composition, relying not on balanced parts with parallels and contrasts but on mathematical proportion and harmony of structure. This was first shown in 1949 by Guy Le Grelle,[5] who analyzed the entire book on the basis of the Golden Mean ratio, according to which the greater part is to the lesser as the whole is to the greater; i.e., with M and m denoting the major and minor parts respectively:

$$M/m = (M + m)/M,$$

the quotient, computed to three decimal places, is 1.618. Likewise, if the smaller part is divided by the larger:

$$m/M = M/(M + m),$$

the quotient to three decimal places is .618. Le Grelle accepts the division of *Georgics* I into "Works" (43-203) and "Days" (204-463a); these parts as minor and major reveal the Golden Section, i.e.,

$$m/M = 161/259.5 = .620;$$

$$M/(M + m) = 259.5/420.5 = .617.[6]$$

The prologue (5b-42) and the epilogue (463b-514) combine to form a major of 89 lines, which, with the astronomical passage ("foyer astronomique," 204-258) as the minor part, produces the exact Golden Mean ratio, as shown on the following page:

$$55/89 = 89/144 = .618$$

We have here a tripartite framework design, with a and c enclosing b, in the pattern b/(a + c). Le Grelle breaks up *Georgics* I into numerous subdivisions, which he terms "chrysodes," and in each of these also he finds a major and a minor part which are in proportion.[7] To cite two instances: 100-117 (18 = m) and 118-146 (29 = M) produce the ratio .617 (29/47), and these two passages combine to form the major of a larger proportion. [I use the term "proportion" not only for the ratio but also for the passage containing the major and minor parts.] Lines 100-146 (47 = M) and 147-175 (29 = m) give the exact ratio .618 (47/76).

Le Grelle believes that the appearance of the Golden Section throughout *Georgics* I cannot be the result of accident, but reveals Vergil's interest in mathematical proportion and harmonious structure.[8] Le Grelle's results are of course valid only if we accept his divisions into major and minor parts. Some scholars, for instance, begin the epilogue at 461 or 466 instead of 463b, but Le Grelle seems correct here; the phrase *sol tibi signa dabit* (463a) marks the conclusion of the passage beginning with *sol . . . signa dabit* (438 f.). Saint-Denis, criticizing Le Grelle's division at 463a, continues the main portion of the book through 497;[9] this is inaccurate, for the many portents connected with the death of Caesar are an integral part of the epilogue; 463b-514 form a unit, and as such the passage was imitated by Horace in *Odes* I, 2.[10]

Actually, there are more examples of the Golden Section in *Georgics* I than Le Grelle lists. The descriptions of the moon (424-437) and the sun (438-460, omitting the conclusion 461-463a) are in proportion; $M/(M+m) = 23/37 = .622$. The first part of the "foyer astronomique" (204-230) is a major, with 187-203 as minor; $27/44 = .614$; the second part of the same astronomical passage (231-258) is a major, with 259-275 as minor; $28/45 = .622$. These ratios prove how closely 204-258 are linked to the passages which precede and follow, in structure as well as in content. Also, Le Grelle in his analysis of the "chrysodes" has only the a/b or b/a types; a restudy of *Georgics* I in the light of the tripartite patterns, which are so frequent in the *Aeneid* and which I have found also in the *Eclogues* and the *Georgics*, might be fruitful in its results.

## THE GOLDEN MEAN RATIO

By discovering the Golden Section in *Georgics* I, Le Grelle was the first to show that Vergil used mathematical proportion as a basic method of poetic composition; he rightly reminds us that ancient poetry was meant to be heard and that the appeal of such harmony and proportion was to the ear and the imagination, as in the case of music.[11] The fact that many hearers (or readers) would not be conscious of the proportions is no argument against their existence. Homer provides us with an interesting parallel; as Whitman points out, a Homeric audience would hardly appreciate the mathematical symmetry of the *Iliad*, e.g., the grouping of days in I (1 - 9 - 1 - 12) and in XXIV (12 - 1 - 9 - 1), or the elaborate correspondence of the parts "in which episodes, and even whole books, balance each other through similarity or opposition."[12] A great poet or artist or musician always puts more into a work than is ordinarily realized.

The Golden Section is important not only in mathematics but in art and architecture as well.[13] The two ratios, 1.618 and .618, added together total 2.236, which is the square root of five; in other words, $1.618 = \frac{1}{2}(\sqrt{5} + 1)$ and $.618 = \frac{1}{2}(\sqrt{5} - 1)$. Weyl says of $\frac{1}{2}(\sqrt{5} - 1)$: "This number is no other but the ratio known as the *aurea sectio*, which has played such a role in attempts to reduce beauty of proportion to a mathematical formula."[14]

It may help the reader, as it has been of assistance to me, to realize that the ratios of the Golden Section (1.618 and .618) are achieved by a very simple mathematical series, known as the Fibonacci series, with each number merely the sum of its two predecessors, i.e., 1, 1, 2, 3, 5, 8, 13, 21, 34, 55, etc. Thompson, after describing this particular series, writes as follows:

The Golden Mean itself is only the numerical equivalent, the 'arithmetization', of Euclid II. 11; where we are shown how to divide a line in 'extreme and mean ratio', as a preliminary to the construction of a regular pentagon: that again being the half-way house to the final triumph, perhaps the ultimate aim, of Euclidian or Pythagorean geometry, the construction of the regular dodecahedron, Plato's symbol of the Cosmos itself . . . . The Golden Mean series is a very curious one; . . . . For the fact is, we may begin it as we please, with 1, 1, or 1, 2, or 1, 3, or *any two numbers* whatsoever, whole or fractional, and in the end it comes always to the same thing! For instance, we may have the series 1, 5, 6, 11, 17, 28, 45, 73, 118, 191, 309 &c., which only agrees with the former in that each number is the sum of its two predecessors: but as before, the fractions soon approximate closely to the Golden Mean; 191/309 = 0.61812 . . . ; and (as a consequence) 309/191 = 1.618 . . . approximately.[15]

This second series described by Thompson (1, 5, 6, 11, 17, etc.) needs more stages to achieve the Golden Section and also involves much higher numbers than does the first (1, 1, 2, 3, 5, 8, 13, etc.). This will be readily apparent if I express each series in the form of fractions with the corresponding decimals approaching .618 (Cols. I and IV). The series beginning 1, 1, 2, 3, 5 and 1, 2, 3, 5 are identical except that the second requires one less stage to reach .618. I shall list, also, for purposes of comparison, the series beginning 1, 3, 4, 7, 11, etc., 1, 4, 5, 9, 14, etc., and 1, 6, 7, 13, 20, etc. (Cols. II, III, and V):

| I | | II | | III | | IV | | V | |
|---|---|---|---|---|---|---|---|---|---|
| 1/1 | 1.0 | | | | | | | | |
| 1/2 | .50 | 1/3 | .333 | 1/4 | .250 | 1/5 | .20 | 1/6 | .166 |
| 2/3 | .666 | 3/4 | .750 | 4/5 | .80 | 5/6 | .833 | 6/7 | .857 |
| 3/5 | .60 | 4/7 | .571 | 5/9 | .555 | 6/11 | .545 | 7/13 | .538 |
| 5/8 | .625 | 7/11 | .636 | 9/14 | .643 | 11/17 | .647 | 13/20 | .650 |
| 8/13 | .615 | 11/18 | .611 | 14/23 | .609 | 17/28 | .607 | 20/33 | .606 |
| 13/21 | .619 | 18/29 | .621 | 23/37 | .622 | 28/45 | .622 | 33/53 | .623 |
| 21/34 | .618 | 29/47 | .617 | 37/60 | .617 | 45/73 | .616 | 53/86 | .616 |
| | | 47/76 | .618 | 60/97 | .619 | 73/118 | .619 | 86/139 | .619 |
| | | | | 97/157 | .618 | 118/191 | .618 | 139/225 | .618 |

In Cols. III ff. not only are more stages necessary to reach .618 than in Cols. I and II, but the total numbers involved become steadily greater.

The relationship of these other series to the Fibonacci series is very close. This will be clearly seen if to multiples of the Fibonacci series we add another Fibonacci series, as follows:

```
           3 (1,  1,  2,  3,  5,  8, 13, . . . )
       =      3,  3,  6,  9, 15, 24, 39, . . .
       +          1,  1,  2,  3,  5,  8, . . .
              ─────────────────────────────────
          (1,) 3,  4,  7, 11, 18, 29, 47, . . .
```

or

```
           4 (1,  1,  2,  3,  5,  8, 13, . . . )
       =      4,  4,  8, 12, 20, 32, 52, . . .
       +          1,  1,  2,  3,  5,  8, . . .
              ─────────────────────────────────
          (1,) 4,  5,  9, 14, 23, 37, 60, . . .
```

or

```
           6 (1,  1,  2,  3,  5,  8, 13, . . . )
       =      6,  6, 12, 18, 30, 48, 78, . . .
       +          1,  1,  2,  3,  5,  8, . . .
              ─────────────────────────────────
          (1,) 6,  7, 13, 20, 33, 53, 86, . . .
```

Similarly,

$$2 (1, 2, 3, 5, 8, 13, 21, \ldots)$$
$$= \quad 2, 4, 6, 10, 16, 26, 42, \ldots$$
$$+ \quad\quad 1, 1, 2, 3, 5, 8, \ldots$$
$$\overline{\quad\quad 2, 5, 7, 12, 19, 31, 50, \ldots}$$

and

$$3 (1, 2, 3, 5, 8, 13, 21, \ldots)$$
$$= \quad 3, 6, 9, 15, 24, 39, 63, \ldots$$
$$+ \quad\quad 1, 1, 2, 3, 5, 8, \ldots$$
$$\overline{\quad\quad 3, 7, 10, 17, 27, 44, 71, \ldots}$$

Thus the different Golden Mean series listed above in Cols. II-V, as well as similar series (2, 5, 7, 12, 19, ...; 3, 7, 10, 17, 27, ...; 4, 9, 13, 22, 35, ...; etc.), are merely variations of the basic Fibonacci numbers.

Let us now return to the Golden Mean ratios in *Georgics* I as given above. The prologue (5b-42) and epilogue (463b-514) are the major of 89 lines, and the astronomical passage (204-258) is the minor of 55 lines, i.e., 55/89 = 89/144 = .618; this is the continuation of the Fibonacci series in Col. I: 21, 34, 55, 89, 144. The ratios in 100-146 (29/47 = .617) and in 100-175 (47/76 = .618) are built on the series in Col. II, beginning 1, 3, 4, 7, 11, etc. The proportions which I suggested adding to *Georgics* I likewise reveal numbers in additive series of this same type. In 424-460 we find 23/37 = .622; these numbers appear in the series 1, 4, 5, 9, 14, etc. (Col. III); in 231-275 we have 28/45 = .622; cf. the series 1, 5, 6, 11, 17, etc. (Col. IV); in 187-230 we have 27/44 = .614, and these two numbers belong to the series beginning 3, 7, 10, 17, 27, 44, etc.; to attain the exact .618 we must continue to 115/186. It is an interesting fact that the proportions in *Georgics* I reveal the numbers of the Fibonacci and other Golden Mean series. We shall find that these same series appear in the *Eclogues* and elsewhere in the *Georgics* and also, to an even more striking degree, in the *Aeneid;* see my "Summary of the Proportions" in Chapter 4.

Throughout the remainder of the book I shall refer frequently, as I have already done, to "the exact Golden Mean ratio" or to "the perfect .618." I realize, of course, that there is no such thing; $\frac{1}{2}(\sqrt{5} - 1)$ is an irrational number which approaches .618034. Since I compute the ratios only to three decimal places, I use the terms "exact .618" and "perfect .618" to distinguish the more exact ratio from approximate ratios such as .610 or .625.

## PROPORTIONS IN THE *ECLOGUES*

The recessed panel construction of *Eclogue* I around a focal point was discussed above in Chapter 2. The central core (40-45) and the two passages which frame it (27-39, 46-58) total 32 verses; this is the minor, in relation to the balanced passages at the beginning and the end of the poem (1-26 and 59-83) which provide the major of 51 verses; M/(M + m) = 51/83 = .614; we have here a framework pattern, b/(a + c). *Eclogue* II is similarly constructed around a focal point, the invitations to Alexis and the gifts offered (28-55); this passage is framed by 6-27, in which Corydon reproaches Alexis, and by 56-68, in which Corydon reproaches himself, and these in turn by 1-5 (introduction) and 69-73 (conclusion). The central core of 28 verses provides the minor part; the remainder, totaling 45 lines, is the major, and here again we have the Golden Mean ratio in the same tripartite arrangement b/(a + c); 45/73 = .616. In both I and II Vergil has combined his concentric panel patterns with the Golden Section in a striking fashion. In *Eclogue* X we find a proportion of a simpler type (a/b); the minor is 1-30 (introduction, nymphs, shepherds, Apollo, Pan), the major 31-77 (song of Gallus, conclusion); 47/77 = .610.

*Eclogue* VI is even more interesting; if we exclude the first twelve verses, the dedication to Varus, the entire poem is constructed on a Golden Mean series.[16] We begin with the short song on cosmology (31-40), which in thought, sentence structure, and meter divides as 4, 2, 4;[17] the central number 2 and the framing passages of 8 verses, totaling 10, provide the beginning of an additive series: 2, 8, 10, 18, 28, 46, 74, leading to the Golden Section. The preliminary description of Silenus (13-30) is 18 lines in length, and this, added to the central core, totals 28; the mythological songs of Silenus (41-86) comprise 46 verses, and the poem as a whole (exclusive of the dedication) contains 74 lines; in other words, $M/(M+m) = 46/74 = .622$. The numbers in this poem are merely two times 1, 4, 5, 9, 14, 23, 37, the series listed above in Col. III. To reach the exact ratio .618, we must continue the series to 97/157.

I described in Chapter 1 the structure of the *Eclogues*, both the reverse parallelism of I-IV and VI-IX about V as the central poem (with X, the poet as a shepherd, added to balance V, the shepherd-poet as a god), and also the arrangement into triads with the central triad (IV-VI) on more cosmic themes and X combining the themes of all three triads. Golden Mean ratios appear in the collection as a whole, with the corresponding poems grouped into major and minor units, as follows:

$$(I + II)/(VIII + IX + X) = m\ (156)/M\ (253); \quad 253/409 = .619;$$
$$VII/III = m\ (70)/M\ (111); \quad 111/181 = .613;$$
$$V/(IV + VI) = m\ (90)/M\ (149); \quad 149/239 = .623.^{18}$$

The existence of these proportions in the *Eclogues*, especially in the two corresponding responsive songs III and VII and in IV-VI, where IV ("the world to come") and VI ("the world that was") enclose V (the death and deification of Daphnis), raises important questions concerning the composition and chronology of the poems; the usual view that X is a later addition to the group of nine pastorals constructed in a framework pattern about V must perhaps be revised, since X fits into so perfect a ratio with VIII and IX in relation to I and II. Were the ten *Eclogues* composed, or at least planned as a unit, or did Vergil do considerable rewriting and mathematical adjustment when X was added to the collection?

The relation between the major and minor poems may be diagramed as follows:

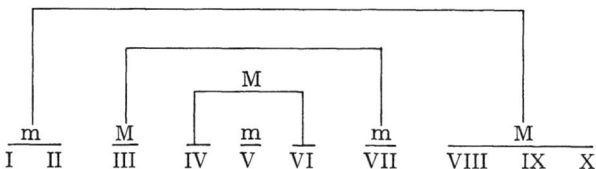

When we turn to Hahn's arrangement of the *Eclogues* in triads, the Golden Section is likewise present. The ratio in IV-VI, given above, places additional emphasis on the central and most significant triad. Also, the first two triads (I-III, IV-VI) form the major (506 lines) and the final triad plus X the minor (323); $506/829 = .610$, and the pattern is tripartite, $c/(a+b)$:

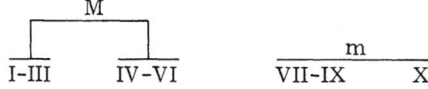

The proportions in the *Eclogues* thus support both the framework structure with Daphnis (V) as the focal poem and the triadic arrangement of the collection, and they illustrate the complexity of Vergil's compositional patterns.[19]

## PROPORTIONS IN THE *GEORGICS*

I was originally of the opinion that the astronomical passage in *Georgics* I, 204-258, which Le Grelle terms "le foyer astronomique" and which is so important for his analysis of the proportions in the book, was perhaps the explanation of Vergil's use of the Golden Section in *Georgics* I; this could be supported by the central position of the song on cosmology in *Eclogue* VI, where Brown has detected a Golden Mean series. The presence of the Golden Section elsewhere in the *Georgics* as well as in the *Eclogues* argues against this view.

It was purely by chance that I found the Golden Section in the *Georgics* as a whole; in 1958 I happened to read Wili's statement that less than half of the *Georgics* is composed of didactic material, the greater part consisting of descriptive passages, of "Bild und Reflexion,"[20] and curiosity led me to check his assertion; Wili is quite wrong, for the technical sections on farming have a much greater extent than the numerous passages which praise Italy and country life and present themes of national or philosophical interest, the ratio being 1352 verses to 835.[21] The total number of verses in the *Georgics* is 2187. It was amazing to find here the exact Golden Mean ratio: $835/1352 = 1352/2187 = .618$.[22] This was my first discovery of the Divine Proportion in any of Vergil's works, and it removed any lingering doubts concerning the soundness and importance of Le Grelle's theory about the structure of *Georgics* I. The fact that Vergil employed the Golden Section in the relation of didactic to descriptive material throughout all four books of the *Georgics* (this cannot be accidental!) gives strong support not only to Le Grelle's work but also to the ratios which Brown and I have detected in the *Eclogues,* my own findings there being subsequent to the discovery of the exact ratio in the *Georgics* as a whole.

Also, the presence of this proportion in the four books of the *Georgics* suggested a tentative look at Books II-IV for additional examples of the Golden Section. The following are the ones which I have noted (I did not examine the books in their entirety), and it is interesting to see how frequently the ratio appears in the famous descriptive passages:

(1) The praise of country life in II, 458-540 (four parts in a chiastic arrangement): country life (458-474) and the poet's ambition (475-489) as minor (32 lines); the *felix-fortunatus* passage (490-494) and the advantages of country life with the new Golden Age in Italy (495-540) as major (51 lines); $51/83 = .614$.

(2-3) Vergil's temple of song in III, 1-48, to honor Octavian; this is tripartite: (a) 1-15 (15), personal; (b) 16-39 (24), Caesar's temple; (c) 40-48 (9), personal; the ratio is double, in the patterns a/b and c/a; $24/39 = .615$ and $15/24 = .625$.

(4-5) Again we have a double ratio; III, 295-321, on the care of flocks in winter, is the major (27) in relation to 322-338, on the care of flocks in summer (17 = m); $27/44 = .614$. Lines 295-321 become the minor in relation to the description of the Libyans and the Scythians in 339-383 (45 = M); $45/72 = .625$. If we designate the three consecutive passages as a, b, and c, we have the patterns b/a and a/c. This relationship is interesting in view of Vergil's careful integration of the descriptions of Libya and Scythia in a chiastic order with the preceding passages on winter and summer.[23]

(6) The plague in III, 478-566, which divides into three stages: 479-502, 503-536, and 537-566 (the third stage beginning with the reversal of nature—almost a Golden Age in reverse); the ratio is tripartite in a framework pattern b/(a + c), with the first and third stages (a total of 55 lines = M) enclosing the second stage (34 = m); $34/55 = 55/89 = .618$, a perfect Golden Section in the Fibonacci series 34, 55, 89.[24]

(7) The Aristaeus and Orpheus stories in IV, 281-558; this is tripartite: (a) Aristaeus, 281-452 (171) as M; (b) Orpheus and Eurydice, 453-529; and (c) the restoration of the bees, 530-558, as m (77 + 29 = 106); the pattern is (b + c)/a, and the ratio is $171/277 = .617$.[25]

(8) IV, 281-452 (part a of the above) also contains a Golden Mean ratio. The major consists of (a) 281-332 (loss of bees and lament of Aristaeus) and (b) 333-386 (Cyrene and Aristaeus), a total of 105 lines (omitting verse 338); the minor is 387-452 (Cyrene sends Aristaeus to Proteus), or 66 lines; $105/171 = .614$, the type being c/(a +b).

(9) The Orpheus and Eurydice story in IV, 453-527; again the pattern is tripartite, c/(a +b), with the death of Eurydice (453-466 = 14) and Orpheus in the Underworld (467-498 = 32) as the major (46 lines), and the lament and death of Orpheus (499-527) the minor (29); 46/75 = .613.[26]

(10) With the speech of Cyrene (IV, 530-547) as major (18) and the restoration of the bees (548-558) as minor (11), we have a b/a pattern; 18/29 = .621.

It is amazing to find in these ten ratios a range from .613 to .625, i.e., only .007 from the perfect .618 or a variation of 1.13 per cent.

*Georgics* II as a whole, like Book I, is difficult to analyze; many find in it two main divisions, but they do not agree whether the second part begins at 259 or 315, while others see a threefold arrangement (1-176, 177-345, 346-542).[27] Richardson detects several units, each with a recessed center pattern.[28] There are two reasons for the difficulty: Vergil has avoided the clear-cut division into two parts which we find in III and IV by refusing to separate his treatment of trees and vines,[29] and enthusiastic descriptive passages are unusually prominent: the glorification of Italy (136-176), the description of springtime (319-345), the ancient festival in honor of Bacchus (380-396), and the epilogue (458-540) praising country life as a time of peace, as a new Golden Age devoted to work. Perret looks upon the apparent disorder in II as intentional and considers it the result of "lyrisme," of Vergil's desire to enrich the technical details of the book with passages of a lyrical nature.[30]

Here too, as in *Georgics* I, the solution of the structure probably lies in Vergil's use of the Golden Mean ratio. In II lines 1-8 and 541 f. are the prologue and conclusion respectively; the book proper falls into three parts: (a) 9-258 (250), devoted chiefly to trees; (b) 259-457 (199), mostly on vines; and (c) 458-540 (83), the praise of country life. We have here a framework pattern with a and c forming the major part of 333 verses and enclosing b, i.e., b/(a + c); the ratio is 333/532 = .626.[31] However, if we include the eight-line prologue and the two-line conclusion in our calculations, we have the exact Golden Mean ratio. The prologue, addressed to Bacchus, is properly reckoned with 259-407, "mostly on vines."[32] We thus have a minor of 207 verses (8 + 199) and a major of 335 (333 + 2); 335/542 = .618.[33]

The Donatus-Suetonius *Vita* (15) gives us the following important information about Vergil: "inter cetera studia medicinae quoque ac maxime mathematicae operam dedit." This especial interest in mathematics is more than confirmed by the poet's use of the Golden Section in both the *Eclogues* and the *Georgics*.

# Notes to Chapter 3

1. Büchner, *P. Vergilius Maro,* col. 252, divides the book into three parts, 1-203, 204-350, 351-514. But the passage beginning with 351 is a resumption, after the advice in 338-350 to worship Ceres, of the earlier treatment of days and weather (259 ff.).

2. See above, Chapter 2, note 18.

3. *The Magical Art of Virgil,* p. 205.

4. *Virgile,* pp. 66 f.

5. *LEC* 17 (1949), pp. 139-235.

6. Le Grelle regularly divides the total of major and minor by 1.618 to derive the major part; i.e., 420.5/1.618 = 259.88, a variation of a fraction of a line from the actual major (204-463a = 259.5). My procedure throughout is to divide the major section by the total lines in the passage in order to derive the Golden Mean ratio, in this case .617, a variation of .001 from the perfect proportion. The range of the proportions in the area of .618 is thus more easily discerned. I shall omit henceforth the ratios resulting from the division of the minor part by the major; this ratio of m/M can always be deduced from the slightly more accurate M/(M + m); see below, Chapter 4, note 7. Also, in the ratio cited above, I have used Le Grelle's figure of 259.5 for the "Days" (204-463a), but the first part of 463 *(sol tibi signa dabit),* consisting of two and one-half hexameters (= 5/12), should actually be valued as .4 rather than .5. I have in Chapters 4 and 5 counted fractional lines of this nature as .4; see below, Chapter 4, note 5.

7. For a summary of the chrysodes, see Le Grelle, *LEC* 17 (1949), p. 184. If, instead of dividing the total of major and minor by 1.618 to produce the approximate major (Le Grelle's method), we divide the major by the total in each instance, the ratios in the chrysodes range from .606 to .630.

8. Le Grelle maintains also that Vergil, as a Neo-Pythagorean, made copious use of numerical symbolism; see below, Chapter 5, "Vergil and Pythagoreanism."

9. *Virgile, Géorgiques,* p. xxiv, note 7. Richter, in his recent analysis of the structure of Book I, makes the division between 465 and 466 *(Vergil, Georgica,* pp. 81 ff.), but this must be an error; elsewhere in his edition he begins the epilogue at 463b; cf. text, p. 31, commentary, pp. 161, 173, 175 f., and especially the outline of Book I on pp. 408 f.

10. See Duckworth, *TAPhA* 87 (1956), p. 292 and note 42.

11. *LEC* 17 (1949), p. 158, note 12. It is significant in this connection that, according to Suetonius, Vergil and Maecenas took turns reading the *Georgics* to Augustus and that Vergil read to Augustus and Octavia three books of the *Aeneid* (II, IV, and VI); cf. the Donatus-Suetonius Life, 27, 32. This indicates clearly that Vergil composed his poetry to be heard.

12. *Homer and the Heroic Tradition,* p. 258.

13. See Ghyka, *Le nombre d'Or,* I, pp. 57-77; *The Geometry of Art and Life,* pp. 124-174; also Ghyka in Northrup, *Ideological Differences and World Order,* pp. 97 f., 108 ff.; von Simson, *The Gothic Cathedral,* pp. 155 and (on the cathedral at Chartres) 208 ff.; Borissavlievitch, *The Golden Number.* Cf. also Le Grelle, *LEC* 17 (1949), pp. 144 f., and the bibliography cited in note 4; Graf, *Bibliographie zum Problem der Proportionen,* I, pp. 27-85.

14. *Symmetry,* p. 72. For the algebraic equation, cf. Le Grelle, *LEC* 17 (1949), p. 143, note 2; for the geometry, see Sarton, *A History of Science,* p. 442, note 38; cf. also Hambidge, *The Elements of Dynamic Symmetry,* pp. 29 ff., on "The Rectangle of the Whirling Squares (1.618) and the Root-five Rectangle (2.236)"; Ogden, *The Psychology of Art,* pp. 181 ff.

15. *Science and the Classics,* pp. 205, 208. For discussion and bibliography on the Golden Section and the Fibonacci series, see Archibald, *AMM* 25 (1918), pp. 232-238, reprinted with additions and corrections in Hambidge, *Dynamic Symmetry: The Greek Vase,* pp. 152-157; cf. Capparelli, *Sophia* 26 (1958), pp. 197-210. On the Fibonacci series in nature, e.g., in the spirals of fir-cones and flowers (phyllotaxis), cf. Thompson, *On Growth and Form,* II, pp. 921-933; Ghyka, *The Geometry of Art and Life,* pp. 87-110.

16. For this analysis of *Eclogue* VI, I am grateful to Edwin L. Brown of the University of North Carolina; see now his Princeton doctoral dissertation, *Studies in the Eclogues and Georgics of Vergil,* pp. 89-95. Lines 1-12, as a prooemium, are properly excluded from the calculations; Brown points out that, if we begin with verse 13, line 28 *(ludere . . . quercus),* the sixteenth verse, is echoed by 71 *(cantando . . . ornos),* the sixteenth verse from the end, and also that lines 47 and 52, each beginning *a, virgo infelix,* are, respectively, 35 lines from the beginning (= line 13) and 35 lines from the end. Brown's analysis of *Eclogue* VI and his discoveries of the Golden Section in the *Eclogues* as a whole (see below) are all the more impressive since he at first shared my original skepticism concerning Le Grelle's work, and his study of Le Grelle and

17. The metrical pattern in this passage created by the alternation of lines with fourth-foot clash of metrical ictus and word accent ("heterodyned," denoted by a) and fourth-foot coincidence of ictus and accent ("homodyned," denoted by b) is the following: a b b a   b b   a b b a. On the importance of these heterodyned and homodyned verses, see Knight, *Vergil's Troy*, pp. 17 ff.; *Accentual Symmetry in Vergil; Roman Vergil*, pp. 240-243; and see below, Appendix F.

18. I am indebted to Edwin L. Brown (see above, note 16) for these three ratios in the corresponding poems of the *Eclogues;* see *Studies in the Eclogues and Georgics*, pp. 99-101.

19. For additional ratios in the *Eclogues*, cf. VII/VIII = m (70)/M (109); 109/179 = .609; IX/VIII = m (67)/M (109); 109/176 = .619.

20. *Vergil*, p. 54.

21. The descriptive passages are the following: I, 5b-42, 125-146, 231-258, 463b-514; II, 1-8, 136-176, 319-345, 380-396, 458-542; III, 1-48, 242-283, 284-294, 339-383, 478-566; IV, 1-7, 116-148, 315-558. The section about the rebirth of bees (IV, 281-314), although actually the beginning of the Aristaeus story, is itself of a technical nature and is not included among the descriptive passages. Also, the introductory statement of the plan of the poem (I, 1-5a) and the final recapitulation (IV, 559-566) are properly classed with the didactic portions.

22. These totals do not include IV, 338, which is bracketed as an interpolation by Hirtzel and other editors. If we retain the verse, the ratio is not affected: 836/1352 = 1352/2188 = .618. See below, note 25.

23. Cf. Duckworth, *AJPh* 80 (1959), p. 231.

24. My divisions here are those of Sabbadini's edition. Most editors (e.g., Hirtzel, Janell, Sabbadini, Richter) begin a new paragraph after 477; so also Saint-Denis, although in his preface to *Georgics* III he lists the plague as 474-566; cf. Richter, *Vergil, Georgica*, pp. 99, 411, where he gives the plague as 478-566. Richter also points out (p. 325) that the third stage of the plague (537 ff.) describes the plight of small animals, etc., whereas the second stage (503-536) had been concerned with horses and cattle, and he comments upon the ironical "Golden Age description" in the third stage.

25. I reject 338, omitted by the best MSS and bracketed by Hirtzel, Janell, Sabbadini, Saint-Denis, and Richter. If we retain 338, the ratio is 172/278 = .619.

26. If we add 528-529, which conclude the Proteus episode and form the transition back to the Aristaeus story, we have in 453-529 a tripartite ratio in a framework pattern b/(a + c), with Eurydice's death and Orpheus in the Underworld (453-498 = 46) plus 528-529 (2) as the major of 48 verses, the grief and death of Orpheus (499-527) as the minor (29); 48/77 = .623. For other ratios in the Aristaeus episode, see Appendix C and Tables XXII and XXIII.

27. Richter favors divisions after 176 and 345; cf. *Vergil, Georgica*, pp. 87 ff., 409 f.

28. *Poetical Theory*, pp. 137 ff.

29. Cf. Rand, *The Magical Art of Virgil*, p. 263: "the two themes are started on a kind of race throughout the whole passage, now one and now the other forging ahead, till in the end the victory is the Tree's."

30. *Virgile*, p. 65.

31. If, with Hirtzel, we reject 433, the ratio is 333/531 = .627. Janell, Sabbadini, Saint-Denis, and Richter accept 433 as genuine.

32. For this arrangement and ratio I am indebted to one of the readers of the manuscript for the University of Michigan Press.

33. This is a four-part interlocked pattern, with a + c the minor and b + d the major, and resembles many ratios which appear in the *Aeneid;* see below, Chapter 4, "Short Passages: Four or Five Parts," and Table IV. The inclusion of the prologue of *Georgics* II is supported by the fact that it is integrated with the beginning of the book in ratios based on the Fibonacci series: with 1-8 as minor and 9-21 as major, the ratio is 13/21 = .619; with 1-21 as major and 22-34 as minor, the ratio is 21/34 = .618. Furthermore, each of these three passages subdivides into Fibonacci units: 1-8 into 1-3 (m = 3) and 4-8 (M = 5) with a ratio of .625 (5/8); 9-21 into 9-16 (M = 8) and 17-21 (m = 5) and 22-34 into 22-29 (M = 8) and 30-34 (m = 5), each with a ratio of .615 (8/13.). Hirtzel does not begin a new paragraph after verse 21, as do most editors, including Forbiger, Ribbeck, Conington, Janell, Sabbadini, and Saint-Denis. Ribbeck, however, places 39-46 after verse 8; this transposition is to be rejected, not only on other grounds (cf. Forbiger on 39 ff.) but for mathematical reasons: (1) the symmetrical naming of Maecenas in each of the four books, in lines 2, 41, 41 and 2 respectively (see above, p. 5), and (2) the presence in 1-34 of ratios based on the Fibonacci series. Ribbeck likewise suggests moving several passages in the *Aeneid* to different contexts, but again the Golden Mean ratios in the passages argue in favor of the traditional text; see below, Chapter 5, "Transpositions."

# Chapter 4
# THE PROPORTIONS IN THE *AENEID*

WE HAVE SEEN above that Le Grelle's discovery of the Golden Section in *Georgics* I has been confirmed by many additional instances in the *Eclogues* and the *Georgics*, both in smaller units and in the two works in their entirety. If Vergil used this type of mathematical composition in his two earlier productions, we might have expected that he would do so likewise in the *Aeneid*. But the elaborate structure of the epic as a whole—the alternation between more serious and lighter books, the parallelism of the books in each half, and his arrangement of the *Aeneid* as a trilogy with two units of four books, each of a tragic nature, framing a central core of four books of a more patriotic and national interest—plus his use of alternating, concentric, and tripartite patterns in the individual books perhaps blinded me to the fact that each book has also received the most careful attention mathematically and has been constructed on the basis of the same Divine Proportion which appears in the *Eclogues* and the *Georgics*.

## THE NATURE OF MATHEMATICAL COMPOSITION

It was only when I had analyzed the structural patterns in the individual books, especially the recessed panels around a focal point and the tripartite divisions (described above in Chapter 2), that the Golden Section appeared. When I detected the proportion in the main divisions of the books, I examined the structure of each individual division and again it was found, and likewise in the smaller sections which combine to form the main divisions; in other words, Vergil composed the smaller units of each book in the approximate Golden Mean ratio and then combined them into larger sections, also in proportion, until he created the main divisions of each book.

We shall see later, when we re-examine the structure of each book of the *Aeneid*, that there is an astounding correlation between the patterns of thought and the proportions both in short passages and in larger narrative units;[1] the large sections which determine the ratio in each book as a whole are in every case identical with the main narrative divisions, and also, in most instances, the component parts determining the ratio in each main division are the three narrative subdivisions of each main section.

If I may anticipate the later discussion, I shall illustrate Vergil's procedure by citing as a typical main division the third and final section of Book X:

689-908 Deaths of Lausus and Mezentius
   (1) 689-746   Aristeia of Mezentius
   (2) 747-832   Death of Lausus
   (3) 833-908   Death of Mezentius

This one main division contains twenty-four proportions, with the smaller passages combining and recombining into larger units until we reach the main division itself. The passages with ratios may be charted as on the following page. The underlined passages are not only the subdivisions of the narrative and the passages with ratios into which the shorter proportions combine, but they are also the component parts which produce the ratio in 689-908. This striking correspondence is added proof of the manner in which Vergil uses the smaller passages as building blocks to form the proportions in the main divisions.

I am certain that I have not discovered every possible instance of the exact or approximate Golden Section in the *Aeneid* (nor have I attempted to do so), but I have found, in short passages,

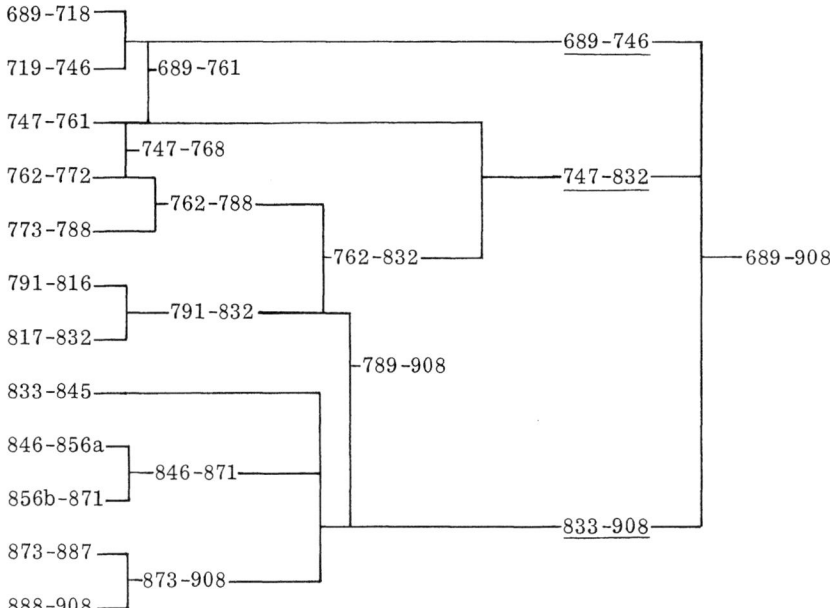

in larger patterns of thought, and in the main divisions of each book, the amazingly large total of 1044 ratios.[2] This seems conclusive proof that Vergil was convinced that the Golden Section contributed to the symmetry of the poetic structure and deliberately used this type of mathematical composition. I say "deliberately" because a subconscious or intuitive feeling for this particular proportion could explain many approximate ratios, but not the almost exact proportions which appear in large quantities everywhere in the *Aeneid*, nor could it account either for the constant construction of longer passages with ratios from the combination of shorter ones, as we have seen illustrated in X, 689-908, or for the almost complete correspondence of these same passages with the narrative units, both small and large.

Even less can the existence of the proportions be attributed to chance. By the law of averages, any writer composing passages of unequal length would hit frequently upon a proportion between .60 and .64, and occasionally on the perfect ratio of .618. But, in a passage of 100 lines, even if we eliminate the extremes of 1-19 and 99-81, we have a possible range for $M/(M + m)$ from .800 (80/100) to .500 (50/100); 61/100 produces a ratio of .610, 62/100 that of .620, and 63/100 that of .630. In a series from 20-80 to 80-20 these three ratios each occur twice; the probability of their appearance is six out of 60 times, or one in ten times. A range from .610 to .626, i.e., .008 from the perfect .618, would thus be extremely difficult to achieve accidentally, and in our hypothetical 100-line passage the exact .618 is impossible. We find, however, that of the 1044 ratios listed below, 622 (59.6 per cent of the total) are within the range from .610 to .626, and 301 (28.8 per cent) are in the almost perfect area of .615 to .621. Furthermore, we have 45 examples of the exact Golden Mean ratio of .618, 33 of them in passages of less than 70 lines! When every book of the *Aeneid* displays these perfect and approximate proportions, both in shorter passages and in all main divisions, we can no longer believe that such proportions are the result of either chance or intuition; Vergil in composing the *Aeneid* purposely arranged the units of his narrative, small and large alike, on the basis of the Golden Mean ratio.

The Golden Section is achieved in the *Aeneid* by means of four different patterns:

(1) We have merely two passages, a and b, and the major is either a or b; i.e., $m/M = a/b$ or $b/a$, and the ratio $M/(M + m) = b/(a + b)$ or $a/(a + b)$.

(2) We have a tripartite arrangement with a + b or b + c forming either the major or the minor; actually, this is only a modification of the first pattern: but if the major or minor in short

passages falls clearly into two sections, e.g., two phases of action, or part speech and part action, it seems advisable to make a threefold division. The ratio m/M = (b + c)/a or c/(a + b) or a/(b + c) or (a + b)/c; the ratio M/(M + m) = a/(a + b + c) or (a + b)/(a + b + c) or (b + c)/(a + b + c) or c/(a + b + c).

(3) The tripartite framework pattern appears with great frequency (276 times, or 26.4 per cent); a and c, as major or minor, enclose the middle portion = b; i.e., m/M = b/(a + c) or (a + c)/b, and the ratio M/(M + m) = (a + c)/(a + b + c) or b/(a + b + c). Proportions of this type seem a natural concomitant of Vergil's fondness for recessed panels around a focal point and for tripartite divisions in general.

(4) We have four or more passages alternating, usually in an interlocked pattern either of four parts, in which m/M = (b + d)/(a + c) or (a + c)/(b + d), or of five parts, in which m/M = (b + d)/(a + c + e) or (a + c + e)/(b + d); the ratio M/(M + m) then = (a + c)/(a + b + c + d) or (b + d)/(a + b + c + d); (a + c + e)/(a + b + c + d + e) or (b + d)/(a + b + c + d + e). These complex patterns are far more numerous than one would anticipate, with the four-part interlocked passages occurring 152 times and the five part interlocked 39 times, and they too are determined by the natural divisions of the narrative; this again rules out the element of chance or intuition.

If I had listed in the following tables only those ratios ranging from .615 to .621, or even from .610 to .626, Vergil's accuracy in the use of the Golden Section would have seemed little short of miraculous, but our picture of the construction of the individual books would have been lacking in clarity. By extending the range of the ratios .018 in each direction from the exact .618, i.e., from .60 (= 3/5 in the Fibonacci series 1, 2, 3, 5, 8, 13 . . .) to .636 (= 7/11 in the series 1, 3, 4, 7, 11, 18 . . .), we shall see that Vergil has included every part of every book in amazingly intricate designs, with very short passages combining into ever larger narrative units until we come to the main divisions of the books and the books as a whole. In the shortest passages the units are sentences; in the larger passages the major and the minor may be determined by divisions of thought within speeches or episodes, or by the alternation of speeches or of speeches and narrative. In the larger proportions, as well as in the main divisions, the major and minor parts are of course each composed of many shorter passages in proportion. But in all this development from the shortest passages to the books as a whole, we have always the ratios centering about the Golden Section (.618), with a range from .60 to .636.[3] The dozens and hundreds of passages in which the proportions occur result in no way from arbitrary divisions of the text but follow naturally the narrative units large and small.

Vergil deliberately introduced the Golden Mean ratios into every part of his narrative. This is not to say that he was conscious of the existence of every one of the 1044 proportions which I shall list in Tables I-VII; some of these may have been, and probably were, the more or less accidental result of the ratios in smaller or larger passages; e.g., if two passages appear side by side each in the pattern a/b, the combined passage may well be in proportion in the familiar interlocked pattern (a + c)/(b + d); conversely, if we have the framework pattern b/(a + c), a and b or b and c may likewise provide a ratio. But the existence of so many perfect or nearly perfect ratios, the overwhelming preference for the Fibonacci series rather than any of the other Golden Mean series,[4] and the correlation of narrative and proportion throughout each book are all the result of Vergil's deliberate plan for the structure of the epic.

The following details are important to note before we turn to the proportions themselves. The numerical totals which accompany the major and minor parts will often display decimal fractions; in many instances these are caused by divisions in the middle of a line, but frequently they result from the much-discussed "half-lines" which have been considered one sign of the lack of revision of the *Aeneid*. These particular verses presented a problem at the beginning of my investigations; since the *Aeneid* was composed for recitation rather than for silent reading, a half-line would not be heard as a whole verse and should be considered a fraction of a line. Also, at an early stage in my examination of the proportions, I discovered that the Golden Mean ratios in many short passages were possible only if the half-lines were counted as fractions, i.e., .2, .3, .4, .6, .7, depending upon the length of the half-line;[5] e.g., in No. 546, with V, 322 (*tertius Euryalus*) counted as .4, the ratio is 7/11.4 = .614, but it drops to an impossible .583 (7/12) if we treat 322 as a whole line; likewise, in No. 646, if we count X, 17 (*pauca refert*) as a whole line, the proportion changes from .617 to .588. In somewhat longer passages also the ratios are more exact with the half-lines counted as fractions; e.g., in No. 595, with VII, 702 (*pulsa palus*) counted as .2, and in No. 836, with VIII, 469 (*rex prior haec*) counted as .2, we have perfect .618 ratios; with VII, 702, and VIII, 469, treated as whole lines, the proportions become .610 and .608 respectively. The fact that the so-called "incompleted" verses seem in many instances very important in achieving more perfect ratios suggests that Vergil may have used them for this purpose; if so, the half-lines, or many of them at least, must have been deliberate, and the generally accepted theory that Vergil, in his final revision of the *Aeneid,* would have replaced all the half-lines by complete verses must be abandoned. I shall return to this problem in Chapter 5.

Also, in computing the numerical total of the major and minor parts, I have omitted the verses which most editors reject on the basis of manuscript evidence.[6] One final point: as in the case of the *Eclogues* and the *Georgics* above, I shall give only the ratio of M/(M + m), since that of m/M can always be derived from the other and so need not be listed.[7]

The tables at the end of the volume give in full the exact and approximate Golden Mean ratios which I have discovered in the *Aeneid*. Tables I through IV contain the short passages in the four patterns described above, and by "short passages" I mean speeches and narrative episodes extending from 10 or 15 lines to 200 or 300; in other words, these four tables, plus Table VII (a supplementary list of short proportions), include all ratios except those in the main divisions of each book. The proportions in the main divisions appear in Table V, and Table VI presents the main divisions in relation to each other. Finally, Table VIII shows the manner in which the books of the *Aeneid* are likewise in proportion. The ratios are numbered consecutively, and these arabic numbers will be useful for cross reference both in the preliminary discussion of selected ratios and especially in Chapter 5, where I shall endeavor to show the significance of Vergil's mathematical composition for many problems of the text—half-lines, interpolations, spurious passages, transpositions, paragraphing, etc. The ratios in Chapter 4 thus provide the raw material for the conclusions to be derived in the final chapter.

## SHORT PASSAGES: BIPARTITE PATTERN

In presenting the ratios I had originally planned a separate category for speeches; two or more speeches, with or without introductory and concluding matter, often combine to form the Golden Section, and frequently a relatively short speech breaks into natural divisions which are in proportion. But I abandoned the attempt to present the speeches apart from the narrative episodes for several reasons: in many proportions the major is composed of a speech, the minor of narrative, or vice versa; long speeches such as the prophecy of Helenus (III, 374-462), the Sibyl's description of Tartarus (VI, 562-627), Anchises' review of Roman rulers and heroes (VI, 756-886), and Evander's story of Hercules and Cacus (VIII, 184-267), in each of which several proportions appear, lack the dramatic quality of the shorter speeches and actually belong with the narrative episodes; also, to have one or more categories for speeches would increase unnecessarily the number of tables.

The bipartite proportions in Table I total 352.[8] The passages in this table (and also in Tables II-IV) fall into four main classifications: (1) ratios within single speeches; (2) two or more speeches in proportion; (3) speech (major or minor) and narrative (minor or major); and (4) narrative or description in both major and minor. I shall mention briefly several typical proportions in each group to illustrate the nature of the bipartite proportions and also to show how the major and minor parts follow the natural divisions of the thought.

(1) Ratios within single speeches: cf. No. 53, Venus' speech to Aeneas in II, 594-620, with the major (604-620) describing the destruction of Troy by the gods; No. 74, Celaeno's words in III, 245-257, with the minor (253-257) containing the prophecy that the Trojans will eat their ta-

bles; No. 296, Drances and his speech in XI, 336-375; the division between major and minor occurs after 359, when Drances ceases speaking to Latinus and addresses Turnus directly.

(2) Typical examples of two speeches in proportion include No. 14: Venus' words in I, 227-253 (m) and Jupiter's prophecy in 254-296 (M); No. 57: Anchises agrees to leave Troy in II, 699-706 (m), Aeneas gives instructions for their departure in 707-720 (M); No. 292: Aeneas' words to the Latin envoys in XI, 106-121 (M) and the speech of Drances in 122-131 (m); No. 297: Drances' speech in XI, 336-375 (m) and Turnus' angry rejoinder in 376-444 (M).

At times a speech may be in proportion with both one preceding and one following: cf. Nos. 94 and 95, where Venus' words in IV, 105-114a provide the minor for Juno's speeches in 90-104 and 114b-128 as major; likewise, in Nos. 344 and 345, Juno's speech in XII, 808-828, is the major, with the minor the words of Jupiter both in 793-806a and in 830-842.

(3) The passages in which the major or minor is a speech and the other part (minor or major) is narrative or description are especially numerous; the following, which I shall list in outline form, are typical:

No. 24:  I, 586-595a (m), Aeneas appears before Dido;
            595b-610a (M), Aeneas' words of gratitude.
No. 25:  I, 613-630 (M), Dido's speech to Aeneas;
            631-642 (m), preparations for the banquet.
No. 51:  II, 567-574 (m), Aeneas sees Helen;
            575-587 (M), his desire to kill Helen.[9]
No. 56:  II, 679-686 (M), the first omen;
            687-691 (m), Anchises' words.
No. 108: IV, 522-533 (m), Dido's last night;
            534-552 (M), Dido's lament.
No. 111: IV, 571-579a (M), Aeneas' words;
            579b-583 (m), departure of the Trojans.
No. 146: V, 799-815 (M), speech of Neptune;
            816-826 (m), Neptune quiets the waves.
No. 159: VI, 295-316 (M), journey with Charon;
            317-330 (m), two speeches: Aeneas and the Sibyl.
No. 167: VI, 477-499 (m), description of heroes;
            500-534 (M), two speeches: Aeneas and Deiphobus.[10]
No. 193: VII, 323-329 (m), description of Allecto;
            330-340 (M), Juno's instructions.
No. 219: VIII, 337-350 (M), description of site of Rome;
            351-358 (m), Evander's words.
No. 253: IX, 590-620 (m), the taunts of Numanus;
            621-671 (M), Numanus slain by Ascanius.

(4) The following illustrate the proportions in which both major and minor are narrative or description:

No. 5:   I, 102-123 (m), the storm and shipwreck;
            124-156 (M), Neptune quiets the storm.
No. 48:  II, 506-525 (m), Priam and Hecuba at the altar;
            526-558 (M), Priam slain by Pyrrhus.
No. 91:  III, 692-706 (M), final stage of the journey;
            707-715 (m), death of Anchises.
No. 97:  IV, 160-164 (m), the storm;
            165-172 (M), the cave and "coniugium."
No. 181: VI, 789b-800 (M), Augustus and the Golden Age;
            801-807 (m), Augustus compared to Hercules and Bacchus.

No. 199: VII, 670-677 (m), Catillus and Coras from Tibur;
678-690 (M), Caeculus from Praeneste.
No. 282: X, 791-816 (M), conflict between Aeneas and Lausus;
817-832 (m), Lausus' death and effect on Aeneas.
No. 285: X, 888-895 (m), Mezentius' horse slain;
896-908 (M), death of Mezentius.
No. 351: XII, 919-939 (M), Turnus is wounded and appeals for mercy;
940-952 (m), Aeneas kills Turnus.

This final passage is of especial interest, in addition to the fact that the ratio is a perfect .618; the division between the major and minor parts occurs after 939, where editors punctuate with a semicolon; a full stop seems advisable here as it places more stress on the *clementia* of Aeneas in 940 f., before he sees the swordbelt of Pallas and slays Turnus (= *iustitia*).[11] I pointed out above (in Chapter I) the close relationship between the conclusion of XII and that of VI; cf. VI, 853: *parcere subiectis et debellare superbos,* where Anchises recommends both *clementia* and *iustitia*.

In a few instances I have listed passages which are not contiguous but which contain speeches or episodes similar in content or important in framing a central passage.[12] The Hector and Panthus scenes in II, 268-297 and 318-335, are in proportion (No. 39), as are the two speeches over the dead body of Pallas in XI, that of Aeneas in 39-58 and that of Evander in 148-181 (No. 289). In No. 32 the two Laocoon scenes in II, 40-56 and 199-227, as minor and major respectively, provide a framework for the Sinon episode. I mentioned above (in Chapter 2) the concentric pattern in VI, 56-123, with two speeches of Aeneas enclosing descriptions of the Sibyl, whose speech in 83-97 is a focal point; this passage (cf. No. 562 in Table III) is framed by 1-55, description of Cumae and the cave of the Sibyl, and 124-155, the Sibyl's instructions, as major and minor respectively of No. 153; cf. No. 202, where VIII, 1-101 as major and 306-369 as minor, enclose two passages with ratios, 102-183 (No. 425) and 184-305 (No. 830) and with them comprise 1-369, the first of the three main divisions of the book (cf. No. 964 in Table V). Another type of nonadjacent proportion appears when three passages are combined in a complex such as b/a and a/c; cf. Nos. 17-18, where I, 402-406, is the major in relation to 407-409 and the minor in relation to 410-417; this same pattern, b/a and a/c, with the same totals for major and minor, is found in II, 437-452 (Nos. 45-46). In the second main division of IX, we have a similar pattern, c/a and b/c, in which IX, 367-449, the third subdivision of the Nisus-Euryalus episode, serves both as minor and major (Nos. 968-969 in Table V).

The ratio in XII, 919-952, the wounding and death of Turnus (No. 351), was cited above as an example of an exact Golden Section, .618; there are fourteen such perfect ratios in Table I, and in most instances they appear at or near the beginnings and the ends of books; e.g., cf. Nos. 1 and 2, in which Juno's lament and appeal to Aeolus in I, 34-75, is the major both to 8-33 (Juno's hostility to the Trojans) and to 76-101 (Aeolus and the storm); No. 113: IV, 672-692, Anna's lament after Dido's suicide (M), and 693-705, Iris releases Dido's spirit (m); No. 187: VII, 37-45a, invocation (m), and 45b-58, description of Latinus and Lavinia (M); No. 310: XI, 768-793, Camilla's pursuit of Chloreus and Arruns' prayer (m), and 794-835, Camilla's death (M).

In the majority of cases, with the smaller proportions combining into larger ones and eventually forming the main divisions of each book, there is no overlapping of ratios from one main division to another; several passages, however, contain what I term "link ratios" and bind together adjacent parts of two main divisions; e.g., No. 69 links together the first two sections of Book III, the major (147-191) concluding the first main division (1-191) and the minor (192-218) beginning the second (192-505); in No. 104, the minor (416-449) concludes the second division of Book IV and the major (450-503) begins the third and final section of the book.[13]

One final note concerning the ratios of Table I (and this applies also to the shorter proportions in Tables II-IV): many of the ratios appear in very short passages of eight, ten, thirteen, fifteen, or more lines; these in most instances appear to be Vergil's smallest narrative and mathematical units; cf., e.g., in Book I, 142-156 (No. 8), 157-169 (No. 9), 170-179 (No. 10), 180-197

(No. 11), 198-207 (No. 12), 208-222 (No. 13); or in Book III, 192-204 (No. 71), 205-218 (No. 72), 219-244 (No. 73), 245-257 (No. 74), 258-273 (No. 75). In most of these passages the same totals appear again and again for major and minor parts, as follows:

| Major | 5 | 6  | 8  | 9  | 10 | 12 | 13 | 15 | 16 | 18 | 21 |
|-------|---|----|----|----|----|----|----|----|----|----|----|
| Minor | 3 | 4  | 5  | 6  | 6  | 8  | 8  | 10 | 10 | 12 | 13 |
| M + m | 8 | 10 | 13 | 15 | 16 | 20 | 21 | 25 | 26 | 30 | 34 |

Here we have the Fibonacci series (1, 1, 2, 3, 5, 8, 13, 21, 34...) or its multiples. With amazing frequency M/(M + m) appears as follows:  3/5 = .60; 5/8 = .625; 8/13 = .615; and 13/21 = .619; an examination of the perfect .618 ratios in Table I will reveal that the majors and minors usually total 21 and 13 respectively (or 42 and 26) and that Vergil has arrived at the exact Golden Section by the quickest possible route: M/(M + m) = 21/34 = .618. The extent to which Vergil has used the Fibonacci series in the *Aeneid* will be discussed in my summary of the ratios.

## SHORT PASSAGES: TRIPARTITE (NONFRAMEWORK) PATTERN

Table II (Nos. 353-469) presents the proportions composed of three parts, with a + b or b + c as major or minor. These are less numerous than the tripartite proportions arranged in a framework pattern (see Table III) and are actually, as I said above, a modification of the bipartite pattern of Table I; a threefold arrangement seems advisable, however, as the major (or minor) divides naturally into two speeches or into speech and narrative. A few typical passages will establish the *raison d'être* for this special category.

(1) A single speech falls into three sections: cf. Aeneas' words in V, 45-71 (No. 405); *nunc ultro* (55) and *praeterea* (64) begin the second and third units of thought, and the third part (64-71), a portion of the major, announces the four contests to be held. In No. 398 (Dido's speech in IV, 416-436), *i, soror* (424) and *quo ruit* (429) begin the second and third parts of the speech; in this instance, b/a and b/c are also in proportion, and the two ratios (Nos. 399 and 400) are listed in Table II rather than in Table I since each is made up of sections of the one speech; 424-428 is the minor both to 416-423 and to 429-436 as major (cf. Nos. 1 and 2 above, where I, 34-75, is the major to both 8-33 and 76-101).

(2) The three parts are frequently composed of three speeches. In No. 359, the major is I, 520-560, the address of Ilioneus, and the minor consists of 561-578, the reply of Dido, and 579-585, the speech of Achates; similarly, in No. 388, the words of Andromache in III, 306-313a and of Aeneas in 313b-319 are the minor, with Andromache's speech in 320-343 as the major. More often, the two speeches comprise the major and the third is the minor: cf. No. 368: the three speeches of Sinon in Book II (words only), with 77-104 and 108-144 the major and 154-194 the minor; No. 393: Juno's speech in IV, 90-104, the minor and Venus' words in 105-114a and Juno's reply in 114b-128 the major (also, the second speech is in proportion as minor with both the first and third speeches; see above, Nos. 94 and 95); No. 390: III, 472-505, with the farewell speeches of Helenus and Andromache the major (cf. No. 750 in Table IV, where the words only of these two speakers form a major) and that of Aeneas the minor; see also No. 430: speeches of Euryalus, Nisus, Euryalus in IX, 197-223 (these form part of a larger unit of four speeches in an interlocked pattern; see No. 848 in Table IV).

(3) The tripartite divisions are part speech, part narrative, and the major (or minor) breaks naturally into speech and narrative; cf. No. 353: I, 34-75 (= M of Nos. 1 and 2) has three sections: 34-49, Juno's lament (m), 50-64, description of Aeolus, and 65-75, Juno's words to Aeolus (M); No. 354: I, 76-80, Aeolus' speech, and 81-91, the storm (M), with 92-101, Aeneas' words to his men, as minor; No. 355: I, 157-197, the landing of the Trojans in Africa after the storm (M), Aeneas' speech in 198-207 and the meal in 208-222 (m); No. 361: I, 586-612, Aeneas appears before Dido and speaks (m), 613-630, Dido's words of welcome, and 631-656, preparation for the banquet and the gifts (M).[14]

(4) All three parts are composed of narrative units, two of which form the major; typical examples include No. 356: I, 297-304, Mercury sent to Carthage, and 305-313, Aeneas and Achates set forth (M), 314-324, meeting with Venus (m); No. 357: I, 402-417, departure of Venus, and 418-440, Aeneas and Achates to Carthage (M), 441-465, Dido's temple and the pictures of Troy (m). In No. 416 we have VI, 756-776, the Alban kings, and 777-787, Romulus (M), in proportion with 788-807, the description of Augustus (m); this part of Anchises' speech constitutes, as is regularly the case, one unit of a larger proportion; cf. No. 417, where VI, 756-807, the Julian portion, forms the minor part of a ratio, with the major consisting of 808-853, the Roman kings and heroes,[15] and 854-892, the famous Marcellus passage.

One of the most important passages in the *Aeneid* is the death of Pallas in X, since, as Vergil points out in 501-509, it is crucial for the fate of Turnus; this episode likewise divides into three parts; see No. 441: the preliminary stage of the fight in 439-478 (m); the death of Pallas in 479-509 and its effect on Aeneas in 510-542 (M). This passage (X, 439-542) contains several shorter proportions which combine to form No. 441, and these may be charted as follows (with the arabic numbers in parentheses referring to the ratios in Tables I-IV):

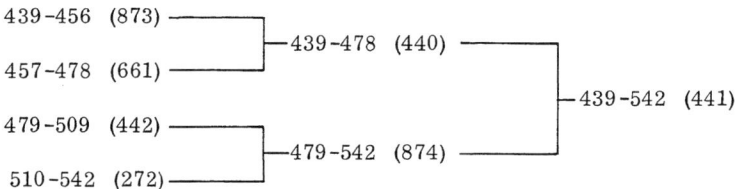

This passage, like the third main division of Book X which was analyzed above, shows the manner in which the shorter proportions constantly combine into larger passages with ratios.[16]

The shortest passages with ratios are subdivided into the natural units of thought, but in the standard editions these divisions are seldom indicated by paragraphs. In the case of somewhat longer passages, e.g., X, 439-542 (No. 441), discussed above, the divisions agree in most instances with the paragraphing in Hirtzel's Oxford Classical Text, which I have used as the basis for my investigations into Vergil's use of the Golden Mean ratio. Exceptions occur, and No. 467 deserves especial comment in this connection: I begin the major part at XII, 725, where Jupiter weighs the scales. Some editors, e.g., Hirtzel and Mackail, add 725-727 to the preceding passage and begin a new paragraph with 728. I have on this point always followed Cartault who gives 725-790 as a unit of composition,[17] and Sabbadini likewise begins a new section at 725. The presence of the Golden Mean ratio in 725-790 (No. 467) supports the paragraphing of Cartault and Sabbadini.[18]

## SHORT PASSAGES: TRIPARTITE FRAMEWORK PATTERN

Table III (Nos. 470-710) presents passages composed of three parts, with the first and third (either as M or m) enclosing the second (m or M); that is, the pattern for m/M is b/(a + c) or (a + c)/b. Vergil's fondness for tripartite divisions and for recessed panel patterns, as described above in Chapter 2, undoubtedly accounts for the frequency with which the Golden Mean ratio is achieved in passages of this type.

We find the proportions appearing in speeches and in narrative, or in a combination of speech and narrative, in the same manner that was pointed out in the preliminary analysis of passages in Tables I and II; e.g., (1) the speech of Venus to Aeneas in I, 335-370a (No. 477) falls into three sections: (a) 335-342, introduction (cf. 338: *Punica regna vides* . . .); (b) 343-364, the story of Dido, Sychaeus, and Pygmalion; (c) 365-370a, conclusion (cf. 365 f.: *ingentia cernes moenia* . . .); a and c combine to form the minor, and b, the central portion of the speech, is the major. (2) This same speech as a whole is the major of a larger unit, with the framing speeches of Aeneas in 326-334 and 372-385a providing the minor (No. 476); for other examples of three speeches in a framework pattern, see, e.g., No. 529: the speeches of Dido, Aeneas, and Dido in IV, 305-387, where Aeneas' defense in 331-361 is the focal point of the book as a whole; No. 562: the speeches of Aeneas, the Sibyl, and Aeneas in VI, 56-123;[19] No. 690: the speeches of Turnus, Latinus, and Turnus in XII, 10-53. (3) A speech may be framed by narrative, as in No. 507, where III, 294-319 and 344-355 as major enclose Andromache's words in 320-343 (minor), or in No. 509, where Aeneas' speech to Helenus in III, 358-368 (M) is enclosed by 356-357 and 369-373 (m). (4) All three parts of the framework pattern may be narrative; e.g., No. 517, where the description of Aetna in III, 570-587 (minor) is framed by Scylla and Charybdis (548-569) and the appearance of Achaemenides (588-595), these two passages providing the major. The description of the shield in VIII, 626-728, provides an excellent illustration of the framework ratio; cf. No. 622: the major part consists of the scenes from the history of Rome in 626-674 and the triumph of Augustus in 714-728; these enclose and emphasize the victory at Actium described in 675-713 (m).

Perhaps the most interesting and important fact about the ratios in Table III is this: the passage b, which is framed by a and c, is regularly a kind of focal point—a significant speech or episode which gains added emphasis from being enclosed by the framing passages. In the case of the proportions cited above, cf. especially the central passages in

No. 476: I, 335-370a, Venus' speech to Aeneas.
No. 477: I, 343-364, the story of Dido, Sychaeus, and Pygmalion in Venus' speech.
No. 507: III, 320-343, Andromache's speech.
No. 529: IV, 331-361, Aeneas' defense.
No. 562: VI, 77-103a, the Sibyl's prophecy (cf. No. 563).
No. 622: VIII, 675-713, Octavian's victory at Actium.

Other episodes of especial significance which as b are enclosed by a and c and thus emphasized by the framework pattern include the following:

No. 471: I, 76-123, Aeolus and the storm.
No. 483: II, 57-198, the Sinon episode.
No. 488: II, 289-295, the words of Hector.
No. 489: II, 318-335, the Panthus episode.
No. 504: III, 219-258, the episode of the Harpies.
No. 519: III, 596-654, Achaemenides and his speech.
No. 524: IV, 90-128, the Juno-Venus episode.
No. 537: IV, 648-671, the final speech and suicide of Dido.
No. 541: V, 45-71, Aeneas' speech announcing the contests.
No. 557: V, 724-740, the words of Anchises.
No. 558: V, 779-826, the Venus-Neptune scene.
No. 564: VI, 136b-148, the Sibyl describes the Golden Bough.
No. 567: VI, 298-304, the description of Charon.
No. 568: VI, 347-371, the speech of Palinurus.
No. 569: VI, 450-476, the meeting with Dido in the Underworld.
No. 572: VI, 679-751, the meeting with Anchises and the doctrine of metempsychosis.
No. 579: VI, 868-886a, Anchises' speech about Marcellus.
No. 585: VII, 212-248, Ilioneus' speech to Latinus.
No. 586: VII, 259b-273, speech of Latinus.
No. 589: VII, 421-434, Allecto's first speech to Turnus.
No. 601: VIII, 126-151, Aeneas' speech to Evander.
No. 607: VIII, 184-267, Evander's story of Hercules and Cacus (cf. No. 606).
No. 609: VIII, 293b-302, song in honor of Hercules.
No. 611: VIII, 337-358, description of the site of Rome.
No. 627: IX, 93-103, Jupiter's words to Cybele.
No. 628: IX, 126-158, Turnus' speech after the metamorphosis of the ships.
No. 633: IX, 324-356, aristeia of Nisus and Euryalus.
No. 638: IX, 691-755, the frenzy of Turnus and the slaying of Bitias and Pandarus.

No. 647: X, 16-62a, the speech of Venus.
No. 652: X, 163-214, the catalogue of ships.
No. 655: X, 228b-245, words of Cymodocea to Aeneas.
No. 661: X, 466-473, Jupiter's speech to Hercules.
No. 663: X, 666-679, the lament of Turnus.
No. 664: X, 719-746, the deaths of Acron and Orodes.[20]
No. 668: X, 821-830a, Aeneas' sorrow at the death of Lausus.
No. 674: XI, 108-131, the speeches of Aeneas and Drances.
No. 675: XI, 148-181, Evander's lament over the body of Pallas.
No. 677: XI, 535b-584a, the story of Camilla (cf. No. 679).
No. 687: XI, 768-835, the death of Camilla (cf. No. 684).
No. 692: XII, 175-215, the vows of Aeneas and Latinus (cf. No. 693, the vows of Aeneas in 175-194).
No. 696: XII, 324-382, the aristeia of Turnus.
No. 697: XII, 313-317, Aeneas' words to the Latins.

These and other passages listed in Table III show how the patterns of Vergil's thought, the framework structure, and the Golden Mean ratio are all integrated in an effective and meaningful manner. It is difficult to conceive how a poet could display more conscious artistry than is apparent in these proportions and in the more complex patterns presented in Table IV.

The following proportions in Table III require additional comment:

In No. 595 (VII, 647-705) and No. 596 (VII, 691-817) the important warriors of the catalogue frame the less important but more picturesque leaders, and the Messapus passage (691-705) appears in both proportions, concluding No. 595 and beginning No. 596. These two ratios give added support to the alternating pattern of the catalogue as described above.[21]

In No. 673 (XI, 39-181) the two laments for Pallas (39-58, 148-181) which combine to form the minor are also in proportion; see No. 289 in Table I.[22]

## SHORT PASSAGES: FOUR OR FIVE PARTS

Table IV (Nos. 711-941) contains the episodes which divide naturally into four or five parts rather than into two or three; usually the parts are interlocked, i.e., a + c or b + d may form either the major or the minor (138 instances) or, with five parts, the major or the minor may be a + c + e or b + d (34 instances). These totals for the interlocking patterns (138 and 34) do not include those in Tables V-VI or the Supplementary List; with these added, the number of four-part interlocking ratios is 152, the five-part 39. The large number of these interlocking proportions again reveals Vergil's interest in alternation, discussed above in Chapters 1 and 2.

In very short passages the component parts are clauses or sentences; cf., e.g., No. 723: I, 544-558, the conclusion of Ilioneus' speech; No. 738 (five-part): III, 1-18, departure from Troy and arrival in Thrace; No. 775: V, 172-182, Menoetes displaced as helmsman; No. 806: VI, 808-818, the Roman kings (Numa, Tullus, Ancus, the Tarquins); No. 840: VIII, 714-731, the triumph of Augustus; No. 850 (five-part): IX, 280b-292a, the speech of Euryalus to Ascanius. In most ratios of this type, however, the units are speeches and short passages of narrative or description.

What is most surprising about the ratios in Table IV is the fact that the parts of the major and/or the minor (e.g., a + c and b + d) are so often linked not only mathematically but by the content as well, by a similarity of theme or idea, by the identity of the speakers, or by the alternation of speech and narrative. To illustrate this striking coincidence of pattern and thought, I shall turn again to the same fourfold classification already utilized in Tables I-III—ratios within single speeches, speeches in proportion, ratios combining speech and narrative, and those composed entirely of narrative.

(1) In Ilioneus' speech to Dido in I, 520-560 (No. 721), the minor (b + d), consisting of 530-538 and 544-550, is concerned with Italy and Aeneas, the remainder of the speech (a + c + e = M) with the fate of the Trojans in Libya. Turnus' speech in XI, 376-444 (No. 909) has four parts: (a) denunciation of Drances, ending at 391; (b) Turnus defends himself in 392-409; (c) in 410-433 he addresses Latinus and suggests either surrender or continuation of the war; and (d) in the final section of the speech, 434-444, he suggests a third possibility—single combat with Aeneas, if that is what the others desire; he, rather than Drances, will gain victory or death. Here the pattern is (a + d)/(b + c) and a and d are linked by the accusation of Drances' inability to fight (a) and Turnus' expressed willingness to fight (d).

(2) A striking example of the interlocking of four alternating speeches occurs in No. 848: the two of Nisus in IX, 184-196 and 207-218 as the major (a + c), those of Euryalus in 197-206 and 219-223 the minor (b + d). In X, 606-632 (No. 879) we also have four alternating speeches—Jupiter, Juno, Jupiter, Juno—and here the pattern is not interlocked, but again (a + d)/(b + c); Juno's request to remove Turnus from the battle (611-620) and Jupiter's consent (621-627) are appropriately linked as b and c to form the major.

(3) The alternation of speech and narrative in interlocking passages appears with great frequency, and usually the major (a + c or b + d) contains the two speeches, with the descriptive or narrative sections providing the minor. I shall give several examples in outline form, and it should be noted how appropriately the speeches are linked together in the ratio by similarity of theme or speaker, or by the second speech being the answer to or development of the first:

No. 736: b + d = M; b = II, 657-670, Aeneas' speech;
             d =      675-678, Creusa's speech.
No. 742: a + c = M; a = III, 84-89, Aeneas' prayer to Apollo;
             c =      94-98, reply of oracle.
No. 746: a + c = M; a = III, 306-313a, Andromache's greeting;
             c =      320-343, her speech.
No. 750: b + d = M; b = III, 475-481, Helenus' farewell;
             d =      486-491, Andromache's farewell.
No. 759: b + d = m; b = IV, 219-237, Jupiter to Mercury;
             d =      265-278, Mercury to Aeneas.
No. 784: b + d = M; b = V, 708-718, advice of Nautes;
             d =      724-740, words of Anchises.
No. 794: b + c = M; b = VI, 388-397, words of Charon;
             c =      398-408a, speech of the Sibyl.
No. 833: b + d = M; b = VIII, 374-386, Venus to Vulcan;
             d =       395-404a, Vulcan to Venus.
No. 849: b + d = M; b = IX, 234-245, Nisus to Trojans;
             d =      257-280a, Ascanius to Nisus.
No. 887: a + c = M; a = X, 846-856a, Mezentius' self-reproach;
             c =      861-866, Mezentius to Rhaebus.
No. 919: b + d = M; b = XII, 142-153, Juno to Juturna;
             d =       156-160, Juno to Juturna.

In Nos. 762 and 763 we find IV, 279-415, in a five-part pattern. This passage is both the second part of the second main division of Book IV and also the central portion of the recessed panel around Aeneas' speech in 331-361 as the focal point; in my analysis of IV above,[23] I commented upon the numerical symmetry in the center of the book, and this symmetry produces a variety of proportions within the passage. The description of Aeneas and Dido in 279-304, Aeneas' speech in 331-361, and the description of Dido and Aeneas in 388-415 are the major, with the two speeches of Dido in 305-330 and 362-387 as the minor; the pattern is thus interlocked (b + d)/(a + c + e), and the ratio is the exact Golden Section (.618). Also, if we take the three central speeches (Dido, Aeneas, Dido) as major and the two passages of description as minor, an approximate Golden Mean ratio appears in the pattern (a + e)/(b + c + d); also Dido's two speeches (as M) frame Aeneas' speech (m) to produce a third ratio (No. 529), and Aeneas' speech is the minor both to 279-330 as major (No. 395) and to 362-415 as major (No. 396). Additional proportions appear within the descriptive passages and the speeches. The many ratios in this passage thus combine with the recessed panel structure to emphasize the significance of Aeneas' speech in the very center of the book.

(4) In passages composed largely or entirely of narrative and description we often find the same close relation between the two parts which combine to form the major (and/or minor). One of the most striking of these is the hunting scene in IV, 129-172 (No. 755); the four interlocking parts are (a) 129-135, preparation for the hunt; (b) 136-150, description of Dido and Aeneas; (c) 151-159, the hunt; (d) 160-172, the storm and the "coniugium." The two Dido and Aeneas passages (b + d) provide the major, the two scenes concerning the hunt (a + c) the minor.

The following passages likewise reveal the union of similar themes in major or minor:

No. 727: b + d = M; b = I, 657-694, the Venus-Cupid episode;
                    d =    712-722, activity of Cupid.
No. 737: b + d = M; b = II, 745-770, Aeneas' reaction to the loss of Creusa;
                    d =    775-795, Creusa's speech and departure.
No. 795: b + d = M; b = VI, 445-476, Dido episode;
                    d =    494-534, Deiphobus episode.
No. 821: a + c = M; a = VII, 406-434, Allecto visits Turnus and speaks;
                    b =    435-444, reply of Turnus;
                    c =    445-457, action and words of Allecto;
                    d =    458-474, effect on Turnus.

Here, as in No. 755, both major and minor are linked by content, the two passages of the major being concerned with Allecto, those of the minor with Turnus.

In No. 932 we have a six-part ratio, also interlocked, of the type (a + c + e)/(b + d + f). The episode presents the activity of Aeneas and Turnus (XII, 505-520, 529-547) in short alternating passages.[24] The minor part is composed of the deeds of Aeneas (505-508, 513-515, 529-534), the major of those of Turnus (509-512, 516-520, 535-547). It is also possible to treat 505-520 and 529-547 as units (a and c) and to add the intervening passage (521-528) as a second part (b); then we have the familiar type a/(b + c); see No. 464 (Table II).

Many of the four-part and five-part passages in Table IV are composed of two or more smaller ratios, e.g.:

No. 713 (I, 157-207) includes Nos. S998 and S999 (157-179 and 180-207), which in turn are
                composed of Nos. 9-12 (157-169, 170-179, 180-197, 198-207);[25]
No. 732 (II, 370-401) combines Nos. 41 and 43 (370-385, 386-401);
No. 739 (III, 1-26) includes Nos. 738 and 62 (1-18, 19-26);
No. 740 (III, 27-68) is made up of Nos. 382 and 741 (27-52, 53-68);
No. 745 (III, 192-218) combines Nos. 71 and 72 (192-204, 205-218).

In several instances the shorter proportions lack self-sufficiency and have been listed under the larger passages; this is especially the case when the major and minor are parts of a longer speech, or when one part is narrative and the other a section of a speech; cf. Nos. 715-716, 722, 768-769, 818-819, etc. These, however, are the normal bipartite or tripartite patterns; i.e., b/c or c/d = a/b, and c/(b + d) = b/(a + c).

A very different problem arises in the case of shorter ratios when the major and the minor are not adjacent. In Table I, I listed a few noncontiguous passages in proportion (Nos. 32, 39, 153, 202, 289); these passages were similar in theme or important in framing a central passage containing one or more ratios. The component but nonadjacent parts of many passages in Table IV are likewise in proportion; these sections are not narrative units in the normal sense; the proportions seem secondary and largely the result of the interlocking of the four or five parts. We find ratios resulting from the patterns a/c (or c/a), a/d (or d/a), a/e (or e/a), b/d (or d/b), also a/(c + d), a/(d + e), c/(a + e), etc. Many proportions of the type a/c or c/a may be found also in the component parts of the passages in Tables II and III. I have eliminated from the tables and charts all noncontiguous ratios of this type (more than 150 in number); many of these rejected proportions, however, display the Fibonacci series and include some perfect .618 ratios.

I have selected, from such secondary ratios in *Aeneid* I and II, several which are close to the exact Golden Section, and these will illustrate the nature of the ratios which I have omitted from the tables. The arabic numbers refer to the passages in Tables II-IV which provide the subordinate parts in proportion.

| Basic no. | m/M: type | Book | Major | Total lines | Minor | Total lines | Ratio: M/(M + m) |
|---|---|---|---|---|---|---|---|
| 714. | d/(a+c) | I | 257-260<br>272-285 | 18 | 286-296 | 11 | .621 |
| 717. | a/d | I | 387-401 | 15 | 326-334 | 9 | .625 |
| 721. | e/(a+c) | I | 520-529<br>539-543 | 15 | 551-560 | 9.2 | .620 |
|  | a/(d+e) |  | 544-550<br>551-560 | 16.2 | 520-529 | 10 | .618 |
|  | e/(b+d) |  | 530-538<br>544-550 | 15.3 | 551-560 | 9.2 | .624 |
| 732. | d/(a+b) | II | 370-378<br>379-382 | 13 | 394-401 | 8 | .619 |
|  | c/(a+d) |  | 370-378<br>394-401 | 17 | 383-393 | 11 | .607 |
| 492. | a/c | II | 453-505 | 52.7 | 370-401 | 32 | .622 |
| 379. | a/c | II | 771-795 | 24.7 | 730-744 | 15 | .622 |

The pattern a/(d + e) in No. 721 gave an exact Golden Mean ratio. Other .618 proportions which I have rejected are the following:

| 759. | d/a | IV | 198-218 | 21 | 265-278 | 13 | .618 |
|---|---|---|---|---|---|---|---|
| 820. | a/d | VII | 385-405 | 21 | 341-353 | 13 | .618 |
| 926. | b/d | XII | 420-440 | 21 | 398-410 | 13 | .618 |

These last three are excellent examples of the Fibonacci series (13, 21, 34) but have been omitted along with the others, inasmuch as the major and minor, being noncontiguous, do not combine to form a single narrative unit.

Before I present the ratios of Table IV, I wish to call attention to No. 800, which deserves special comment. Although Nettleship, Hirtzel, and others indicate a lacuna after VI, 601, there is no MS evidence for an additional verse, and other editors (e.g., Janell, Mackail, Sabbadini) join 602 to 601 without a break; I have departed from Hirtzel's text in this respect and disregarded the lacuna. Although the ratios in Nos. 172 and 801 are more accurate with a verse added (.617 and .621 instead of .609 and .607), No. 800 is a perfect Golden Mean ratio, .618 (Fibonacci series, 34, 55, 89), with the text as transmitted; this suggests that the lacuna is ill-advised. The reading of Madvig in 601 (*Pirithoumque et/quo*) is accepted by Mackail, who says that the emendation "solves the difficulty satisfactorily."

## THE MAIN DIVISIONS OF THE BOOKS

We turn now to the proportions in the main divisions of the books of the *Aeneid*. As I pointed out above (in Chapter 2), every book divides naturally into three sections.[26] There are thus 36 main divisions in the twelve books. It is an amazing fact that the approximate Golden Mean ratio appears in each of these 36 divisions. In Table V we find again the familiar patterns already listed in Tables I-IV: two parts, three parts, three parts in a framework design, and four or more parts usually interlocked. The tripartite patterns are by far the most numerous, occurring 30 times, and 16 of these are of the framework type, $b/(a + c)$ or $(a + c)/b$.

The two most perfect ratios, the exact .618, appear in the first two divisions of *Aeneid* IV. In No. 951, 1-53 (Dido and Anna) and 160-172 (the storm and the "coniugium") frame 54-159 (the development of Dido's love, the Venus-Juno speeches, and the hunting scene); in the second division (No. 952) we have the pattern $a/(b + c)$, the minor consisting of 173-278 (Fama, Iarbas, Jupiter, Mercury), the major of 279-415 (Aeneas and Dido) and 416-449 (the failure of the attempted reconciliation).

In these and the other main divisions the natural development of the narrative determines the major and minor parts; this may be illustrated from Book X, where all three divisions have the framework pattern $b/(a + c)$. In the first division (No. 971), 1-117 (the council of the gods) and 256-361 (Aeneas' landing and aristeia) enclose 118-255 (catalogue of ships and night journey of Aeneas).[27] The second division (No. 972) likewise has three parts, with the central portion, 479-605 (the death of Pallas and its effect on Aeneas), framed by 362-478 (the aristeia of Pallas) and 606-688 (the removal of Turnus from battle). In the final portion of the book (No. 973), 689-746 (the aristeia of Mezentius) and 833-908 (the death of Mezentius) enclose 747-832 (the death of Lausus). In each main section of X the three parts which combine in the framework pattern to produce the Golden Mean ratio are the same three subdivisions which I listed in my analysis of the book in Chapter 2. This close correlation between the tripartite subdivisions of the narrative and the component parts which determine the ratios in the main divisions is a characteristic feature of Vergil's composition and will be seen more clearly when we refer to the Chart-Index in Table IX.

The ratio in the second main division of Book IX is double, in the pattern $c/a$ and $b/c$; cf. Nos. 968-969: the third part of the Nisus-Euryalus episode (capture and death in 367-449) is the minor in relation to 176-313 and the major in relation to 314-366; for a similar pattern ($b/a$ and $a/c$), see above Nos. 17-18 and 45-46 in Table I.

No. 942 gives the first main division of Book I, and lines 1a-1d, the four verses preceding *arma virumque cano*, are included in this proportion; these are the verses which were deleted by Varius after Vergil's death (Donatus-Suetonius Life, 42). With the four verses omitted, the ratio is a less accurate .640; this supports Hirtzel who accepts the four preliminary verses. Many editors ignore, bracket, or italicize 1a-1d (cf., e.g., Janell, Sabbadini, Mackail). The verses are not in the important MSS and *arma virumque* was viewed in antiquity as the beginning of the *Aeneid;* this was the natural result of the deletion of the verses by Varius. The four verses should not be rejected, however.[28]

For the proportions in Table V, I have used the divisions of the books which I had established prior to the discovery of the Golden Mean ratios in the *Aeneid*. I treat the conclusions of two books, I and XI, as epilogues and do not include them as a part of the third and final division. In the case of I, the third division really ends at 747, the last nine lines serving as an introduction to Aeneas' narrative in II and III; cf. Nos. 944 and (in Table VI) 981.[29]

In XI the death of Camilla, the climax of the book, ends at 867, and 868-915 provide a transition to XII. No. 977, an alternate to No. 976, carries the final division through 915, but No. 976 is to be preferred because the second and third divisions of XI are in proportion only if the third division stops at 867; cf. No. 995 in Table VI. Mackail ends the third division with 867.[30]

Not only do we have two different proportions for the third main division of Book XI, but the third main division of Book VII (the catalogue of Latin warriors in 641-817) has both a framework pattern, b/(a + c), with 641-690 and 761-817 as M and 691-760 as m (No. 962), and a six-part interlocking pattern (No. 963); this last has as major 655-690 (the first group of three unimportant warriors), 706-743 (the second group of three unimportant warriors), and 783-817 (Turnus and Camilla); the minor consists of 641-654 (invocation; Mezentius and Lausus), 691-705 (Messapus), and 744-782 (the third group of three unimportant warriors).[31]

## THE MAIN DIVISIONS IN PROPORTION

Not only does the approximate Golden Mean ratio appear in each main division of each book but the main divisions themselves are in proportion in each book. From my earlier analysis of the tripartite nature of the books, the fact emerged that the three parts were of surprisingly uneven length, with the first division usually the shortest; this suggested to me the possibility that some sort of mathematical ratio might exist in the relation of the parts to each other. I must repeat that no reader of this work will be more astonished at Vergil's use of mathematical proportions than I have been in tracing the presence of the Golden Mean ratio from the proportions of the main divisions in relation to each other (Table VI) to the proportions within the main divisions (Table V), and down to the ratios in the shorter passages of speeches and narrative episodes (Tables I-IV, VII), and in discovering the various patterns which Vergil used to produce the major and minor parts in the proportions.

In six books of the *Aeneid,* two of the main divisions are in proportion, the first part as minor and the second as major, as in III (No. 983) and IX (No. 992), or the first part as major and the third as minor, as in VII (No. 990); in the case of I (No. 981) and XI (No. 995) the second division is the minor and the third the major, and for XII (No. 996) the third is the minor and the second the major.

In the remaining six books all three divisions combine to form the Golden Section, either of the type c/(a + b), in VI (No. 988) and VIII (No. 991), or (b + c)/a, in V (No. 986), or a/(b + c), in X (No. 993), or in the framework pattern b/(a + c), in II (No. 982) and IV (No. 984). In four of these books, IV, V, VI, and X, two of the three divisions are also in proportion; cf. Nos. 985, 987, 989, and 994 respectively.[32] With one exception (Book XII) the more significant books with even numbers (cf. Chapter 1, "The Alternating Rhythm") reveal in their main divisions a greater attention to mathematical symmetry than appears in the lighter, odd-numbered books.

The Donatus-Suetonius Life (32) states that Vergil read Books II, IV, and VI to Augustus and Octavia. It is interesting to note that these three books which Suetonius describes as having been completed (*perfectaque demum materia*) are in this second group where all three divisions are used to produce the Golden Section. This suggests the possibility that Book XII, in which only two of the three divisions are utilized for the ratio, might have undergone revision which would have brought it in line in this respect with Books II, IV, VI, VIII, and X.

In Table VI, as in Table V, I have listed the ratios by order of books rather than by patterns. In describing the patterns in Table VI, I use the letters a, b, and c to denote the three main divisions respectively.

## SUPPLEMENTARY LIST OF RATIOS

After the ratios were tabulated and numbered consecutively from 1 to 996, with the numbers listed (in parentheses) in the Chart-Indices in Table IX, several additional proportions in short passages came to my attention. To insert these in Tables I-IV under the appropriate pattern and book would have meant a complete renumbering of all the ratios in both the tables and the charts, and the changing of hundreds of cross references in the descriptive material in Chapters 4 and 5;

the revision of so many cross references in the final stage of the work undoubtedly would have created many additional errors.

It seemed advisable, therefore, to present the new ratios in a separate list as Table VII; they are included in the statistical summary which follows and have been added also both to the later discussion and to the charts where they fill in a few obvious gaps in the first or second column of ratios. The new ratios are numbered from S997 to S1044, the capital S being prefixed to call attention to the fact that these ratios, when they appear in the text and the charts, are to be found in this supplementary list ( = Table VII).

## SUMMARY OF THE PROPORTIONS

As we review the 1044 proportions listed in Tables I-VII, several significant facts emerge. Vergil's interest in tripartite divisions, framework patterns, and alternating patterns is confirmed to an astounding degree. One would expect the ratios in most instances to be derived from two passages, a/b or b/a, but such is not the case; the bipartite pattern occurs only 394 times, or 37.7 per cent of the total number. 414 ratios, 39.7 per cent of the total, are composed of three parts and of these exactly two-thirds (276) are of the framework type, b/(a + c) or (a + c)/b; this pattern thus occurs 26.4 per cent of the total. The ratios derived from passages falling into four or more parts are also surprisingly numerous (236 instances); most of these are interlocked patterns, such as (a + c)/(b + d) or (b + d)/(a + c + e), and reveal the poet's especial fondness for an alternating rhythm. As I pointed out above, the four-part interlocked pattern appears 152 times, that with five parts 39 times, and we find three instances of a six-part interlocked pattern (Nos. 932, 963, and S1036).

In Chapter 1, I discussed the alternating rhythm of the *Aeneid* as a whole; the books of a more serious and tragic nature are those with even numbers, the lighter books which provide a relief from tension are the ones with odd numbers. One might expect that Vergil would devote more attention to the significant books than to the lighter, odd-numbered books. The proportions listed in Tables I-VII support this conclusion. The number of ratios in the even-numbered books varies from 79 (Book IV) to 107 (XII), with an average of 89.3 per book, that in the odd-numbered books from 74 (VII) to 93 (XI), an average of 84.8. Another noticeable difference between the books with even numbers and those with odd numbers appears in Table VI. All the even-numbered books except XII produce the Golden Section by the use of all three main division, but in the case of the books with odd numbers only two of the three divisions combine to form the ratio (Book V is the exception among the odd-numbered books, in that the entire book is in proportion whether it is divided into three parts, as seems correct, or into only two).

I stated early in the chapter that 622 proportions (59.6 per cent of the total) are within .008 of the perfect Golden Mean ratio, i.e., a range from .610 to .626. If we extend the range .002 and include the ratios from .608 to .628, the total rises to 697; in other words, two thirds of the ratios are within .01 of the perfect .618. On the other hand, 301 passages are in the almost perfect range of .615 to .621, and the exact .618 ratios total 45. The percentage for the group ranging from .615 to .621 is slightly higher in the shorter passages than in the main divisions; this might seem surprising, as the mathematical series leading to the Golden Mean ratio become more accurate as the numbers grow larger, but it indicates that Vergil worked with small narrative units and gave special attention to the ratios in these passages. The totals and percentages are as follows:

THE PROPORTIONS IN THE AENEID

|  |  | Total proportions: Tables I-VII | Shorter passages: Tables I-IV and VII | Main divisions: Tables V-VI |
|---|---|---|---|---|
|  | Totals | 1044 | 989 | 55 |
| .618 | Totals | 45 | 42 | 3 |
|  | Per cent | 4.3 | 4.2 | 5.5 |
| .615-.621 | Totals | 301 | 286 | 15 |
|  | Per cent | 28.8 | 28.9 | 27.3 |
| .610-.626 | Totals | 622 | 585 | 37 |
|  | Per cent | 59.6 | 59.2 | 67.3 |
| .608-.628 | Totals | 697 | 652 | 45 |
|  | Per cent | 66.8 | 65.9 | 81.8 |

The 45 examples of the perfect .618 ratios are very revealing and again prove Vergil's deliberate devotion to mathematical proportion. Of these 45 proportions 33 are in short passages of less than 70 verses, and in passages of this length it is not likely that the exact Golden Section would appear by accident. Seventeen of these perfect ratios are of particular interest, for the major is 21 lines and the minor 13, or the major is 42 (21 x 2) and the minor 26 (13 x 2); for 21 (M) and 13 (m), see Nos. 103, 113, 255, 345, 351, 364, 390, 539, 678, 778, 894, 906, S1041; for 42 (M) and 26 (m), see Nos. 1, 2, 310, 381.

The Golden Mean series were described above (in Chapter 3) as various additive series with each number merely the sum of its two predecessors; we may begin with any two numbers and sooner or later the ratio .618 appears: for instance, if we take the series 1, 5, 6, 11, 17, 28, 45, 73, 118, 191,..., 45/73 = .616, 73/118 = .619, and 118/191 = .618. To reach .618 most quickly, we must begin with the Fibonacci series 1, 1, 2, 3, 5, 8, 13, 21, 34,...; 21/34 = .61764 = .618. In order to achieve a perfect .618 in very short passages, Vergil has used the numbers of the Fibonacci series, 13 and 21, whereby the exact ratio, $M/(M + m) = 21/34 = .618$, can be most swiftly attained. If this happened two or three times only it might be considered accidental, but it occurs in 17 of the perfect .618 ratios. In addition to these 17 exact proportions constructed with 13 (or 26) as minor and 21 (or 42) as major, we have in Nos. 293 and 458 a minor of 21 and a major of 34 (M + m = 55) and in Nos. 748 and 800 a minor of 34 and a major of 55 (M + m = 89); thus, in 21 perfect .618 ratios we find the Fibonacci series 13, 21, 34, 55, 89.

An even more astounding fact is this: in seven instances of the exact Golden Section the majors and minors are composed of totals with fractions which reveal these same numbers of the Fibonacci series. In Nos. 128, 184, 701, and 914 we have short passages of 22 verses, with a minor of 8.4 and a major of 13.6; these two numbers are 21 and 34 times .4; in No. 402 the minor is 29.4 (= 21 x 1.4) and the major is 47.6 (= 34 x 1.4); in No. 690 the minor is 16.8 (= 21 x .8) and the major is 27.2 (= 34 x .8); in No. 187 the minor is 13.6 (= 34 x .4) and the major is 22 (= 55 x .4). The presence of the Fibonacci series in these ratios where the major and minor contain decimals seems an added proof of Vergil's deliberate use of the numbers 13, 21, 34, 55 to achieve the exact Golden Section.[33] Of the 45 perfect .618 ratios, the Fibonacci series appears in 28, or 62.2 per cent of the total.

That Vergil has purposely used these numbers to produce the exact Golden Mean ratio is confirmed by the places in the *Aeneid* where these perfect proportions appear; they are in many instances at or near the beginning and the end of the books, as follows:

Book I, 8-75    No. 1
I, 34-101   No. 2
I, 81-156   No. S997

| | |
|---|---|
| Book I, 723-756 | No. 364 |
| II, 634-804 | No. 376 |
| III, 1-68 | No. 381 |
| IV, 1-172 | No. 951 (main division) |
| IV, 672-705 | No. 113 |
| V, 1-34 | No. 539 |
| VII, 37-58 | No. 184 |
| VII, 45b-80 | No. 187 |
| VII, 647-705 | No. 595 |
| VIII, 1-101 | No. 599 |
| X, 1-62a | No. S1036 |
| XI, 768-835 | No. 310 |
| XI, 794-815 | No. 914 |
| XII, 10-53 | No. 690 |
| XII, 19-45a | No. 317 |
| XII, 808-842 | No. 345 |
| XII, 919-952 | No. 351 |

Vergil has apparently placed in key positions many of his perfect ratios in order that the attentive hearer might be conscious of the mathematical symmetry and perhaps detect more easily the approximate ratios elsewhere in the books.[34]

We saw above (in Chapter 3) that *Eclogue* VI is constructed on the series 2, 8, 10, 18, 28, 46, 74,... (actually, this is merely a doubling of 1, 4, 5, 9, 14, 23, 37,...) and also that the major and minor parts of the proportions in *Georgics* I show the numbers of several different series: 1, 1, 2, 3, 5, 8, 13, 21,...; 1, 3, 4, 7, 11, 18,...; 1, 4, 5, 9, 14, 23,...; 1, 5, 6, 11, 17, 28,...; and 3, 7, 10, 17, 27, 44,....[35] We now find that in the *Aeneid* 28 of the 45 perfect .618 ratios are based on the Fibonacci series, 1, 1, 2, 3, 5, 8, 13, 21,....

It will be of interest to see the extent to which the other proportions in the *Aeneid* (as given in Tables I-VII) are constructed on these series and others of the same type. I shall give below those series which appear most frequently, disregarding for this purpose the proportions which contain decimal fractions in either major or minor part. As in the case of the .618 ratios above, the numbers in the Fibonacci series are often doubled, and frequently they are multiplied by three, four, five, etc.; e.g., minor of 15 and major of 25 (= 3 and 5 times 5), cf. Nos. 232, 422, 440, 527, etc.; minor of 10 and major of 16 (= 5 and 8 times 2), very frequent (22 times), cf. Nos. 26, 132, 210, 261, 292, 309, 348, 354, etc.; minor of 16 and major of 26 (= 8 and 13 times 2), cf. Nos. 86, 282, 306, 353, etc.; minor of 32 and major of 52 (= 8 and 13 times 4), cf. No. 177; minor of 150 and major of 250 (= 3 and 5 times 50), cf. No. 976. Multiples of other series likewise appear with considerable frequency; e.g., a minor of 14 and a major of 22 belong to the series 1, 3, 4, 7, 11,..., cf. Nos. 78, 93, 159, 663, 716, 728, etc.; 27 (m) and 42 (M) belong to 1, 4, 5, 9, 14,..., cf. Nos. 541, 542; 22 (m) and 34 (M) fall in the series 1, 5, 6, 11, 17,..., cf. Nos. 608, 904.

In listing the occurrences of the Golden Mean series, I have disregarded 373 passages which contain decimals in either major or minor part, or in both; although these series appear likewise in the totals with decimals,[36] the 671 passages with whole numbers will give an adequate idea of the relative frequency with which the more familiar series occur. The one used by far the most times is of course the Fibonacci series, the simplest of all, and in 21 passages (excluding the seven with decimals) the poet uses this series to achieve the exact .618 ratio, in a majority of instances in passages of 34 verses or less. The total occurrences of this series (with decimals disregarded) are as follows:

$$M/(M+m) = 21/34 \ (34/55, \text{etc.}) = .618 \quad 21 \text{ times}$$
$$13/21 = .619 \quad 39 \text{ ''}$$
$$8/13 = .615 \quad 94 \text{ ''}$$
$$5/8 = .625 \quad 85 \text{ ''}$$
$$3/5 = .60 \quad 79 \text{ ''}$$

This gives us the surprisingly large total of 318 instances of the Fibonacci series in the *Aeneid*, 47.4 per cent of the 671 ratios under consideration. The relative frequency of the other Golden Mean series which appear 10 or more times is as follows:

| | | |
|---|---|---|
| 92 times: | | 1, 3, 4, 7, 11, 18, 29,... |
| 27 | " | 1, 4, 5, 9, 14, 23, 37,... |
| 24 | " | 2, 5, 7, 12, 19, 31, 50,... |
| 23 | " | 1, 5, 6, 11, 17, 28, 45,... |
| 21 | " | 3, 7, 10, 17, 27, 44, 71,... |
| 15 | " | 4, 9, 13, 22, 35, 57, 92,... |
| 14 | " | 1, 6, 7, 13, 20, 33, 53,... |
| 11 | " | 6, 13, 19, 32, 51, 83, 134,... |

We thus have a total for the Fibonacci series and these other eight series of 545, or 81 per cent of the 671 ratios in which both major and minor are whole numbers. Vergil's use of the various Golden Mean series is now apparent, as is his decided preference for the Fibonacci series which occurs more than three times as often as the one next in order of simplicity (1, 3, 4, 7, 11, 18,...), and this in turn is more than three times as frequent as any of the others.

The close relationship between the numbers in these series and the same numbers as they appear in the major and minor parts of the proportions indicates again that neither chance nor intuition can be an adequate explanation of the phenomena. In all the passages in which these series appear, Vergil is purposely using numbers from different types of the same mathematical series to produce the exact or approximate Golden Section.

The large number of proportions in which the Fibonacci series and its multiples appear and also the great variety of the other Golden Mean series which recur so many times are important in another connection. Thompson points out the curious fact that in the history of Greek mathematics there is no account of the Golden Mean series nor any allusion to it; it is often thought to have been discovered by Leonardo Pisano, called Fibonacci, in the early thirteenth century, but Thompson himself believes that the series was known to the ancient mathematicians; he says:

> The Greeks were familiar with the series 2, 3: 5, 7: 12, 17, etc.; which converges to $\sqrt{2}$, as the other does to the Golden Mean; and so closely related are the two series, that it seems impossible that the Greeks could have known the one and remained ignorant of the other. [37]

The fact that Vergil uses the numbers of the Fibonacci series to such an extent and reveals a knowledge of many other varieties of the Golden Mean series seems conclusive proof that this series was known to the ancient Greek and Roman mathematicians. The poet's acquaintance with the series was probably gained in the days when he was "giving especial attention to mathematics," as the Donatus-Suetonius Life (15) tells us. Perhaps at that time his interest was aroused in composing poetry by mathematical symmetry, although such a method of composition was not original with him, as we shall see below. [38]

## THE *AENEID* AS A WHOLE

To summarize the architecture of the *Aeneid*, which was described in full detail in Chapter 1, the structural patterns are three in number: the alternation between the less serious books with odd numbers and the more significant and tragic books with even numbers; the division of the poem into two halves with the books of the second half paralleling those in the first half (both similarities and contrasts); and the tripartite division of the poem, with the tragedy of Dido (I-IV) and the tragedy of Turnus (IX-XII) enclosing the more national and Augustan portion (V-VIII).

Here also the approximate Golden Section appears; Books I-VI may be divided into three parts, each composed of a lighter book followed by a more serious book (I-II, III-IV, V-VI), and likewise the second half (VII-VIII, IX-X, XI-XII);[39] thus in each half the regular pattern c/(a + b), i.e., (V-VI)/(I-II + III-IV) and (XI-XII)/(VII-VIII + IX-X), produces the ratios .627 and .636 respectively (Table VIII, Nos. 1045 and 1046).[40] The pattern in each half is thus not unlike that of the *Eclogues* when we group the poems in triads: (VII-IX, X)/(I-III + IV-VI).[41] Since each half of the *Aeneid* reveals the Golden Mean ratio, the entire poem falls inevitably into proportion in the fourfold interlocked pattern seen frequently above, i.e., (V-VI + XI-XII)/(I-IV + VII-X), or (b + d)/(a + c); the ratio is .632 (No. 1047). The over-all structure of the *Aeneid* may thus be presented in the following diagrams:

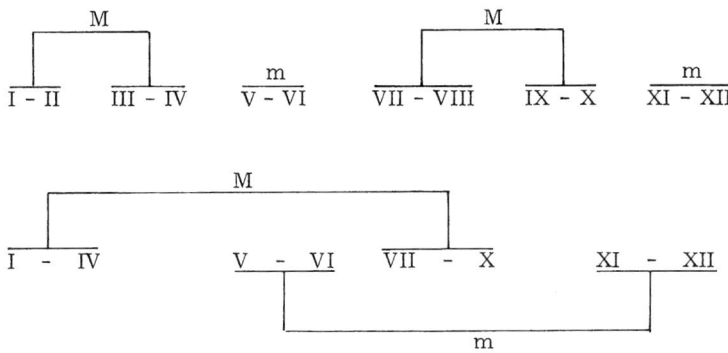

But Vergil also arranged his epic as a trilogy, and in the three units of four books each we again have the c/(a + b) pattern:

The ratio here is .637 (No. 1048). Thus, even in the architecture of the poem as a whole, the structural designs and the approximate Divine Proportion go hand in hand. The fact that these ratios are somewhat less accurate than most of those in the individual books is interesting and perhaps indicative of the unfinished state of the poem; this seems the more likely when we recall that in the completed *Georgics* the ratio of the descriptive to the didactic passages is the exact Golden Section, .618.[42]

Table VIII gives the four proportions for the *Aeneid* in its entirety. As in dealing with the individual books, I have established the totals by counting the half-lines as fractions and by omitting the verses rejected by editors on manuscript evidence. *Aen.* I, 1a-1d, have been included, but not III, 204a-204c, or VI, 289a-289d.[43]

# Notes to Chapter 4

1. Table IX contains a chart-index in which every proportion in every book is listed, with arabic numbers in parentheses keyed to Tables I-VII where the details concerning the ratios are presented. The charts enable the reader to see at a glance the relation of the shorter to the longer passages, as well as the agreement of the ratios with the structural outlines of each book; cf. also Chapter 5, "Vergil's Method of Composition."

2. This total does not include between 150 and 200 "secondary" ratios where the major and minor parts are separated; these noncontiguous proportions seem to be the accidental result of the interlocking patterns of Table IV and have been excluded from my calculations; see above, pp. 56 f.

3. A few proportions ranging from .585 to .599 and from .637 to .650 also appear; these have not been listed, but they perhaps suggest that Vergil in certain passages had aimed at the more correct ratio but had not achieved it. Was an even more perfect mathematical structure part of the revision which he had planned for the *Aeneid*?

4. See above, pp. 61-63. For these additive series in general, cf. above, pp. 37-39.

5. I have, somewhat arbitrarily, assigned the following values to the "incompleted" verses: .2 for verses composed of one metrical foot or of one and one-half; .3 for two feet; .4 for two and one-half feet; .6 for three and one-half feet; and .7 for four feet. There is no such thing as a true "half-line," i.e., a verse composed of three metrical feet. For divisions of thought within a verse, the same values are given to the part-lines. (The following fractions would be slightly more accurate: 1/6 and 1/4 instead of .2; 1/3 instead of .3; 5/12 instead of .4; 7/12 instead of .6; and 2/3 instead of .7; the resultant ratio is the same except for the very short passages, but for these we have evidence that .4 and .6, for example, seem more accurate and more in accord with Vergil's regular procedure than 5/12 and 7/12; see above, p. 61. and below, note 33.)

6. These include II, 76; III, 230; IV, 273, 528; V, 595 (half-line); VI, 242; VIII, 46; IX, 29, 121, 151, 529; X, 278, 872; XI, 391 (half-line); XII, 218 (half-line), 612-613; see below, Chapter 5, "Interpolations."

7. If $M/(M + m)$ is .621, i.e., .003 from the perfect .618, the ratio $m/M$ will show a variation about two and one-half to three times as great in the opposite direction, i.e., .007 to .009, or .613 to .611; if $M/(M + m)$ is .614, $m/M$ will be about 628. It is therefore unnecessary to list both ratios.

8. A few other bipartite proportions are listed in Tables II-IV, where they seem part of a larger unit (cf., e.g., Nos. 399-400, 617, 722, 744, 768-769), and in Table VII, a supplementary list of ratios.

9. This passage (II, 567-587) is the minor of No. 52; cf. also No. 50 and the ratio in the third main division (No. 947); for the importance of the ratios in this much discussed passage, see below, pp. 85 f.

10. Nos. 159 and 167 could be listed equally well in Table II, since the two speeches in M or m would justify a tripartite division.

11. A full stop at the end of verse 939 appears in almost every edition of Vergil prior to that by Heyne in 1775; see Duckworth, *TAPhA* 91 (1960), p. 202, note 34.

12. Many noncontiguous proportions which I consider secondary have been excluded from the tables; cf. above, note 2, and see above, pp. 56 f.

13. For other link ratios, see Nos. 170, 217, 277, 339, and, in later tables, Nos. 357, 553, and 869.

14. For instances in the later books of the *Aeneid* of the combination of speech and narrative in the major part, see, e.g., Nos. 439, 447, 453, 458.

15. On the tripartite divisions in VI, 760-807 and 808-853, see above, Chapter 2, note 21.

16. The chart-index in Table IX presents the ratios for all twelve books of the *Aeneid*; see above, note 1.

17. *L'Art de Virgile*, p. 855.

18. With 725-727 removed, the ratio in No. 467 changes from .621 to .603. Cf. also Nos. 702 (Table III) and 940 (Table IV), where lines 725-727 likewise belong to what follows rather than to what precedes; these, being shorter passages, are even more decisive; if we omit 725-727, the ratios in Nos. 702 and 940 move from .60 to .533 and from .610 to .579 respectively. On the importance of the ratios in general for the proper division of the text, see Chapter 5, "Paragraphing."

19. Both IV, 305-387 (No. 529), and VI, 56-123 (No. 562), were cited in Chapter 2 as illustrations of "The Framework or Recessed Panel Pattern"; see above, pp. 23 f.

20. The deaths of Acron and Orodes in X, 719-746 (= No. 665) as minor is framed by fighting in 689-718 (= No. 279) and in 747-761 (= No. 666) which together form the major. Several proportions begin with 762; cf. 762-772 (No. 883), 762-788 (No. 882), and 762-832 (No. 280). These passages all indicate that a new narrative unit begins after 761 and thus support the paragraphing of Sabbadini and Mackail against Hirtzel, who begins a new section at 769. See Chapter 5, "Paragraphing."

21. See Chapter 2, "Alternation and Contrast." For 641-817 as a main division of Book VII, cf. Nos. 962 and 963 in Table V.

22. Actually we have here a double proportion, (a + c)/b and a/c, a being Aeneas' praise of Pallas and c the lament of Evander; see above, p. 50. In most instances I have ignored the noncontiguous secondary proportions such as a/c or c/a which often appear in connection with the ratios in Table III, and even more frequently in Table IV, where we find also a/d, a/e, b/d, etc.; cf. above, note 2.

23. See Chapter 2, "The Framework or Recessed Panel Pattern."

24. See above, Chapter 2, note 1.

25. For ratios with arabic numbers preceded by a capital S, see below, "Supplementary List of Ratios," and Table VII.

26. On Book V, see above, pp. 25 f.

27. Cf. Cartault, *L'Art de Virgile*, pp. 725 and 727, who indicates a break after line 255. Hirtzel, Durand, and Sabbadini begin a new paragraph after 259, but Mackail agrees with Cartault in ending the section at 255.

28. Cf. also Nos. 711 and 712, where the ratios, with 1a-1d omitted, change from .623 to .592 and from .631 to .663. See below, Chapter 5, "Spurious Passages."

29. See above, Chapter 2, note 23. Some proportions include 748-756; cf. Nos. 362, 364, and 482.

30. See above, pp. 22 f. The ratios in the third main division of Book XI are numerous and confused. There appear to be two series of proportions, one ending at 867, the other going through 915, as if 868-915 had been added and a second series of ratios had been superimposed on those ending at 867; cf. the Chart-Index of Book XI in Table IX.

31. On the effective alternation in VII, 641-817, see above, pp. 20 f. A five-part alternating pattern with the major including the three groups of unimportant warriors (655-690, 706-782) and the minor composed of the more important characters (641-654, 691-705, 783-817) produces a less accurate ratio of .640 (112.6/175.8).

32. Book V (No. 987) differs from the others in that the second division (545-603) combines with part of the third, 604-745, to form the major. This is an added argument to show that, whether we view V as composed of two or of three main divisions, the *ludus Troiae* (545-603) is to be separated from the contests ending at 544; see above, pp. 25 f.

33. This appearance of the Fibonacci series in majors and minors totaling 13.6 and 8.4, or 22 and 13.6, or 47.6 and 29.4 is all the more surprising since my totals are an approximation to the nearest decimal. With the hexameter verse composed of twelve units (each consisting of one long or two short syllables) the hephthemimeral caesura would produce most accurately, e.g., instead of 13.6 and 8.4, a major of 13 7/12 and a minor of 8 5/12. But then the Fibonacci series disappears and the ratio is .617, not .618. Does the presence of the Fibonacci series in these passages indicate that Vergil used a decimal system? The exact .618 ratio is also achieved by totals with decimals which fall into other Golden Mean series; see No. 24: major of 15.2 and minor of 9.4 = 76 and 47 times .2; No. S1036: major of 37.6 and minor of 23.2 = 47 and 29 times .8; these numbers are in the series 1, 3, 4, 7, 11, 18, 29, 47, 76, . . . (cf. No. S997); Nos. 317 and S1034: major of 16.2 and minor of 10 = 81 and 50 times .2, in the series 2, 5, 7, 12, 19, 31, 50, 81, . . . (cf. No. 869).

34. The distribution of the exact .618 ratios is interesting; two-thirds of them occur in five books: VII has four, I and XI each have six, IV has seven, as does XII, the book which Mackail, *CJ* 26 (1930-31), p. 17, praises as marking "the utmost of what poetry can do."

35. These and other series appear also in *Georgics* II-IV. For the Fibonacci series, or multiples thereof, cf. the minors, majors, and totals in III, 1-48 (9, 15, 24, 39 = 3, 5, 8, 13 times 3); III, 295-321, 339-383 (27, 45, 72 = 3, 5, 8 times 9); III, 478-566 (34, 55, 89); see above, Chapter 3, "Proportions in the *Georgics*."

36. We saw above in the case of the .618 ratios that different Golden Mean series appear in the totals with decimals; this is true also of the approximate ratios: for the Fibonacci series, cf., e.g., No. 31: minor of 3 and major of 4.8 (= 5 and 8 times .6); Nos. 33 and 337: minor of 7 and major of 11.2 (= 5 and 8 times 1.4); No. 92: minor of 9 and major of 14.4 (= 5 and 8 times 1.8); No. 121: minor of 11 and major of 17.6 (= 5 and 8 times 2.2); No. 266: minor of 9.6 and major of 16 (= 3 and 5 times 3.2). For examples of the other series in the totals with decimals, cf. No. 53: minor of 10 and major of 16.4 (= 25 and 41 times .4) and No. 58: minor of 5 and major of 8.2 (= 25 and 41 times .2); these

are both in the series 2, 7, 9, 16, 25, 41, . . .
No. 61: minor of 4.6 and major of 7.4 (= 23 and 37 times .2) and No. 95: minor of 9.2 and major of 14. 8 (= 23 and 37 times .4); these are both in the series 1, 4, 5, 9, 14, 23, 37, . . .

37. *On Growth and Form,* II, p. 923, note. See also Thompson, *Science and the Classics,* pp. 207 f.

38. Cf. Chapter 5, "Vergil and Pythagoreanism," and Appendices A and B.

39. This grouping of books in the second half of the *Aeneid* accords well with Vergil's arrangement of his material: VII-VIII, preparation for the conflict; IX-X, to the end of the first great battle; XI-XII, to the final defeat of Turnus by Aeneas; cf. Heinze, *Virgils epische Technik,* p. 180. On the close union of V and VI, see Heinze, p. 456.

40. Books V and XI are relatively unimportant books; they are considered the lightest of all the odd-numbered books, and properly so, since they stand between most tragic and significant books. Yet, surprisingly enough, they are considerably longer than the other books with odd numbers. In the first half of the poem Book V (871 lines) is exceeded only by VI (901 lines), and XI, with 915 lines, is the longest book of the *Aeneid* with the exception of XII (952 lines). I have never understood this feature of V and XI, but perhaps the combination of V with VI and of XI with XII as the minor parts in the over-all proportion in each half of the poem explains their extra length.

41. See above, p. 40.

42. See above, p. 41.

43. See below, Chapter 5, "Spurious Passages."

# Chapter 5
# THE VALUE OF THE PROPORTIONS

WE TURN NOW to a number of topics suggested by Vergil's use of mathematical proportions. These concern Vergil's method of poetic composition in general (both *how* and *why* he composed by means of the Golden Section) and the text of the *Aeneid* in particular. The presence of hundreds of passages throughout the epic containing exact or approximate Golden Mean ratios throws important new light on many aspects of the text: half-lines, interpolations, spurious passages, transpositions, and even the proper paragraphing of the text. I shall attempt to show that the ratios in the books may serve as a useful control against the fanciful conjectures of many editors. It will also be necessary to consider the proposed revision of the *Aeneid* in the light of its mathematical structure, and, finally, to discuss the authenticity of the *Appendix Vergiliana*, where many instances of the Golden Section likewise appear.

## VERGIL'S METHOD OF COMPOSITION

The Donatus-Suetonius Life (23) gives us most important information concerning Vergil's procedure:

> Aeneida prosa prius oratione formatam digestamque in XII libros
> particulatim componere instituit, prout liberet quidque, et nihil in
> ordinem arripiens.

This preliminary prose outline presumably contained the tripartite divisions and subdivisions outlined above in Chapter 2 and probably much detail about each. Even the smaller units of the narrative were undoubtedly marked out. Then Vergil began to write *particulatim*, i.e., he composed the short sections just as he pleased and taking nothing in order. But the small units, as well as the larger, were all written on the basis of the exact or approximate Golden Section.

We have seen in Chapter 4 the various patterns for m/M used to achieve the Golden Mean ratios: two parts, i.e., a/b or b/a; three parts, e.g., a/(b + c) or (a + b)/c; three parts with the central portion framed, i.e., b/(a + c) or (a + c)/b; and the more elaborate interlocked patterns, e.g., (b + d)/(a + c) and (b + d)/(a + c + e). Here we find tripartite, framework, and alternating patterns very similar to those appearing in the architecture of the *Aeneid* as a whole (Chapter 1) and in the structure of the individual books (Chapter 2). To reach the Golden Mean ratio, Vergil used again and again the various Golden Mean series, and especially the Fibonacci series (and its multiples) which appears in the tables more than 300 times.[1]

The shortest passages with ratios (many of them only 10 or 15 lines in length) regularly combine to form larger proportions until we reach the main divisions of the books. Also, in most instances, each proportion becomes the major or minor (or a part of either) of the next larger proportion until we come to the final ratio in the book. In order to illustrate this aspect of Vergil's procedure we may take, for instance, *Aen.* IV, 54-89 (No. 93), where the major (68-89) combines with the major (105-128) of No. 393 and the minor (54-67) combines with the minor (90-104) of No. 393 to produce the ratio in 54-128 (No. 754); 54-89 combines with 129-159 (No. 394) to form the major of 54-159 (No. 524), and 90-128 (No. 393) becomes the minor; 54-159, in turn, is the major of 1-172 (No. 951), the first main division; this is the minor of 1-449 (No. 985) and also part of the major of 1-705 (No. 984), where all three divisions combine in a Golden Mean ratio. This combination of the smaller units into larger proportions will be more easily understood when presented in the form of a diagram:

THE VALUE OF THE PROPORTIONS

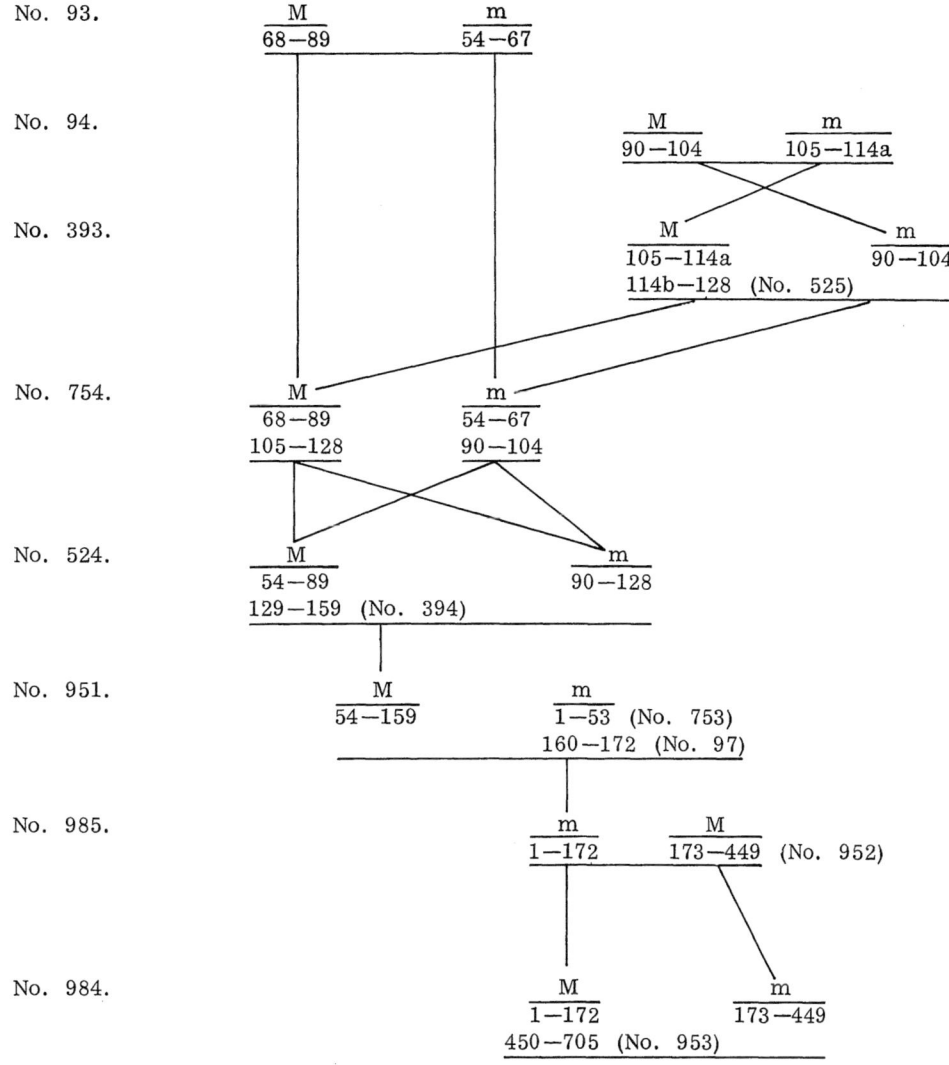

In Book VII, 135-169 (No. 190), the major (148-169) is No. 582 and the minor (135-147) is No. 581; the major combines with 117b-134, the major of No. 188, and the minor combines with 107-117a, the minor of No. 188, to produce the ratio in 107-169 (No. 814); this passage and 170-285 (No. 815) form the major of No. 960, the minor being 1-106 (No. 809), and 1-285 is the major of the exact .618 ratio (No. 990) composed of the first and third main divisions, 1-285 and 641-817. This is diagramed on the following page.

Equally intricate, perhaps even more so, are the relationships of the various proportions in Book IX if we begin with 176-183 (No. 629) and trace the combinations up to the main divisions. 176-183 is the minor of 176-196 (No. 240), the major being 184-196 (No. 241); the major and minor combine with the component parts of 197-223 (No. 430) to produce the ratio in 176-223 (No. 846); 176-223 combines with 224-313 (No. 631) to create the proportions in 176-313 (No. 847); this passage is the major of No. 968, the minor being 367-449 (No. 244), but 367-449 is also the major of No. 969, and the minor is 314-366 (No. 633); finally, 176-449, the second main division, is the major of No. 992, the minor being 1-175 (No. 967), the first main division. The diagram on page 71 will make this clearer.

The various ratios and the manner in which they combine in each book as major or minor of larger units could be traced by similar diagrams, but such diagrams do not show clearly the

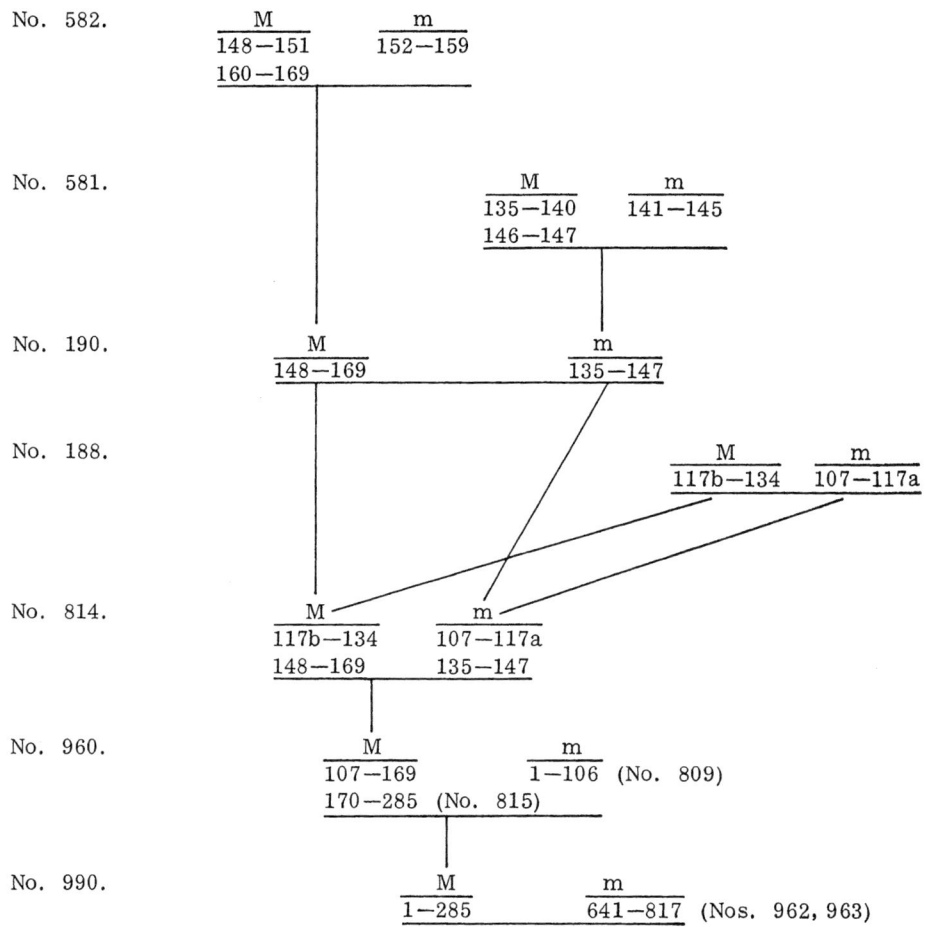

amazing correlation between the tripartite divisions and subdivisions of the books, as originally determined and listed in Chapter 2, and the proportions subsequently discovered in these same books; this can be presented more effectively by means of charts. The diagram given above for Book VII, for instance, shows that the component parts which determine the ratio in No. 960 are 1-106, 107-169, 170-285; these are also the three narrative subdivisions of 1-285, the first main division of VII. Likewise, in Book IX, where the second main division (in the patterns c/a and b/c) divides into 176-313, 314-366, 367-449 (Nos. 968 and 969), we have the three parts of the Nisus-Euryalus episode, the first in the Trojan camp, the second in the camp of the enemy, the third their departure and death. I discussed above the perfect correlation between the narrative subdivisions of Book X and the component parts determining the ratio in each of the three main divisions.[2]

This correlation is so typical of Vergil's procedure and so important a part of the composition of the *Aeneid* that some method seems necessary to present to the reader at a glance not only the manner in which the shorter passages combine into ever larger proportions but also this striking correspondence between the mathematical ratios and the subject matter of the poem. For this purpose I have devised as Table IX a Chart-Index for each book of the *Aeneid*. I give first the three main divisions of each book as established in Chapter 2 and include the subdivisions, which are usually tripartite; the chart for each book lists all the ratios in that book and reveals their combination into larger units; the proportions which form the component parts of the ratio in each main division are underlined, and a reference to the outline of the book facing each chart will show how frequently the tripartite subdivisions are identical with these component parts. Finally, by adding in parentheses arabic numbers to each passage, I provide for the reader a con-

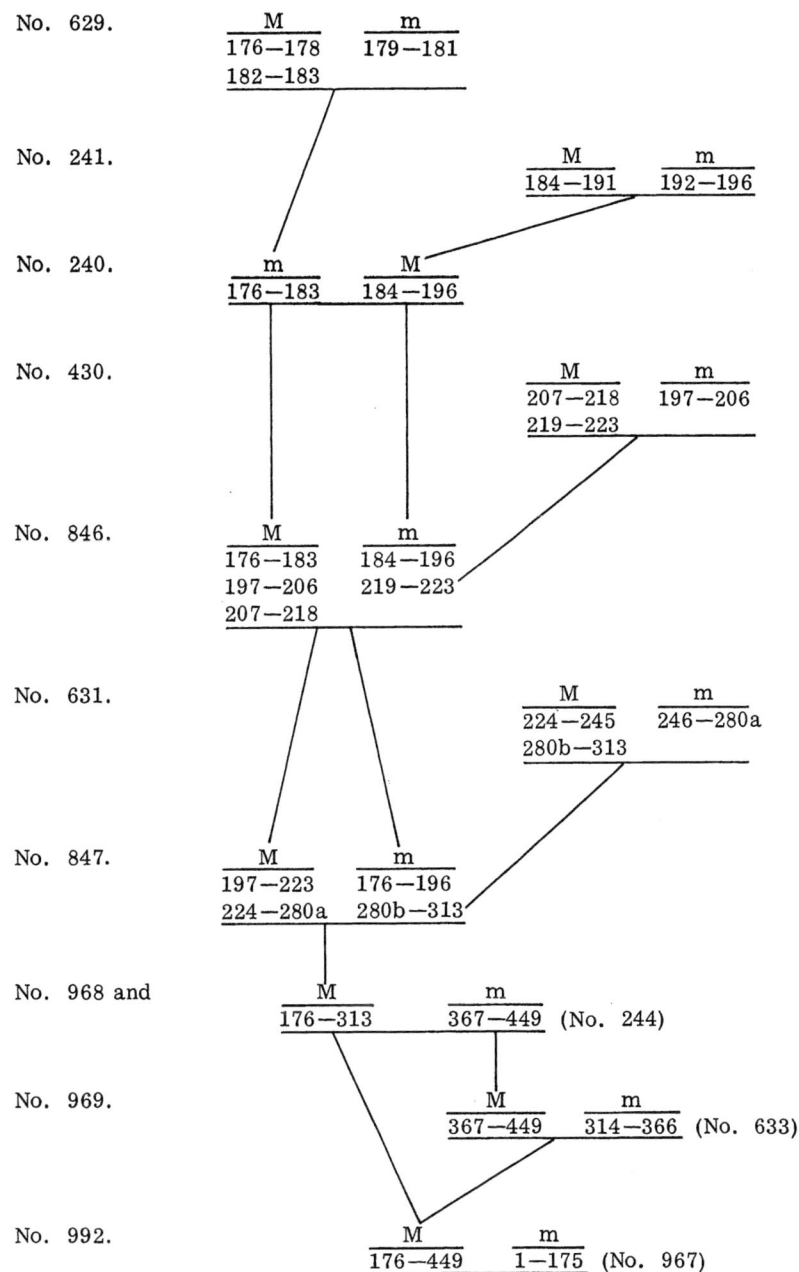

venient means of cross reference to Tables I-VII where the details of each ratio are to be found; this index seems all the more necessary since the proportions appearing in each book of the *Aeneid* are scattered throughout all seven tables.

The charts presented in Table IX reveal how in each book of the *Aeneid* the dozens of small ratios combine as building blocks into ever larger proportional units until we reach the three passages with ratios which in every instance are identical with the three main narrative divisions, as determined in Chapter 2.

These main divisions of the narrative, 36 in number, in most instances are subdivided into three parts. The exceptions are as follows:

Three main divisions too short to have the usual tripartite structure:
    V, 545-603: *ludus Troiae;*
    VII, 641-817: catalogue of Latin warriors (alternating groups of leaders);
    VIII, 370-453: Venus and Vulcan; making of armor (a bipartite division?);
One main division, I, 223-417, clearly bipartite:
    223-296, Venus and Jupiter; 297-417, Venus and Aeneas;
Two main divisions, both in Book V, composed of four or more parts: 1-544, six parts (arrival, preparation for the contests, the four contests), and 604-871, four parts (Iris and the Trojan women, burning of the ships and farewell, Venus and Neptune, death of Palinurus).

As apparent exceptions the following may be noted:

Two main divisions where two or three parts are in a recessed panel pattern:
    II, 1-249, with the three parts arranged (1) (2) (3) (2) (1);
    VI, 1-235, with the arrangement (1) (2) (3) (2);
One main division, VI, 548-901, where a double tripartite structure appears: 548-755, the Underworld, and 756-901, Anchises' forecast of Roman history.
Two main divisions, where the three parts are preceded or followed by a fourth:
    VIII, 1-369: the three parts concerning Aeneas at the site of Rome (107-379) are preceded by 1-106, Aeneas at his camp;
    XI, 468-915: the three parts concerning Camilla (468-867) are followed by 868-915, the epilogue or "coda."

Actually, these "apparent exceptions" belong with the examples of tripartite structure; when we include them, we have a total of 30 main divisions which contain three subdivisions each; in other words, Vergil's fondness for tripartite structure is shown by the fact that it appears in over 83 per cent of the narrative divisions.[3]

The ratios in the main divisions (described in Table V) are determined in almost every instance by three component parts, with m/M derived either from a/(b + c) or c/(a + b), etc., or from the framework pattern b/(a + c) or (a + c)/b. I, 223-417, has two parts corresponding to the twofold division of the narrative, and two main divisions have four parts determining the ratio: VIII, 1-369, where the component parts are identical with the preliminary scene of Aeneas at his camp (1-106) and the three subdivisions concerning Aeneas at the site of Rome (107-369); and IX, 1-175, the first and fourth parts (1-76, 123-175) corresponding exactly to two of the three subdivisions. V, 545-603, and VIII, 370-453, are too short as main divisions to be considered here, and if we exclude them we discover that the ratio is determined by three component parts in 31 main divisions out of a possible 34; i.e., the ratio in the main divisions has a tripartite structure in 91 per cent of the cases. Also, in every instance the component part is itself a passage containing a ratio.

When we add to these 93 component parts (31 times 3) the ten from I, 223-417, VIII, 1-369, and IX, 1-175, we have a total of 103 parts (underlined in the charts), of which 76, or 74 per cent, are absolutely identical with the subdivisions of the narrative. This astonishing correlation between the narrative units and the mathematical structure is best seen in the charts for Books I, II, III, IV, VII, VIII, IX, X, and XI, where six or more of the component parts of the ratio in the main divisions are the same as the subdivisions of the narrative in each book. The most perfect correlation appears in X and in XI, where 868-915 is best considered an epilogue (for the ratio in 868-915, see No. 454).[4] Furthermore, 30 narrative subdivisions which do not correspond to the underlined passages in the charts likewise contain ratios; cf., e.g., I, 520-656 (No. S1001), 657-756 (No. 362); III, 192-273 (No. S1006), 274-293 (S1007), 294-505 (No. 387); IV, 1-89 (No. S1010), 90-128 (No. 393), 129-172 (No. 755), etc. In V, where exact correlation is lacking, two or more narrative units combine to form the component parts of the proportions in the first and third main divisions.

When I first discovered the ratios in the main divisions of the *Aeneid,* I had no idea that so many proportions would appear in each book, with the larger proportions regularly developing from the smaller units, or that the passages with ratios were so closely related to the narrative, from the shortest episodes and speeches to the tripartite subdivisions and main divisions of each book. It was only when I had assembled the ratios of Tables I-VII in the charts (Table IX) and had compared them with the structural analyses of the books of the *Aeneid* (from Chapter 2) that the true nature of Vergil's composition became evident: he wrote the short passages with mathematical symmetry and combined them into increasingly larger units, with every part of every book containing exact or approximate Golden Mean ratios ranging from .60 to .636. Furthermore, each of the 36 main divisions of the poem (three in each book) contains a ratio, and in 74 per cent of the cases the component parts of the ratios are identical with the subdivisions (usually tripartite) of the narrative. It is staggering to think of the planning and the labor involved in working out so elaborate a correlation between the mathematical structure of each book and the divisions of the story, but such was Vergil's method of composition.

## VERGIL AND PYTHAGOREANISM

The first section of this chapter, in a sense a recapitulation and expansion of the description of Vergil's procedure in Chapter 4, reveals *how* Vergil composed the books of the *Aeneid* in Golden Mean ratios, and the chart-index in Table IX shows at a glance the correspondence between the proportions and the divisions of each book. This close correlation between the proportions and the narrative episodes (both small and large) supports my conclusion, based also on the large number of almost exact ratios and the use of the Fibonacci and other series,[5] that Vergil did not achieve the Golden Section in hundreds of instances by means of intuition or poetic instinct but by deliberate planning. But all readers of this work will ask the more difficult question: *why* did he compose the *Aeneid* in this fashion? Was he so obsessed by mathematical ratios that he desired every part of the *Aeneid* to reveal the Divine Proportion? We have seen also (in Chapter 3) the extent to which this same ratio appears in both the *Eclogues* and the *Georgics.* Was Vergil convinced that poetic beauty was enhanced by the use of the Golden Section? Was he, as many Vergilian scholars have maintained for other reasons, a Neo-Pythagorean? If so, perhaps he believed that mathematical relations reflect not only the essence of reality but the essence of poetry as well.

Even if space should permit, it would be unnecessary to describe in full detail the many arguments advanced by various Vergilian scholars to support their view that Vergil belonged to the Neo-Pythagoreans who, under the leadership of Nigidius Figulus, were active at Rome in the first century B.C. This revived Pythagoreanism stressed the symbolism of numbers and was, like most Roman schools of thought, eclectic, borrowing elements derived from Platonic and Stoic sources. Vergil's interest in astronomy and cosmology has usually been ascribed to the influence of Lucretius and Epicureanism, while his attitude toward the gods and immortality has been considered Stoic; it is equally possible that Vergil accepted the scientific and philosophical teachings of Neo-Pythagoreanism as providing a more unified doctrine best suited to his own interests and beliefs. The doctrine of reincarnation in *Aeneid* VI may be Pythagorean as well as Platonic, and the numerous references to Orpheus throughout Vergil's works are at least striking.[6]

One of the many problems connected with the much discussed Fourth *Eclogue* is the background of the poem; its basic source has been variously given as Messianic prophecy, the Sibylline oracles, Egyptian and Eastern religious and philosophical thought, Pythagorean doctrine, and more purely Roman ideas—the desire for a new age of peace; many features of the poem, such as the reference to Apollo, the song of the Sibyl, and the new age to come, are political as well as religious and can be paralleled by Roman coinage of the first century B.C.[7] But the Pythagoreans of Vergil's day had a system of cyclical world-ages with the recurrence of events (as in *Ecl.* IV, 31-36), and Suetonius' statement that Vergil gave especial attention to mathematics could mean that he studied Neo-Pythagorean doctrine as expounded by his contemporaries. Carcopino finds

Pythagorean mysticism everywhere in the Fourth *Eclogue*.[8] This seems an exaggeration; the *Eclogue* is basically Roman in that it reflects the political ideas current in Vergil's day, but it seems indebted to a degree both to Sibylline oracles and to Pythagorean doctrine, as well as to literary sources, Hesiod, Catullus, and others.

In *Ecl.* VI, 31-40, a passage usually considered Lucretian in content and style, Vergil reveals his interest in cosmological themes, and in *Georg.* II, 490-494, he writes as follows:

> felix qui potuit rerum cognoscere causas,
> atque metus omnis et inexorabile fatum
> subiecit pedibus strepitumque Acherontis avari.
> fortunatus et ille deos qui novit agrestis
> Panaque Silvanumque senem Nymphasque sorores.

This passage in its context refers back to 475-489, where Vergil expresses his desire to know the workings of nature (the pathways of heaven, the stars, the eclipses of the sun, the *labores* of the moon, etc.), but if this is not possible his delight will be the country with its valleys, streams, and forests; 475-482 is balanced by the *felix* passage in 490-492, 483-489 by the *fortunatus* passage in 493 f. Most commentators and writers on Vergil have taken line 490, *felix qui potuit rerum cognoscere causas,* as a reference to Lucretius, but several scholars point out that the victory over Fate and Acheron implies, not the Lucretian acceptance of annihilation upon death, but the belief in the immortality of the soul which results from the study of astronomy and cosmology; the man who is *felix* is therefore not Lucretius but Pythagoras.[9] In this case, Vergil here is employing Lucretian phraseology to present beliefs which are the very opposite of Epicureanism. The Pythagorean tone of the passage is strengthened by the reference in II, 536 ff., to the end of the Golden Age and the impious race now feasting on slain bullocks. It is significant in this connection that Ovid combines in one and the same passage (*Metam.* XV, 60-175) Pythagoras' knowledge of the nature of the universe (cf. *rerum causas* in 68 and *Georg.* II, 490), his doctrine of the immortality of the soul (158-172), and his condemnation of the *nefas* committed when man eats the flesh of animals; such a crime did not exist in the Golden Age (75 ff.). Whether Ovid echoes Vergil here or, as seems more probable, both poets go back to a common source is unimportant; Pythagoras' words in the *Metamorphoses* give support to the view that Vergil in *Georg.* II, 475 ff., is expressing Pythagorean ideas.[10]

Le Grelle argues from the Golden Section in *Georgics* I that Vergil is a Pythagorean poet, and he finds in the book much numerical mysticism;[11] e.g., the "Works" (43-203) and the "Days" (259-463a) total 365.4 lines, the number of the days of the year (cf. the final words in 463a: *sol tibi signa dabit*), and the exordium (5b-42) and epilogue (463b-514), with the "foyer astronomique" (204-258), add to 144, the square of twelve, the months of the year. Also, the number 36 (or fractions or multiples of 36) appears throughout the book; this is the Pythagorean Grand Tetractys, considered the source of eternal Nature and the essence of the Universe.[12] The descriptions of the temperate zones (the *terra habitabilis*) in 50-203 and 259-437 total 333 verses, half of 666, the triangular number of 36.

The numbers 333 and 666 had earlier been discussed by Maury in his lengthy treatment of the architecture of the *Bucolics*.[13] Maury likewise stresses Vergil's Pythagorean symbolism; he looks upon the corresponding poems, I-IV and VI-IX, as the columns of a "bucolic chapel" leading to V, the shrine of Daphnis-Caesar, and the verses in each group of four pastorals (with very minor adjustments recommended by earlier scholars) total 333 verses; thus in the eight *Eclogues* framing V we find the mystic number 666, which also has the value (by Greek numerical count) of the names of Caesar (*Eclogue* V) and Gallus (*Eclogue* X) and, incidentally, is the number of the Beast in *Revelation* XIII, 18.

Maury's views have been highly praised and as severely condemned.[14] If Vergil were a Neo-Pythagorean and as interested in the symbolic use of numbers as Maury and Le Grelle maintain, the Golden Section would certainly have appealed to him; although Maury did not detect the

Golden Mean ratios in the *Eclogues,* we now know that these ten poems reveal the use of mathematical proportions just as do the *Georgics* and the *Aeneid.* Vergil is definitely interested in numerical symmetry; this we have seen above in Chapter 2, particularly in the balanced structure of *Eclogues* I and IV and in *Aeneid* IV and VI, and also in his frequent use of the perfect number 28. As I mentioned above in Chapter 1, Vergil refers to Maecenas in each of the four *Georgics:* in line 2 in I and IV, and in line 41 in II and III; this is a curious example of numerical symmetry which, to the best of my knowledge, remains unexplained, except as a means of linking I to IV and II to III.

The many mathematical proportions in the *Aeneid* obviously do not weaken the view that Vergil was a Neo-Pythagorean or at least was interested in Pythagorean doctrines, but neither do they provide conclusive proof for the theory. I have found no evidence of numerical symbolism in the *Aeneid* in connection with the Golden Section beyond the occasional use of the perfect number 28 and the appearance of 333 in Jupiter's prophecy in I, 261-274: three years for Aeneas' rule, 30 years for Ascanius, 300 years for the Alban kings. Vergil's use of the Golden Mean ratio in all three major works explains Suetonius' statement in the Donatus-Suetonius Life (15) that Vergil gave especial attention to mathematics, and I am convinced that he employed mathematical proportions to such an extent because of his love for numerical symmetry, but—and this is the main reason why I do not believe that the presence of the Golden Section everywhere in the *Aeneid* proves his Pythagoreanism—*he was not the only poet of his day to compose in Golden Mean ratios.*

When I described in Chapter 2 the framework or recessed panel structure such as we find in several *Eclogues,* in *Georgics* IV, in *Aeneid* IV and in various passages in other books, I referred to the similar structure of Catullus LXIV. This poem, with its story of Ariadne enclosed by the wedding of Peleus and Thetis, suggests particularly the Aristaeus story in the second half of *Georgics* IV. The contrast between the happy wedding ceremonies of Peleus and Thetis and the unhappy story of Ariadne in Catullus corresponds to the contrast in Vergil between Aristaeus' success in having a swarm of bees brought to life and Orpheus' inability to restore Eurydice to life. We saw above in Chapter 3 that the Golden Section appeared in the relation of the Orpheus story to that of Aristaeus, and also within the central episode and the two parts of the framing story. When we turn to Catullus LXIV we find the same situation: the central story of Ariadne (52-250) breaks the poem into three parts, and these are in proportion in the pattern c/(a + b), with the ratio M/(M + m) = 250/408 = .613 (I add one verse for the lacuna after 253 and omit 378). Each of the three main divisions breaks into smaller units which are likewise in proportion:

| m/M: type | Minor | Total lines | Major | Total lines | Ratio: M/(M + m) |
|---|---|---|---|---|---|
| b/(a + c) | 1-11<br>31-51 | 32 | 12-30 | 19 | .627 |
| (a + c)/b | 116-237 | 122 | 52-115<br>238-250 | 77 | .613 |
| b/(a + c) | 251-322<br>384-408 | 98 | 323-383 | 60 | .620 |

The framework pattern here is the same as that used so frequently by Vergil.[15]

The influence of Lucretius on the thought and poetic technique of Vergil is well known. The structure of the *Georgics* is not unlike that of the *De Rerum Natura* in certain respects. Lucretius composed his poem in three units of two books each: I-II, the structure and the properties of the atom; III-IV, the soul and the senses; V-VI, celestial and terrestrial phenomena; and the endings of the even-numbered books are of a pessimistic nature: II, 1131 ff., on the decay of the world; IV, 1058 ff., on the evils and pain of love; and VI, 1090 ff., on plague and death, the passage which inspired the conclusion of *Georgics* III. Vergil's outlook was far less pessimistic than that of his predecessor. He composed the *Georgics* in two units of two books each, but he placed the more gloomy parts at the ends of I and III and made the conclusions of II and IV more cheerful. When we view the four books as a whole, we see that the conclusions contribute largely in making the *Georgics* an epic of universal significance: I, War; II, Peace; III, Death; IV, Rebirth.[16]

In another way Vergil may have been indebted structurally to Lucretius as to Catullus. If we take the six books of the *De Rerum Natura* in relation to each other, with I-II and III-IV as the major and V-VI as the minor,

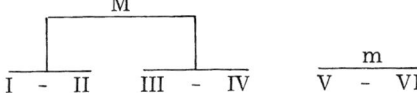

the ratio M/(M + m) = 4675/7422 = .630. This is the same c/(a + b) pattern which we have seen in the *Eclogues*, in the two halves of the *Aeneid*, in the *Aeneid* as a whole, and also in Catullus LXIV. Book V is by far the longest book of the *De Rerum Natura* and its length makes possible the c/(a + b) ratio; is it accidental that *Aeneid* V and XI are also conspicuous for their unusual length?[17]

Each book of the *De Rerum Natura* falls into main divisions which combine to form the Golden Section in the book as a whole:

| Book | m/M: type | Major | Total lines | Minor | Total lines | Ratio: M/(M+m) |
|---|---|---|---|---|---|---|
| I | (a+c)/(b+d) | 146-634<br>921-1117 | 687 | 1-145<br>635-920 | 434 | .613 |
| II | (c+d)/(a+b) | 1-332<br>333-729 | 732 | 730-1022<br>1023-1174 | 448 | .620 |
| III | (a+b)/(c+d) | 417-829<br>830-1094 | 677 | 1-93<br>94-416 | 417 | .619 |
| IV | (a+d+e)/(b+c) | 26-215<br>216-822 | 790 | 1-25<br>823-1057<br>1058-1287 | 490 | .617 |
| V | (b+d)/(a+c+e) | 1-90<br>509-770<br>925-1457 | 887 | 91-508<br>772-924 | 571 | .608 |
| VI | b/(a+c+d) | 1-42<br>535-1089<br>1090-1286 | 798 | 43-534 | 491 | .619 |

The ratios in Lucretius can be only approximate because of the uncertainty of the text; I have followed the text of Bailey's edition of 1947, adding one line for each indicated lacuna and deducting the bracketed passages. Even so, these and other proportions in the *De Rerum Natura* are surprisingly accurate.[18]

When I discovered the Golden Section in the *Aeneid*, and also in various *Eclogues* and in *Georgics* II-IV, I did not then realize that it would likewise appear to such a degree in Catullus and Lucretius. Horace does not mention the Golden Section in his *Ars Poetica*, but if the use of mathematical symmetry was common poetic procedure we should expect to find it in his poetry, and we do—in the structure of the *Ars Poetica* itself. The poem is usually divided into three parts, *poesis* (1-152), *poema* (153-294), and *poeta* (295-476),[19] and, in any case, the division between poetry and the poet and critic is clearly marked at 294; thus, either in the pattern c/(a + b) or b/a, we have the exact Golden Mean ratio: M/(M + m) = 294/476 = .618, and, surprisingly enough, this is also the Fibonacci series; m = 182 = 13 x 14; M = 294 = 21 x 14; M + m = 476 = 34 x 14; thus we have multiples of the usual series 13, 21, 34, so prominent in Vergil's poetry, and perhaps an indication that Horace knew more about Vergil's poetic methods than he revealed in his references to his friend's achievements.[20]

The fact that so many other Roman poets of the first century B.C. used the same Golden Mean ratio indicates that the Golden Section played an important part in Roman poetic theory (perhaps going back to Ennius?[21]) as well as in ancient mathematics, music, art, and architecture, and this strengthens my conviction that the extent to which we find the Golden Mean ratio in Vergil's poetry is strong evidence of his interest in mathematical symmetry but does not necessarily prove that he was a Pythagorean. If for other reasons, such as those listed above, we accept Vergil's Neo-Pythagoreanism, his use of mathematical ratios can serve as a supporting argument and also can be explained more readily.

But if Vergil was not a Neo-Pythagorean, how are we to account for his interest in—almost an obsession with—Golden Mean ratios? Perhaps the answer lies in the formal beauty innate in the Golden Section. Sarton points out that a number of artists and mystics in modern times have believed that the Golden Section is one of the secrets of beauty.[22] Almost a century ago Fechner demonstrated, by requesting opinions from a large number of persons of both sexes, that a rectangle constructed on the Golden Section, with its sides as 21:34, has a far greater appeal than any other rectangle; it had absolutely no rejections and received 35 per cent of the preferences.[23] The result of this poll has been brushed aside as inconclusive by many writers, who apparently ignore the fact that the two rectangles closest to the Golden Rectangle each received over 19 per cent of the preferences; in other words, the Golden Rectangle and the two approximating it received about 74 per cent of the total number of votes; viewed in this light, the poll seems significant and implies that the Golden Mean ratio produces the aesthetically perfect rectangle.[24]

Did Vergil likewise believe that poetic passages or groups of passages bearing this same ratio, exact or approximate, had a mathematically formal beauty which would contribute to the perfection of the structure of his epic? I can think of no other explanation which will account for the presence of the Golden Section everywhere in the *Aeneid* and elsewhere in his poetry.

## THE PROBLEM OF THE HALF-LINES

The hemistichs, or half-lines, of the *Aeneid* have been the subject of lengthy discussion. No other Greek or Roman poet who wrote in dactylic hexameters has verses composed of less than the usual six feet. In the *Aeneid* there are 59 verses which vary in length from one to four dactylic feet.[25] These hemistichs are often included among the *tibicines,* or temporary supports, mentioned in the Donatus-Suetonius Life (24); these lines then would have been replaced by complete verses, had Vergil lived to give to the poem its final revision.

Belling argues that the *quaedam imperfecta* of Suetonius cannot refer to the half-lines, since, also according to Suetonius (32), Vergil read Books II, IV, and VI to Augustus and Octavia in a completed form *(perfectaque demum materia)*. These three books contain 17 hemistichs, almost one-third of those in the entire *Aeneid,* and Book II has ten, more than any other book.[26]

Büchner looks upon the *tibicines* as roughly sketched-in passages rather than as incompleted verses, but he believes that Vergil would not have been the one and only ancient poet to employ half-lines, and he is convinced that the poet would have replaced them by whole lines in his revision of the poem.[27] As a supporting argument, Büchner refers to "der sinnlose Halbvers 3, 337"—apparently an inaccurate allusion to III, 340, *quem tibi iam Troia—*, where Andromache, asking about Ascanius, thinks of the fate of her own son Astyanax (cf. 489-491); overcome by emotion, she is unable to continue. This is the only hemistich in the entire *Aeneid* actually unfinished, where the sense as well as the metrical line is incomplete. But this half-line is not truly "sinnlose"; for Andromache in this instance to break off in the middle of the verse is most effective and suggests that the hemistich was intentional.

Many other half-lines are also effective and may well have been deliberate; Sparrow points out that these occur in spoken passages, appear in an emotional or rhetorical context, and embody, or immediately precede, an exclamation, question, or command. He lists as effective 16 half-lines;[28] these are, in addition to III, 340, the following:

| | | |
|---|---|---|
| II, | 66: | disce omnes. |
| | 346: | audierit! |
| | 640: | vos agitate fugam. |
| | 787: | Dardanis et divae Veneris nurus. |
| III, | 316: | ne dubita, nam vera vides. |
| | 640: | rumpite. |
| IV, | 361: | Italiam non sponte sequor.' |
| VI, | 94: | externique iterum thalami. |
| | 835: | proice tela manu, sanguis meus!— |
| VII, | 129: | exitiis positura modum. |
| VIII, | 536: | laturam auxilio. |
| IX, | 467: | Euryali et Nisi. |
| X, | 728: | ora cruor— |
| XI, | 375: | qui vocat.' |
| | 391: | semper erit? |

Sparrow shows that other half-lines occur in contexts where they have no artistic effect and many appear in passages that suggest incompleteness, and he believes that these would not have been retained in the final revision; those listed above as effective may have been deliberate on Vergil's part. Sparrow's conclusion is noncommittal: "Virgil may in the course of his revision have noticed their effectiveness, and intended to make still further use of the device, or he may never have decided whether or not any of them should be eventually retained."[29]

Fowler is far more definite: he considers that once or twice Vergil may have left a line unfinished because he had not worked out the whole passage, but looks upon the great majority of the half-lines as intentional; they indicate a pause and relieve the listener's ear from the monotony of the hexameter;[30] thus, according to Fowler, Vergil is employing an artistic device to secure variety. On the other hand, Walter asserts that the half-lines are not a means of increasing the poetic effect but rather reveal the incomplete state of the *Aeneid*.[31] *Quot homines tot sententiae!*

The proportions listed in Tables I-VII were established before I had examined in detail the views of Sparrow and the other writers on the half-lines. But the problem of the hemistichs arose at the very beginning of my calculations: should they be treated as whole lines or as fractions of lines? The latter seemed the more correct procedure, since the *Aeneid* was meant to be heard; the listeners would not count lines, but they would receive through the ear a sense of harmonious proportion; also they would hear each incomplete verse as a fraction of a line. Then, when I began to count the half-lines as fractions (.2, .3, .4, .6, or .7), I discovered that in many instances the ratio was more accurate and that in others it was possible only by the use of fractions. I cited above as illustrations V, 322 (No. 546), and X, 17 (No. 646), with ratios of .614 and .617; if in each instance the hemistich is treated as a whole line, the proportions become .583 and .588 respectively and the Golden Section disappears.[32]

In many of the proportions containing hemistichs, especially in longer passages, the mathematical variation is too slight to be of any value (e.g., .619 vs. .620), or the differences cancel out (e.g., .616 vs. .620). The shortest passages with ratios throw the most light on the mathematical value of the half-lines. Some proportions are more accurate if the half-line is treated as a whole verse; but since these are less numerous than those where the use of fractions gives a greater accuracy, I have adopted as my standard procedure the counting of all half-lines as fractions.

THE VALUE OF THE PROPORTIONS

It is now necessary to re-examine the ratios in order to determine the half-lines which seem necessary for Vergil's composition by mathematical proportions and also those which, if completed, would provide better ratios. I cannot state as a certainty that Vergil in his final revision would have retained the half-lines in the first group and filled out the verses in the second group, but the importance of so many half-lines for the mathematical structure of the poem at least makes plausible the assumption that these half-lines were deliberately introduced into the epic.

Table X, the first of the two tables presenting the mathematical evidence for the value of the hemistichs, gives the 41 half-lines where the ratio is better when the half-line is treated as a fraction; the final column contains the less accurate ratio which results when the half-line is calculated as a full verse. I have avoided, as much as possible, the inclusion of proportions containing two or more hemistichs, and I list in both Tables X and XI only the shorter proportions (usually from columns one and two in the charts = Table IX), since these produce the most decisive results and, in case of conflict, are to be preferred.

In this list of 41 half-lines, involving 64 proportions, the variation in some instances is slight, but the cumulative effect is impressive. The table contains five perfect ratios of .618, and in 28 proportions of the 64 listed the ratio ranges from .615 to .621; these exact and almost exact examples of the Golden Section are possible only if we count the half-lines as fractions; when we treat them as whole lines, the ratio moves away from the area of .618, often as far as .643, or in the other direction as far as .582 or .583. The final column of Table X contains no ratios in the range from .615 to .621 and only twelve (out of 64) in the range from .610 to .626. This would seem to indicate that Vergil deliberately employed many half-lines not only for their dramatic or rhetorical effect (as Sparrow and Fowler claim) but also as an aid to more exact numerical ratios. Of the 16 half-lines which Sparrow looks upon as effective and possibly intentional, eleven appear in Table X, where the use of the hemistich as a fraction makes for a more exact ratio. But since the 41 hemistichs in Table X comprise over two-thirds of the total number, the presence in this list of eleven of Sparrow's 16 effective half-lines seems merely a normal distribution and neither proves nor disproves his conclusions concerning effective half-lines.

Table XI gives the shorter list of half-lines which produce less exact ratios when counted as fractions, and these are perhaps the hemistichs which would have been changed to full lines had Vergil lived to revise the *Aeneid*. There are two reasons for this conclusion: (1) the ratio in each instance is more accurate if the line is counted as a complete verse, and (2) the Fibonacci or other Golden Mean series appear in the numbers of the major and minor in almost every case. The final column of Table XI gives the series, with the numbers of minor, major, and total underlined (often reduced from multiples; e.g., 6, 10, 16 are listed as 3, 5, 8 in No. 497).

When ratios are in conflict, the shorter proportion is to be preferred, e.g., in the case of II, 767.[33] *Aen.* V, 574, has been added to the list with some hesitation.[34]

Table XI contains 14 half-lines and 20 ratios, only one of which is in the .615 to .621 range when the hemistichs are treated as fractions; but if these same hemistichs are counted as complete verses, eleven of the 20 ratios move into the area from .615 to .621 and two perfect .618 ratios appear. Also, and this seems the more convincing argument, in 18 of the 20 passages the change from hemistichs to whole lines produces in the totals of minor and major the various Golden Mean series used so frequently by Vergil.[35] The series appearing in Table XI are the following:

    1, 1, 2, 3, 5, 8, . . . (Fibonacci)        10 times
    1, 3, 4, 7, 11, 18, . . .                    3 times
    1, 4, 5, 9, 14, 23, . . .                    3 times
    1, 5, 6, 11, 17, 28, . . .                  2 times

We have here, I believe, strong evidence that Vergil would have replaced these 14 hemistichs by complete lines in his final revision of the poem.

The half-lines listed in Tables X and XI total 55; the four remaining half-lines require special consideration:

1, 560 (.2, as fraction): this hemistich appears in no proportion by itself. I, 534, is better as a fraction (cf. Nos. 722 and 22 in Table X). Nos. 721 and 359 both include I, 534, and I, 560; with 534 as a fraction and 560 as a whole line, the ratios move from .613 to .60 and from .612 to .617 respectively. No. 721, as the slightly shorter passage, might favor the addition of I, 560, to Table X, but in view of the conflicting evidence I prefer to make no decision.

II, 623 (.4, as fraction): this hemistich likewise does not appear by itself, but in combination with II, 614, which is better as a fraction (cf. No. 53 in Table X); if we treat 614 as a fraction and 623 as a complete line, the ratio of No. 735 moves from .632 to .621; this suggests that II, 623, should be included in Table XI.

V, 294 (.6, as fraction): if this verse is treated as a whole line and not a fraction, the ratio in No. 121 moves from .615 to .621, a change which is meaningless. But V, 322, must be treated as a fraction (cf. Nos. 546 and 124 in Table X). No. 122 contains both V, 294 and 322; with both as fractions the ratio is .619, but with 294 as a whole line we have .615; the difference is slight but favors the addition of V, 294, to Table X; this conclusion is supported by the similarity of the hemistich (*Nisus et Euryalus primi*) to IX, 467 (*Euryali et Nisi*), which is better as a fraction (cf. No. 248 in Table X).

V, 595 (.6, as fraction): this apparently interpolated hemistich does not appear by itself in a proportion but with V, 574 (listed in Table XI); with both as whole lines, the ratios move as follows:

| No. 134 | .608 ⟶ .610 |
| 133 | .627 ⟶ .627 |
| 955 | .633 ⟶ .627 |

With 574 as a complete verse and 595 as a fraction, the change is this:

| No. 134 | .608 ⟶ .616 |
| 133 | .627 ⟶ .632 |
| 955 | .633 ⟶ .625 |

The evidence here favors the treatment of V, 595, as a fraction and its addition to Table X; this is supported by the omission of the conclusion of the verse from the better manuscripts.

If we add V, 294 and 595, to Table X and II, 623, to Table XI, we thus have 43 half-lines which as fractions produce better mathematical ratios, 15 where the ratios would be improved and the usual mathematical series would appear if the hemistichs were replaced by complete verses; one half-line (I, 560) can not be assigned to either table.

As I said above, some scholars, e.g., Walter and Büchner, believe the half-lines to be a sign of the incomplete state of the *Aeneid* and assume that all would have been changed to full lines by the poet had he lived. Such an assumption is unwise when we find that 41 (or 43) hemistichs out of 55 (or 58), in other words 74 per cent, are necessary as fractions to make possible the usual Golden Mean ratio; these seem intentional and would probably not have been revised. On the other hand, the substitution of the hemistichs in Table XI by complete verses would produce more accurate ratios, and the totals of the majors and minors would then fall into the normal mathematical series which Vergil favored. These are the half-lines which he probably would have reworked in the final revision of the poem.

## INTERPOLATIONS

Another difficulty in establishing the proportions discussed in Chapter 4 and listed in Tables I-VII was caused by interpolations—verses repeated from another context and bracketed by modern editors. In the shorter passages the retention or rejection of a single verse could affect the ratio very materially; e.g., in a passage of 13 lines, with a major of 8 and a minor of 5, the ratio of $M/(M + m) = 8/13 = .615$; if one verse is subtracted from the minor, we have 8/12, or .666; if the verse is taken from the major, we have 7/12, or .583. Let me illustrate by means of two verses, selected in part because the passages involved contain no half-lines to confuse the issue:

IX, 85, which Hirtzel, following Ribbeck, brackets as an interpolation; if we retain the verse (as I have done), the ratios in the following proportions are those in column 2; if we reject IX, 85, the ratios change to those in column 3:

| No. 627 | .633 | → | .621 |
| 237 | .630 | → | .638 |
| 238 | .60 | → | .595 |
| 428 | .625 | → | .630 |

IX, 121, bracketed not only by Hirtzel but by other modern editors (Janell, Sabbadini, Mackail, Durand); I have rejected the verse but if, on the contrary, we include it in the count, the ratios of Nos. 844 and 843 change as follows:

| No. 844 | .613 | → | .575 |
| 843 | .622 | → | .630 |

My reason for retaining IX, 85, and rejecting IX, 121, is not an arbitrary desire to produce more accurate ratios. IX, 121, is bracketed by editors because it is lacking in the better MSS, whereas there is no MS authority for the rejection of IX, 85, and other modern editors do not consider it an interpolation.

My procedure throughout has been to reject those verses which are missing in the better MSS and which editors agree in viewing as interpolations. These verses are the following (with notations from Hirtzel's critical apparatus):

| II, 76 (= III, 612): | *om.* $M^1 P \gamma b$ |
| III, 230 (= I, 311): | *sine dubio ex* i. 311 *interpolatus*[36] |
| IV, 273 (cf. IV, 233): | *om.* $M P \gamma^1 a^1 b^1$ |
| IV, 528 (cf. IX, 225): | *om.* $M P \gamma^1 a^1 b^2 c^1 m$ |
| VI, 242: | *om.* $FM$ (*in textu*) $P a^1 cm$, *habent* $R \gamma b$ *et* $M$ *in margine* |
| VIII, 46 (= III, 393): | *om.* $M P \gamma^1 a^1$ |
| IX, 29 (=VII, 784): | *om. codd. omnes boni* |
| IX, 121 (= X, 223): | *om. codd. optimi* |
| IX, 151 (= II, 166): | *om. editio Parmensis*[37] |
| IX, 529 (= VII, 645): | *om. codd. praeter deteriores* |
| X, 278 (= IX, 127): | *om.* $M P \gamma m$ |
| X, 872 (= XII, 668): | *om.* $M P R \gamma^1 a^1 b$ |
| XII, 612-613 (cf. XI, 471-472): | *om.* $M P R \gamma b$[38] |

In general, the rejection of the verses listed above, most of which are found elsewhere in the text, is supported by the proportions in which they appear. I give below in Table XII the verses under discussion and the variation in the ratios if the verse is retained. In the case of IV, 528, VIII, 46, X, 278, and X, 872, several proportions include also the hemistichs IV, 516, VIII, 41, X, 284, and X, 876, which are listed in Table XI, and I have included the results if these half-lines are treated as complete verses. As in the case of Tables X and XI, I list only the shorter proportions since they produce the more decisive changes in the ratios.

The manner in which the ratio in almost every instance moves away from the area of the Golden Section if the interpolated verse is included in the proportion gives strong support to the evidence of the manuscripts and confirms the rejection of the verses by modern editors. The only conflicts (six in a list of 50 ratios) appear in III, 230 (No. 73), IV, 273 (No. 760), IV, 528 (No. 108), VIII, 46 (No. 204), IX, 29 (No. 626), and IX, 529 (No. 434), and the majority of the ratios for these verses favors their rejection. It is significant that the treatment of the four hemistichs (IV, 516, VIII, 41, X, 284, and X, 876) as full lines argues also for the omission of IV, 528, VIII, 46, X, 278, and X, 872, and the rejection of these verses as interpolations in turn supports the inclusion of the four half-lines in Table XI.

Hirtzel brackets three other verses: IX, 85 (already mentioned), IX, 363, and XI, 404; there is no MS authority for their rejection, and they are retained by most modern editors. The evidence of the ratios argues for the retention of IX, 85, as we saw above. In the case of XI, 404, the omission of the verse likewise produces a less accurate ratio:

| | | | |
|---|---|---|---|
| No. 909 | .616 | → | .610 |
| 676 | .616 | → | .612[39] |

The ratios thus favor the view that IX, 85, and XI, 404, are not interpolations. The results for IX, 363, are less decisive, but in two instances the ratio is materially improved if we reject the verse:

| | | | |
|---|---|---|---|
| No. 242 | .60 | → | .625 |
| 633 | .623 | → | .635 |
| 969 | .610 | → | .615 |

IX, 363, although accepted by Janell, Sabbadini, Mackail (but cf. *ad loc.*), Durand, and others, was criticized by Servius as one of the most obscure passages in the *Aeneid* and was rejected by Wagner, Forbiger, and Ribbeck; Hirtzel is possibly correct in bracketing this verse.

Mackail retains these three verses, but he in turn italicizes others as interpolations; these include IV, 126 (= I, 73), 256-258,[40] 285-286 (= VIII, 20-21); VI, 901 (= III, 277); IX, 651 (cf. IV, 559); and X, 20-21 *(feratur . . . tumidusque).*[41]

If we reject the six passages italicized by Mackail, the ratios in the shorter proportions change as follows:

| | | | | |
|---|---|---|---|---|
| IV, 126 | No. 525 | .608 | → | .565 |
| | 95 | .617 | → | .60 |
| | 393 | .615 | → | .605 |
| IV, 256-258 | No. 102 | .610 | → | .544 |
| | 527 | .625 | → | .595 |
| | 760 | .610 | → | .659 |
| | 759 | .60 | → | .584 |
| IV, 285-286 | No. 761 | .615 | → | .583 |
| | 395 | .630 | → | .633 |
| VI, 901 | No. 579 | .621 | → | .613 |
| | 803 | .622 | → | .625 |
| IX, 651 | No. 254 | .60 | → | .580 |
| | 255 | .618 | → | .606 |
| X, 20-21 | No. 648 | .619 | → | .610 |
| | S1036 | .618 | → | .612 |

In every instance the omission of the verses italicized by Mackail produces in the shorter proportions an inferior ratio, and in many cases any approximation to the Golden Section disappears; we have here an effective argument against Mackail's rejection of these particular verses, and in general the mathematical structure of the poem provides a means of curbing the desire of various editors and commentators to look upon lines or groups of lines as interpolations.[42] The evidence of the ratios favors only the omission of the verses which are lacking in the best MSS, also III, 230 and IX, 151; and possibly IX, 363, criticized by Servius and condemned by several editors; the other verses suspected by Hirtzel and Mackail are to be retained in the text. As fantastic as Vergil's use of the Golden Mean ratio everywhere in the *Aeneid* may seem, it points in the direction of a conservative text, argues for the essential soundness of the best MSS, and tends to disprove many of the suggestions and criticisms of modern scholars.

The mathematical structure of the poem will be even more effective as a control in connection with larger portions of the text, especially when editors condemn passages as spurious or in need of revision, or when they transpose passages from one context to another. I shall discuss these topics in the two sections which follow.

## SPURIOUS PASSAGES

It is sometimes difficult to make a definite distinction between interpolations and spurious passages. Both are considered the work of a later hand or hands. The interpolations discussed above are passages of one or two lines believed to have been copied or adapted from a similar context elsewhere in the poet's work; the so-called spurious passages are those which create difficulties in the minds of editors and commentators and which therefore are rejected as non-Vergilian. Fortunately, scholars are far more conservative today than they were 75 years ago, when the detection of spurious passages was a favorite pastime of editors and commentators. For example, Leo's text of Plautus (1895-96) bracketed many passages as the result of *retractatio*, or reworking of the text after the death of Plautus, when, as he believed, many alterations and additions were made for later productions of the plays; Lindsay and Ernout retain as Plautine almost all the passages bracketed by Leo.

It is not my intention to go back into nineteenth-century scholarship and re-examine the many passages suspected by various editors of Vergil. My purpose rather is to point out the value of the mathematical structure of the *Aeneid* as a possible guide for scholars today in their treatment of individual passages. A few illustrations will suffice to show that the more recent editors who have rejected the views of older scholars should find substantial support for their position in the Golden Mean ratios of the poem.

Ribbeck, following earlier writers (Heyne, Wagner, Nauck, etc.) bracketed as spurious III, 690-691, VI, 893-896, VII, 624-627, VIII, 42-49a, and XI, 179b-181. The shorter proportions in which these passages appear are listed below, with the new ratios which result if the suspected passages are deleted.

| Passage | No. of proportion | Ratio | Ratio with passage omitted |
|---|---|---|---|
| III, 690-691 | No. 90 | .632 → | .610 |
| | 522 | .609 → | .571 |
| | 521 | .619 → | .640 |
| | 520 | .606 → | .60 |
| | 519 | .606 → | .60 |
| VI, 893-896 | No. 579 | .621 → | .586 |
| | 803 | .622 → | .632 |
| VII, 624-627 | No. 422 | .625 → | .583 |
| | 420 | .604 → | .629 |
| VIII, 42-49a[43] | No. 205 | .613 → | .500 |
| | 204 | .612 → | .550 |
| | 598 | .617 → | .586 |

| | | | |
|---|---|---|---|
| VIII, 42-49a (cont'd) | No. 599 | .618 ⟶ .591 | |
| | 202 | .608 ⟶ .592 | |
| XI, 179b-181 | No. 896 | .615 ⟶ .769 | |
| | 894 | .618 ⟶ .586 | |
| | 293 | .618 ⟶ .599 | |

These five passages, suspected by Ribbeck, are accepted by Hirtzel, Janell, Mackail, Sabbadini, and Goelzer-Durand. If the passages are rejected as spurious, the ratios in almost every instance are less exact and in a majority of the cases the approximate or exact Golden Section vanishes entirely. This gives added support to the twentieth-century editors who agree that Ribbeck was wrong in bracketing the passages.[44]

Ribbeck had no MS evidence to justify his rejection of the passages listed above. We turn now to certain passages which are lacking in the manuscripts but concerning the authenticity of which we have ancient testimony; these are (1) I, 1a-1d; (2) III, 204a-204c; (3) VI, 289a-289d; and (4), most important of all, the famous Helen episode in II, 567-588.

(1) I, 1a-1d:

> Ille ego, qui quondam gracili modulatus avena
> carmen, et egressus silvis vicina coegi
> ut quamvis avido parerent arva colono,
> gratum opus agricolis, at nunc horrentia Martis

These verses appear in none of the early MSS and, according to the Donatus-Suetonius Life (42) and the *Vita* of Servius, were removed from the beginning of the *Aeneid* after Vergil's death; they sound truly Vergilian (cf. *Georg.* IV, 563-566) and are accepted by Hirtzel, who comments in his apparatus that most editors have wrongly rejected the lines as non-Vergilian; Janell omits the verses entirely, Sabbadini and Mackail print them in italics, and Goelzer cites them in his apparatus. I follow Hirtzel in accepting the verses, and the evidence of the ratios supports their authenticity:

No. 711          .623, but .592 with the verses omitted
      712          .631, but .663 " " " "
      942          .628, but .640 " " " "

(2) III, 204a-204c:

> hinc Pelopis gentis Maleaeque sonantia saxa
> circumstant, pariterque undae terraeque minantur.
> pulsamur saevis et circumsistimur undis.

These verses are not in the MSS, but are cited by Servius Danielis with the comment: "hi versus circumducti inventi dicuntur et extra paginam in mundo." Mackail prints the three lines in italics and notes as follows: "In themselves they are quite Vergilian, both in rhythm and in diction, and there need be little if any doubt that they were tentatively written down by Virgil for consideration whether he should incorporate the first two, or the third, or all three." Hirtzel, Sabbadini, and Goelzer give the verses in the apparatus, and Janell fails to mention them. I have excluded them from my calculations, following Hirtzel; cf. the following proportions, which favor the rejection of the verses:

No.   71          .615, but .500 with 204a-204c added
     745          .611, but .548 " " " "
      69          .632, but .606 " " " "
   S1006         .601, but .585 " " " "
     503          .630, but .612 " " " "
     385          .622, but .629 " " " "

(3) VI, 289a-289d:

> Gorgonis in medio portentum immane Medusae,
> vipereae circum ora comae, cui sibila torquent
> infamesque rigent oculi, mentoque sub imo
> serpentum extremis nodantur vincula caudis.

These verses are cited by Servius Danielis, who says: "sane quidam dicunt versus alios hos a poeta hoc loco relictos, qui ab eius emendatoribus sublati sint." Hirtzel and Janell ignore the verses in their editions; Sabbadini and Goelzer each give the passage in the critical apparatus, and Mackail prints it in his text in italics. Sabbadini suggests that it was an early draft replaced by 282-289; Mackail thinks that the verses "may quite well be a piece of Vergilian drafting, meant to follow on the Chimaera, and then struck out and replaced by the single magnificent line 289"; cf. Goelzer: "sed satis liquet illos versus non esse Vergili sed discipuli cuiusdam sese in poetica arte exercentis." Whether the passage is Vergilian or not, the following proportions indicate that it does not belong in the present *Aeneid*:

| No. 158 | .615, but .706 with 289a-289d added to M (289a-289d belong with 282-289 and not with 290-294) |
| 792 | .630, but .677 with 289a-289d added to M |
| 408 | .617, but .633 with 289a-289d included |
| 791 | .617, but .625 "    "    "    " |

If, with Sabbadini, we remove the present 282-289 and substitute 289a-289d, or if, with Mackail, we consider 289a-289d the original version of 289 and substitute for it the four verses, the ratios change as follows:

| No. 158 | .615 ⟶ .555 (Sabbadini) ⟶ .688 (Mackail) |
| 792 | .630 ⟶ .565 " ⟶ .666 " |
| 408 | .617 ⟶ .60 " ⟶ .629 " |
| 791 | .617 ⟶ .608 " ⟶ .623 " |

These results likewise show that VI, 289a-289d, if written by Vergil, are to be rejected as a preliminary draft.

(4) We come now to the very important and much discussed episode, II, 567-588, in which Aeneas sees Helen and contemplates killing her. This passage appears in none of the ancient MSS but is quoted by Servius in his *Vita* as one of the passages removed by Tucca and Varius after Vergil's death; cf. Servius Danielis on II, 566, and see Servius on II, 592. Mackail *(ad loc.)* writes as follows:

> There is no reason to doubt the truth of this; but there is no direct evidence for the reasons which weighed with them in doing so; whether because the passage (1) was ringed or only marginally inserted in the autograph, or (2) was inconsistent with vi. 511-30 where the ghost of Deiphobus describes Helen as having acted in concert with the Greeks all through that night, or (3) offended sentiment by making Aeneas propose to kill a helpless woman, or (4) was too obviously unrevised; or for some or all of these reasons in combination. . . . All that we can say certainly on the internal evidence is that the passage is Virgilian, that it is an incomplete and unrevised draft, and that its rhythm and phrasing are those of Virgil's later, not his earlier manner.

Mackail himself prints the passage in italics, but Hirtzel and Goelzer print it without brackets; Sabbadini italicizes the passage, saying: "Hos versus Vergilius ipse delevit, qui morte impeditus est quominus alios in eorum locum sufficeret." Janell brackets it, obviously as spurious, since he cites in his apparatus Leo, Heinze, and Norden, and these three scholars agree that the passage is non-Vergilian.[45]

The attempt of Leo, Heinze, and Norden to prove that the passage was spurious did not go unchallenged. Those who have defended its authenticity include, among others, Fairclough, Gerloff, Crump, Shipley, Knight, Pease, Palmer, and (most recently) Büchner and Hatch.[46] Some of these scholars, e.g., Fairclough and Shipley, believe that Vergil wrote the passage, that he was dissatisfied with it and planned changes or a substitute passage (especially in view of his treatment of

Helen in Book VI), that his dissatisfaction was known to his friends Varius and Tucca, and that they accordingly deleted the lines. Palmer's interpretation seems more satisfactory: sacrilege committed at altars is a *leitmotif* of Book II; the extraordinary emphasis given to *arae* (e.g., 154 f., 202, 425 ff., 501 f., 512 ff., 523, 550 ff.) becomes more meaningful in the light of Aeneas' temptation to commit murder at an altar. Palmer concludes "not merely that the Helen episode is authentic, but that it constitutes the spiritual crisis of the second book. The incident is the first great temptation which Aeneas with the help of the Gods has to overcome."[47]

The views of those who defend the authenticity of II, 567-588, derive effective confirmation from the Golden Mean ratios in this section of Book II:

| No. of proportion | Type of passage | Ratio | Ratio with 567-588 omitted |
|---|---|---|---|
| No. 51 | Short | .619 | No ratio; entire proportion disappears |
| 52 | Short | .624 | No ratio; 567-587 = minor |
| 50 | Short | .607 ⟶ | .848 |
| 947 | Main division | .610 ⟶ | .571 |
| 982 | Main division in proportion | .614 ⟶ | .603 |

With the deletion of 567-588, No. 51, composed of 567-587, necessarily vanishes; this proportion, however, is typically Vergilian, containing a minor of eight verses (567-574: Aeneas sees Helen) and a major of thirteen verses (575-587: his desire to kill Helen); this combination of 8 and 13 (in the Fibonacci series) appears 39 times, exclusive of fractions and passages with half-lines such as III, 640, and VIII, 536 (which would probably have been changed to full lines in the final revision). In the case of Nos. 52 and 50, the Golden Mean ratios likewise disappear, since 567-587 is the minor of No. 52 and almost all the minor of No. 50. Also, as I pointed out above (in Chapter 4, "The Main Divisions of the Books"; cf. Table V), the approximate Golden Section appears in every main division of every book of the *Aeneid,* but we do not have it in the third main division of Book II (No. 947) unless the Helen episode is retained, or unless we have a substitute of approximately the same length. Even in No. 982, where the totals are large, with all three main divisions creating a ratio of .614 in the usual framework pattern $b/(a + c)$, the deletion of the passage produces a less exact ratio.

We have here the strongest possible argument to support the view that the Helen episode was composed by Vergil. The mathematical structure of the poem proves the authenticity of II, 567-588; that Vergil was dissatisfied with the passage and planned to replace it by a revised version is possible, but the substitute version would necessarily have been similar and of the same approximate length.

## TRANSPOSITIONS

Just as the Golden Mean ratios are helpful in throwing new light on passages thought to be spurious, so they are likewise of value in determining whether editors are correct in transposing lines from one part of the *Aeneid* to another. If the transposed passage remains in the major or minor of the same proportion, there is of course no information to be derived, but otherwise the ratios provide decisive evidence. When a passage is transferred from one series of proportions to another, several more or less perfect ratios may disappear; in other words, the removal of the passage from one context and its addition to another often may disrupt completely the mathematical symmetry of both sections. We have here a useful control against editorial whim and wild conjecture.

In the catalogue of warriors in Book VII, Fowler suggests transferring six verses, 664-669, from Aventinus to Ufens, i.e., after 749.[48] His arguments are not convincing: e.g., he says that there are too many lines here for an unimportant character—but more lines are devoted to Clausus and Virbius; and certainly Aventinus, son of Hercules, should wear the lion skin of Hercules (cf. 666 ff.) rather than Ufens. The proposed transposition affects a number of ratios, and the results are as follows:

Change in ratios with VII, 664-669, removed from its context:

| No. 594 | .609 ⟶ .529 |
| Nos. 592, 824 | .622 ⟶ .548 |
| No. 825 | .636 ⟶ .579 |
| No. 595 | .618 ⟶ .575 |
| No. 593 | .620 ⟶ .568 |

Change in ratios with 664-669 added after 749:

| No. 423 | .616 ⟶ .685 |
| No. 597 | .615 ⟶ .653 |
| No. 201 | .630 ⟶ .652 |
| No. 826 | .615 ⟶ .634 |
| No. S1030 | .608 ⟶ .639 |

The mathematical structure of the book thus provides a most effective argument against Fowler's suggested transposition.

Other proposed transpositions are the following (those of Ribbeck being printed in his edition):

IV, 548-549, after IV, 418 (Ribbeck)
VI, 826-835, after VI, 807 (Ribbeck)[49]
IX, 146-147, after IX, 72 (Ribbeck)
IX, 176-181, after V, 294 (Crump)[50]
IX, 717-719, after IX, 755 (Mackail, p. 336)
X, 747-754 = unplaced fragment (Mackail, pp. 373 f.)
XI, 636-645, after XI, 835 (Mackail, p. 417)
XI, 891-895, after XI, 835 (Sabbadini)
XII, 801-802, after XII, 832 (Ribbeck)

The removal of these passages from their contexts and their insertion elsewhere destroys 48 Golden Mean ratios and produces proportions ranging from .503 to .590 and from .650 to .780. Many examples of the Fibonacci series disappear, including some exact .618 ratios; e.g., if with Mackail we place XI, 636-645, after 835, the ratio in No. 310 changes from .618 to .666; if with Sabbadini we join XI, 891-895, to 835, No. 310 moves from .618 to .644. The .618 ratio of No. 345 changes to .583 if we follow Ribbeck and place XII, 801-802, after 832. In other words, the mathematical structure of the poem gives no support to the suggested transpositions of modern editors and scholars but confirms the text of the poem as it has come down from antiquity.

## PARAGRAPHING

In determining the ratios in the *Aeneid,* I followed as closely as possible the division into paragraphs which appear in Hirtzel's edition. In the shortest proportions, where the units are sentences or groups of sentences, there were naturally no paragraphs to serve as guides, and in the longer passages it was often necessary to make a division where Hirtzel failed to indicate a new paragraph. Mackail comments that overparagraphing which disturbs the continuity is to be avoided,[51] but Hirtzel and other editors often err in the opposite direction and fail to indicate properly the different narrative units.

When I had established the Golden Mean ratios described above in Chapter 4 and listed in Tables I-VII, I turned to three other editions of the *Aeneid* to compare the paragraphing with that of Hirtzel and with the divisions which I had made on a combined basis of sense and mathematical structure. The editions selected were those by Mackail, by Sabbadini, and the nineteenth-century edition by Ladewig and Schaper, revised by Deuticke and (for *Aeneid* I-VI) by Jahn. All three editions contain more frequent breaks in the text than does that by Hirtzel, and the comparison of their paragraphing with my own divisions produced surprising results.

In 99 instances, where Hirtzel contains no paragraphing, Sabbadini, or Mackail, or Deuticke-Jahn (or some combination of the three) begin a new paragraph *at the very point* where I indicate a division between passages containing Golden Mean ratios. In 88 additional instances one or more of these same editors indicates a paragraph where I make internal divisions between major and minor (or between parts of major and minor) and where likewise no paragraphing appears in Hirtzel's edition. The striking agreement between the mathematical structure of the poem and the paragraphing in these three editions provides mutual support: my divisions establishing the proportions are confirmed as natural narrative units, and the paragraphing of the editors is shown to be correct by its correspondence with the mathematical ratios. Among recent editions, those by Mackail and Sabbadini impress me as the most readable; certainly their paragraphs (especially Mackail's) reproduce the units of Vergil's narrative more closely than do those of Hirtzel and Janell.

The following list gives for each book the lines which mark the ratio divisions[52] and after which new paragraphs are indicated in one or more of the editions. The initials in parentheses signify whether the paragraphing is that of Deuticke (D), Jahn (J), Mackail (M), or Sabbadini (S).

I, 409 (M), 465 (J), 558 (M), 688 (M), 747 (S)

II, 227 (J), 346 (S), 385 (J), 436 (M), 485 (JMS), 587 (JS), 593 (M), 620 (M), 720 (S), 729 (JMS), 744 (S), 770 (JS)

III, 101 (M), 131 (M), 153 (M), 171 (JMS), 208 (JMS), 244 (JS), 257 (M), 343 (JM), 380 (M), 395 (M), 409 (M), 432 (M), 440 (M), 460 (M), 612 (M), 668 (S)

IV, 8 (M), 29 (M), 205 (M), 222 (M), 258 (JS), 304 (JM), 387 (M), 647 (J), 671 (JS)

V, 326 (J), 339 (JMS), 408 (J), 420 (MS), 576 (JM), 595 (J), 640 (JMS), 798 (JM)

VI, 55 (JM), 330 (M), 346 (J), 371 (JMS), 439 (JS), 455 (J), 508 (JM), 561 (JM), 665 (M), 787 (MS), 807 (MS), 835 (S)

VIII, 212 (S), 267 (S), 358 (DM), 468 (DS), 625 (DMS), 662 (S), 713 (S)

IX, 125 (M), 158 (MS), 196 (DM), 256 (MS), 341 (DMS), 568 (DMS), 597 (M), 658 (D), 787 (M)

X, 62a (M), 255 (DM), 323 (D), 456 (DM), 463 (M), 665 (DS), 761 (MS), 788 (M), 845 (D), 887 (D)

XI, 13 (M), 485 (DMS), 815 (DS)

XII, 45a (M), 288 (M), 323 (DMS), 504 (DM), 630 (DM), 724 (DS), 807 (M), 828 (DM)

The 99 verses listed above, verses where Hirtzel's text continues without a break, reveal that the edition in which the paragraphs correspond most closely with the divisions into mathematical ratios is that by Mackail (66 instances, or exactly two-thirds of the verses cited). I shall not list the 88 verses where the paragraphing in one or more of the three editions agrees with the divisions within the proportions; e.g., for I, 91 (JS), cf. Nos. 354 and 472; for I, 630 (M), cf. No. 25. Here again, however, the paragraphs inserted into the text by Mackail greatly outnumber those by Sabbadini and Jahn-Deuticke. The paragraphing in Mackail's edition thus not only agrees most closely with the mathematical structure of the poem but also reveals a sensitive feeling for the natural divisions into units of speech and narrative.

Future editors of the *Aeneid* would be well advised to study the paragraphing in the three editions analyzed above and the correspondence of the paragraphs with the ratio units (as seen in the chart-index of each book); this seems all the more necessary since, as I implied above, most editors are guilty of underparagraphing rather than of overparagraphing.

Also, there are a number of places in the text where editors disagree as to the beginning and end of narrative units. Here, I believe, the Golden Mean ratios can help editors to decide where the paragraphs are to be introduced. I list in Table XIII the passages where such disagreement occurs and give the verse where I end the narrative unit, after which, in my opinion, the new paragraph should begin. As above, J = Jahn, D = Deuticke, M = Mackail, S = Sabbadini, and H = Hirtzel. The ratios listed before the first semicolon are those which end with the verse recommended; those which follow the first semicolon are the ratios which begin with the following verse; when I write "cf. Nos...," the divisions within the proportions support the proposed paragraphing.

It is an interesting and curious fact that, although Mackail's divisions seem correct when he introduces new paragraphs (with many of which Sabbadini agrees), he seems less accurate when he is in conflict with Hirtzel's paragraphing. In Table XIII, he appears in column 1 (Paragraphing *not* after) 47 times, but in column 2 (Paragraphing *better* after) only 13 times. The figures for Hirtzel are as follows: column 1, 7 times; column 2, 39 times. Hirtzel, when he does paragraph, thus seems far more accurate, and my praise of Mackail's paragraphing expressed above must be qualified to this extent.

And now, finally, I shall list additional verses after which I should favor a paragraph, but where the editors cited above do not break the text. Two of these seem especially important:

III, 273, which marks the end of the first subdivision (192-273: Strophades) of the second main division of Book III; 274, rather than 278 (JHS) marks the beginning of the Actium episode, the second subdivision; cf. *Leucatae montis* (274), *Apollo* (275), *parvae urbi* (276) [ = Actium]. The break at 273 is clearly indicated by Nos. 75, 504, S1006; S1007, 386, 506; cf. Nos. 503, 385.

VIII, 596, which closes the first subdivision of the third and final portion of the book. It is strange that most editors fail to paragraph here;[53] 596 ends the departure from Pallanteum; 597 describes the arrival near Caere, many miles and certainly many hours on horseback from the site of Rome. The division after 596 is supported by Nos. 620, 221; 227, 228; cf. No. 966.

The editors cited above do not paragraph after the following verses, but a break seems advisable in each instance and might well be incorporated into future editions of the *Aeneid*. As in Table XIII, I shall give the ratios which favor the paragraphing.

| | |
|---|---|
| I, 706, not 708 (S) | Cf. No. 363. |
| III, 492, not 491 (JM) | Nos. 750; 514; cf. No. 390. |
| V, 562 | Nos. ——; 134; cf. Nos. 779, 955. |
| 853 | Nos. 149, 786; 560; cf. No. 148. |
| VI, 312 | Nos. 567; 793. |
| 449[54] | Nos. 796, 163; 164, 165; cf. Nos. 569, 797, 798. |
| 607 | Nos. 801, 172; 173, 174, 802; cf. No. 800. |
| VII, 384 | Nos. 196; 588; cf. No. 820. |
| VIII, 312, not 313 (M) | Nos. S1031; 218, 611; cf. No. 610. |
| 394, not 393 (D) | Nos. 220, S1032; 614; cf. Nos. 834, 833. |
| 422, not 423 (DH) | Nos. 613; S1033; cf. Nos. 835, 965. |
| 674, not 670 (S) | Nos. 623, 839, 621; 624; cf. No. 622. |

| | |
|---|---|
| IX, 280a | Nos. 632, 849; 850, 431; cf. Nos. 631, 847. |
| 398, not 394 (D) | Nos. 243; 634, 245; cf. No. 244. |
| 805 | Nos. 863; 645; cf. Nos. 644, 862. |
| X, 17, not 15 (DHMS) | Nos. 646; 648. |
| 490, not 489 (M) | Cf. Nos. 442, 874. |
| 600, not 601 (DM) | No. 878; —; cf. No. 877. |
| XI, 38, not 41 (M) | Nos. 288, 671; 889, 446, 673, 289; cf. No. 287. |
| 71 | Nos. 290; 891; cf. No. 890. |
| XII, 480 | Nos. 332; 334; cf. No. 930. |

## THE PROBLEM OF THE REVISION

Vergil intended to devote three years to the revision of the *Aeneid*, a plan rendered impossible by his unexpected death in 19 B.C. (Donatus-Suetonius Life, 35). Various editors and commentators have criticized numerous passages, both small and large, and have suggested what Vergil would have done had he lived to revise the epic to his complete satisfaction. To review these many theories in the light of the mathematical structure of the poem would be tedious as well as unnecessary; a few selected examples will suffice to show that most of the proposals are impossible, or at least highly improbable.

The presence of so many Golden Mean ratios in the *Aeneid* and the tightly knit structure of each book with its tripartite divisions and subdivisions all in proportion (cf. the chart-index of each book) makes it difficult to introduce most of the changes which have been recommended. The mathematical symmetry of the poem in general is such that we can assert with more confidence what Vergil would not have done than what he would or might have done. I have shown above that a majority of the hemistichs probably would not have been changed to full lines in the final revision; on the other hand, there is strong evidence that fourteen (or fifteen) hemistichs would have been replaced by complete verses. In the case of suggested transpositions, we have seen that the ratios argue strongly for the text of the poem as we have it, and against the view that Vergil would have introduced such changes at a later date. In like manner, the revisions proposed by scholars receive little or no support from a study of the ratios involved.

The passages selected for examination are the following:

(1) II, 40-56 and 199-227, the Laocoon episode. Mackail writes about Book II:

> It is remarkable not only for magnificence, but, with one exception, for complete structural finish in the succession and connexion of its episodes. That exception is the Laocoon episode, in two parts, ll. 40-56 and 199-233. Both are insertions, which have only a slight organic connexion with the main current of the story. Either, or both, could be omitted without leaving any gap, and in fact with some added continuity; and the latter and longer of the two has a touch of overloadedness which suggests a comparatively early date of composition, and the absence of final revision after it had been inserted in its present place.[55]

Mackail is quite wrong here. The two parts of the episode are structurally necessary for the recessed panel pattern of the first main division (cf. the separation of the death of Misenus from his burial in the first section of Book VI), and one could not be omitted without the other; the two parts, although separated, are in proportion; cf. No. 32: 40-56 = minor, 199-227 = major.[56]

The Laocoon story is firmly embedded in the symmetrical structure of the book; 40-56 comprise the minor of No. 366 and with the omission of the passage the proportion disappears; if we remove 40-56 from Nos. 728 and 729, the ratios change from .611 and .628 to .789 and .706 respectively. In like manner, 199-227 form the minor and almost the entire major of No. 485, and also almost all the major of No. 370; with the passage rejected these ratios vanish. Nos. 729 and 370 are component parts of No. 945, the first main division of the book; without the Laocoon episode the ratio in No. 945 moves from .621 to .680. The importance of the Laocoon story to the structure of the book becomes still more evident when we recall that every one of the 36 main divisions of the *Aeneid* contains an approximate Golden Mean ratio.

(2) Mackail says that IV, 261-264 "are in effect a rather infelicitous parenthesis and might also have disappeared in the final revision of the whole paragraph."[57] With the deletion of the passage, the changes in the ratios are as follows:

| No. S1013 | .630 ⟶ .836 |
|---|---|
| 760 | .610 ⟶ .566 |
| 759 | .60 ⟶ .579 |
| 526 | .619 ⟶ .604 |

(3) VII, 553-554. Mackail (p. 256; cf. *ad loc.)* looks upon these two lines as "pretty certainly alternatives set down for consideration."

(4) VIII, 268-272. Durand *(ad loc.)* considers these lines marginal additions wrongly placed in this position by Varius and Tucca,[58] thus implying that Vergil would have deleted them in his revision or placed them elsewhere.

(5) VIII, 457-464. Mackail (p. 297) sees here two alternate drafts. I assume he means that Vergil would have rejected either 457-460 or 461-464 in the final revision.

(6) IX, 176-181. Crump, discussing the problem of the revision, says that Book V "needed no correction, except that perhaps the account of Nisus and Euryalus in IX. 176-181 would have been put into its natural place here."[59]

(7) IX, 266, 272-274. Cf. Mackail (p. 336): "The list of rewards offered by Ascanius to Nisus, ll. 263-274, would probably have been compressed on revision; we may be inclined to think that in any case ll. 266 and 272-4 would have been struck out entirely, to no little advantage."

(8) XII, 593-611. Mackail suggests that "Virgil contemplated discarding the whole passage, and leaving Amata's end in silence. While its introduction gives an additional note of tragedy, . . . it is of doubtful value; if it had been cancelled by Virgil himself or by his editors, we should hardly feel anything to be missing."[60] Again I disagree. Amata's suicide is a necessary part of the tragedy of Turnus; his failure to face Aeneas leads her to believe him dead, and she carries out her threat of lines 61-63.[61] We should miss the episode structurally also; the Golden Mean ratios Nos. 934, S1044, and 931 vanish, as does No. 929 (.614 ⟶ .566), the second component part of the second main division.[62]

I have omitted the details for the passages cited from Books VII, VIII, IX, and XII, inasmuch as the results resemble those for the Laocoon episode and IV, 261-264; either the proportions disappear entirely, or the ratios change to .550, .709, .731, and the like. In other words, such proposed revisions disrupt entirely the mathematical structure of the poem.

In his discussion of half-lines, Mackail writes as follows (pp. li f.):

> In no less than seventeen cases the hemistich forms part of a passage of a line and a half which is similarly detachable. These are i. 559-60, ii. 232-3, 345-6, 622-3, iii. 660-1, iv. 399-400, 515-16, v. 573-4, vi. 93-4, 834-5, vii. 128-9, 438-9, 454-5, 759-60, viii. 40-1, ix. 166-7, 720-1. It seems not unlikely that some at least of these passages were *marginalia* which the editors decided to incorporate into the text.

I understand Mackail here to mean that Vergil would have rejected some or all of these passages in his final revision, and therefore I comment briefly upon the passages at this point. In most instances the ratios, especially those in the shorter passages, move away from the area of the Golden Section if we remove the half-line and the preceding verse; e.g., cf. IV, 515-516, where the ratios change as follows:

| No. 106 | .625 ⟶ .688 |
|---|---|
| 107 | .630 ⟶ .652 |

| No. 105 | .626 ⟶ .648 |
| 401 | .606 ⟶ .597 |
| 402 | .618 ⟶ .610 |
| 765 | .614 ⟶ .608 |

The evidence is less conclusive for VI, 93-94, and VII, 128-129,[63] and in the case of II, 345-346, the proportions, small and large alike, move closer to the exact Golden Section with the half-line and the preceding verse removed:

| No. 491 | .603 ⟶ .630 |
| 489 | .627 ⟶ .617 |
| 490 | .604 ⟶ .614 |
| 731 | .612 ⟶ .619 |
| 38 | .623 ⟶ .621 |
| 946 | .613 ⟶ .615 |

These three half-lines, II, 346, VI, 94, and VII, 129, are listed above in Table X, since the ratios are more accurate with the half-lines as fractions. The ratios are even more exact if II, 345-346, are deleted. In this instance Mackail may be correct, and perhaps also concerning VI, 93-94, and VII, 128-129. It is worth noting, however, that the ratios with the lines retained are all within the accepted range and also that all three half-lines are included among those which Sparrow considered emotionally or rhetorically effective.[64]

We turn now from short passages to an entire book, to Book III, which Mackail, in full accord with many earlier scholars, considers "the most incomplete and the least coherent in the whole Aeneid."[65] The injustice of such a sweeping denunciation has been pointed out recently by Lloyd, whose careful analysis of the book[66] leads him to the conclusion that the various episodes are in concept and structure truly Vergilian and executed with meticulous care. An examination of the chart-index of III reveal clearly the tripartite divisions and subdivisions of the book (composed in no other book with greater care) and also the close correlation between the narrative units and the mathematical ratios; the component parts of the ratios of the first and third main divisions (Nos. 948 and 950) are identical with the narrative subdivisions (the stops on the journey), but the great length of the Helenus-Andromache episode (294-505) renders such correspondence impossible in the second main section. Even so, the correlation in III as a whole is as accurate as in I, IV, and VII, and more so than in V, VI, and XII, and most of these books Mackail praises highly.[67] The structural analysis of III does not support the criticisms of Mackail, Crump, and others, but on the contrary confirms Lloyd's high opinion of the book.

We shall never know what changes in subject matter, wording, metrical patterns, poetic effect, and structural organization Vergil might have introduced into the final and completed version of the poem, but the coherence and consistency of his composition by Golden Mean ratios and the close agreement of narrative units and mathematical proportions argue against any sweeping alterations in the books. The suggested revisions, especially those based upon supposed flaws or inconsistencies, cannot stand.

Since Vergil has provided for the *Aeneid* so surprising and so complete a symmetrical structure built by mathematical proportion, it seems plausible to suggest that the revision might have been devoted, at least in part, to perfecting this aspect of the poem. The following possibilities occur to me:

(1) An even closer correlation between the narrative subdivisions and the component parts of the ratios in the main divisions would be possible, especially in Books V, VI, and XII. Perhaps in the case of XI, where the many ratios in the third main division give the impression of a new ending (through 915) superimposed upon the original conclusion,[68] Vergil would have brought the revised final section into proportion with the first or second main division.

(2) In the more significant books (those with even numbers) all three main divisions combine into a final ratio for the book as a whole with the exception of XII.[69] It is perhaps possible that XII might have been brought into line in this respect.[70] Of the odd-numbered books, only V

has a ratio derived from all three main divisions; whether the other odd-numbered books would have been revised to the point where all three divisions would form a single ratio remains uncertain but seems highly improbable in view of the many structural and mathematical changes involved. It is significant, however, that in the *De Rerum Natura* all the main divisions of each book combine into a ratio for the book as a whole.[71]

(3) We saw above that the books of the *Aeneid* are also in proportion, with the ratios ranging from .627 to .637. In the finished and polished *Georgics*, where the didactic passages total 1352 lines and the descriptive passages total 835, we find a ratio of .618, the exact Golden Section. Had the *Aeneid* been completed, the architecture of the poem as a whole might have been revised slightly to produce more accurate ratios between the books.[72]

The changes suggested above may or may not have been part of Vergil's plan for the final revision, and certainly attention to more exact ratios and to a closer correspondence between narrative units and the mathematical structure would have been only a part, at most, of the proposed revision. The Golden Mean ratios in the poem make it easier to assert what Vergil would not have done than to affirm what he would have done; all too often this second category becomes what editors and commentators believe he should have done. We must not forget that in his revision he had planned to meet the criticisms of his contemporary *obtrectatores* (Donatus-Suetonius Life, 46), not those of modern scholars.

## THE AUTHENTICITY OF THE MINOR POEMS

It seems improbable that Vergil wrote no poetry until about 28 years of age, at which time he began the *Bucolics* and chose as his model the Alexandrian poet Theocritus; some form of literary apprenticeship seems inevitable to explain the maturity of expression and the poetic perfection displayed in the *Eclogues*. A youthful poet in the middle of the first century B.C., under the influence of Catullus and the other poets of the period, would naturally have tried his hand at various Alexandrian forms such as elegy, epigram, and the short epic (epyllion).

A collection of such poems, known as the *Appendix Vergiliana* and ascribed to Vergil, has come down from antiquity. The manuscripts are late and the textual difficulties are unusually numerous. The *Appendix*, although a small collection, has been the source of endless discussion (and little agreement) for the past sixty years.

Briefly, the problems are these: are the poems by Vergil and, if so, what do they tell us about his life and early poetry? Can it be shown from the study of their language, meter, and style that they are not by Vergil? If they are not, are they earlier or later? How are the many similarities between passages in these poems and Vergil's authentic poetry to be explained—is Vergil imitating the work of earlier poets, or are later poets imitating Vergil? Or is Vergil in his later poetry using material from his own earlier unpublished work? If some of the poems in the collection are obviously not genuine, should we reject the others also? Is the fact that the poems are different from and inferior to the *Bucolics, Georgics,* and *Aeneid* a sufficient reason for rejecting them?

The Donatus-Suetonius Life (17-18) vouches for the fact that Vergil composed poetry as a young man. After quoting a two-line epigram about Ballista who was stoned to death, it lists the following poems: *Catalepton* (both *Priapea* and *Epigrammata*), *Dirae, Ciris, Culex* (with a brief summary), and *Aetna,* with an expression of doubt concerning its authenticity *(de qua ambigitur).* The Servius *Vita* names the same poems in a different order, with the addition of the *Copa (Ciris, Aetna, Culex, Priapea, Catalepton, Epigrammata, Copa, Dirae),* and omits the expression of doubt concerning the *Aetna.*[73] These poems have come down in a variety of mediaeval manuscripts (ninth century or later) along with several other poems which are demonstrably non-Vergilian; these last include two elegies on Maecenas, who died eleven years later than Vergil, and other short poems believed to date from the fourth century. The *Elegiae in Maecenatem* may have been added to the collection because they commemorated Vergil's patron, and the later poems because their contents were considered appropriate.

Donatus-Suetonius and Servius thus provide the ancient external evidence to support the Vergilian authorship of the major portion of the collection. In the first century Quintilian *(Inst. Orat.* VIII, 3, 28) quotes *Catal.* II as Vergil's, and we have additional testimony from the same period for the existence of a *Culex* by Vergil; Statius *(Silvae* I, praef.; II, 7, 73-74) and Martial (VIII, 56, 19-20; XIV, 185) each alludes to it twice, and Suetonius mentions it also in his Life of Lucan (ed. Rostagni, p. 144). These references may not prove that the extant *Culex* is the one composed by Vergil, but they are evidence for the belief in the first century that Vergil in his early years composed a poem entitled *Culex*.[74]

One might suppose that the combination of the statements in the Lives and the supplementary evidence supporting the authenticity of the *Culex* would be a strong argument that certain of these poems, especially the *Culex, Ciris, Dirae,* and *Catalepton,* were in reality the work of the youthful Vergil. This was not the case in the nineteenth century, when all the poems were pronounced non-Vergilian; such trifling and inferior poems could not have been written by the author of the *Bucolics, Georgics,* and *Aeneid*. Gudeman at the turn of the century gave a typical expression of the nineteenth-century view when he listed the poems as *pseudo-Virgiliana* and said that "their spuriousness is established by incontrovertible proofs."[75]

The first quarter of the twentieth century saw a rapid change in the attitude of many scholars toward the poems. Several German and British writers argued for the authenticity of the *Culex* and the *Ciris*, while others stated their conviction that the collection contained the genuine youthful works of Vergil; in America, Rand, Frank, DeWitt, and others accepted the poems as Vergil's and, using them as a source of biographical material, created in a sense a new Vergil for twentieth-century readers.[76] Other scholars were more cautious; cf., e.g., the words of Prescott:

> These poems, if by Virgil, offer a rich supply of material for the understanding of his early environment and of his apprenticeship in the art of poetry. . . . Few, if any, of these poems can be proved not to be from Virgil's hand. But whether Virgil's or not, many of them reveal, more clearly than any other evidence, the influences that affected the poetry of his boyhood and youth.[77]

At the present time, however, the scholarly trend is to deny Vergilian authorship to most of the minor poems. Two of the most exhaustive studies of the *Appendix* in recent years are those by Bickel and Büchner.[78] Bickel examines the *Culex, Ciris,* and *Aetna* on the basis of the poet's knowledge of events, his imitation of other works, his use of language and meter, and concludes that the *Culex* is a spurious work of the late Augustan period, the *Ciris*, with its imitations of the *Aeneid,* is to be dated about 18 B.C., and the *Aetna* was composed after the time of Manilius, either late in Tiberius' reign or during the reign of Claudius; only *Catalepton* I, V, VII, and VIII were written by Vergil. Büchner also considers the *Culex* and the *Ciris* to be later than the publication of the *Aeneid*, and he dates the *Aetna* between 65 and 79 A.D.; of the entire *Appendix* he accepts as genuine only *Catalepton* V and VIII. It is important to note that these scholars maintain not only that the poems were not written by Vergil, but also that they were not composed in the days of Vergil's youth and therefore do not reflect the literary interests of the period.

It is not my purpose here to summarize the various poems of the *Appendix* or to examine the validity of the arguments already advanced for or against the genuineness of the poems. The presence of so many Golden Mean ratios in the *Eclogues, Georgics,* and especially the *Aeneid* suggests a new approach—an examination of the longer of the "minor works" to see if they are also composed on the basis of mathematical proportion and also whether the structure of the poems resembles in any way that which we have found in Vergil's authentic works. Again the results are more than surprising.

I begin with the *Culex,* since the external evidence from antiquity argues strongly for the existence of a *Culex* composed by Vergil, and since themes such as praise of country life, description of a serpent, portrayal of the Underworld with a generous listing of heroes, both mythological and Roman, could reveal Vergil's youthful interest in the same topics which he later developed with far greater success in the *Georgics* and the *Aeneid*. Many modern scholars view the presence of these topics in the *Culex* as proof that the poem is a deliberate forgery which echoes in an awkward fashion themes from Vergil's published works.[79] But it is difficult to believe that a forger, attempting to pass off his composition as an early poem of Vergil's, would have assumed that Vergil's youthful work would have been so different and so imperfect. It is even more difficult to believe that the forger understood Vergil's composition by Golden Mean ratios and his correlation of the narrative sections with the mathematical structure so well that he composed the *Culex* in the same pattern.

I have discovered in the *Culex* 35 Golden Mean ratios ranging from .60 to .636; those in small passages combine into larger units and finally into three ratios which represent the tripartite structure of the poem, and these three ratios are also the component parts of the ratio for the poem as a whole. This is very similar to the structure of the books of the *Aeneid*. Furthermore, the *Culex* resembles the *Georgics* in this respect: the narrative of the shepherd, the serpent, and the gnat is expanded by descriptive passages (prooemium, 1-41; praise of country life, 58-97; catalogue of trees, 123-156; the heroes and heroines seen in the Underworld, 231b-371), just as the didactic sections in the *Georgics* are enriched by many passages of a more universal nature. With the didactic passages as major and the descriptive passages as minor, we find in the *Georgics* as a whole the exact Golden Section, .618.[80] The descriptive passages in the *Culex* mentioned above total 255.4 lines; if we take this total as major and the remainder of the poem (158.6 lines) as minor, $M/(M + m) = 255.4/414 = .617$. This too resembles Vergil's procedure in the *Georgics* in a most striking fashion. I give this ratio below in Table XIV as No. A1, after which I list the other proportions in the *Culex* and number them consecutively from A2 to A35.

The *Ciris* and the *Aetna* also contain Golden Mean ratios, as do, to a lesser degree, the *Moretum* and the *Dirae*. The ratios in these poems are listed in Table XIV after those in the *Culex* and are numbered as follows: *Ciris*, Nos. A36 to A72; *Aetna*, Nos. A73 to A111; *Moretum*, Nos. A112 to A123; *Dirae*, Nos. A124 to A134. No. A135 gives a ratio in the *Copa* and No. A136 one in *Cat*. IX, the longest of the *Catalepton*. I have used for text and paragraphing the Oxford text of the *Appendix* by Ellis, and, for the *Aetna*, the Loeb text by J. W. Duff and A. M. Duff.[81]

The ratios in the poems of the *Appendix* listed in Table XIV are derived from the same bipartite, tripartite, and alternating patterns which produce the ratios in Vergil's authentic works. Also, as in Vergil, a large number of the proportions are built on numbers in the Fibonacci series. If Vergil were the only Roman poet to compose by Golden Mean ratios, the presence of so many similar proportions in the *Culex*, *Ciris*, and *Aetna* would be a most convincing argument for Vergilian authorship. But both in Lucretius and in Catullus LXIV numerous ratios appear and, as in the *Aeneid*, there is a close correlation between the narrative units and the passages, both large and small, which contain the ratios.

In some respects the mathematical structure of these minor poems resembles that of Catullus LXIV or Lucretius almost as closely as it does that of the books of the *Aeneid*; this could indicate a date of composition contemporary with or earlier than the writing of the *Eclogues* and the *Georgics*; is this an indication of Vergilian authorship? Most scholars today look upon the *Culex*, *Ciris*, and *Aetna* as post-Vergilian, but would a later imitator or forger have understood or appreciated the subtleties of the mathematical composition which appear in Catullus, in Lucretius, in Vergil's *Eclogues*, *Georgics*, and, especially, in his *Aeneid*?

The relative frequency in the *Aeneid* of the most accurate ratios, those ranging from .615 to .621, is 301/1044, or 28.8 per cent; in the *Culex* ten proportions appear in this same range out of 35, or 28.6 per cent; Catullus LXIV is similar, with 29 per cent. The figures for the *Ciris* and the *Aetna* are higher (35.1 and 38.5 per cent, respectively), but for Lucretius, Book I, somewhat lower (21 per cent). Again, the tripartite structure of the *Culex* and the *Moretum* resembles that of the twelve books of the *Aeneid* (and also Catullus LXIV), but both the *Ciris* and the *Aetna* have four main divisions and in this respect they are similar to the majority of the books of the *De Rerum Natura*. Could those scholars be correct who have maintained that these poems, even if not by Vergil, were written under the influence of Catullus and Lucretius and reflect the literary environment of Vergil's youth?

In Table XV, I compare the ratios in the minor poems with those in the *Aeneid*, and likewise with those in Catullus LXIV, Lucretius I, and *Georg*. IV, 281-558 (the Aristaeus episode).[82] I give the relative frequencies of the ratios as I have found them, also the percentages for the occurrence of the more exact ratios (.618, from .615 to .621, and from .610 to .626) and the percentages for the two Golden Mean series which appear most frequently—the Fibonacci series, and 1, 3, 4, 7, 11, etc. I realize, of course, that the smaller number of ratios in the *Moretum* and the *Dirae*, as well as in *Georg.* IV, 281-558, makes the percentage lists of less value for these poems.

I present in Table XVI chart-indices to show the correspondence between the structure of the *Culex, Ciris, Aetna, Moretum,* and *Dirae* and the proportions which appear in these poems. These charts may be compared both with those of the books of the *Aeneid* (Table IX) and with those of Catullus LXIV, Lucretius I, and *Georg.* IV, 281-558 (Tables XVIII, XXI, XXIII). The component parts of the final ratio are underlined, and these are identical with the main divisions of the poem in the case of the *Culex, Ciris, Aetna,* and *Moretum;* also, every main division of each poem, with the exception of the *Dirae*, contains a ratio.[83]

The chart-index of the *Dirae* shows that there is a ratio in the poem as a whole (No. A124 in Table XIV) only if 104-183, the so-called *Lydia*, are included with the *Dirae*. My outline is based on that of Van der Graaf, who divides the *Dirae* after 47 and 81 (1-47, imprecation by fire; 48-81, imprecation by water).[84] The division of the *Dirae* into two poems *(Dirae,* 1-103; *Lydia,* 104-183) has been accepted by many scholars, but the mathematical structure of the poem supports the view of Van der Graaf and others who maintain that the poem is a unity. The so-called *Lydia* is the fourth main division and the ratio is derived from the alternating pattern $(a + c)/(b + d)$; cf. the similar alternation of the main divisions of the *Ciris* and the *Aetna* (Nos. A36 and A73 in Table XIV).

The Golden Mean ratios in these five minor poems and the striking correspondence between subject matter and mathematical structure (e.g., the complete agreement between the main divisions and the component parts of the final ratio) do not prove Vergilian authorship, but they certainly do not disprove it, and they suggest a date of composition shortly after Catullus and Lucretius, at the very time when Vergil was devoting special attention to mathematics and also undoubtedly engaging in early poetic efforts. The symmetrical structure of the poems seems not unworthy of him.

The results from Tables XIV-XVI are more decisive for the *Culex* than for the other poems in the collection. Not only are the three main divisions of the story identical with the three component parts of the final ratio (No. A2), in the familiar framework pattern $b/(a + c)$, but the distribution of the ratios, especially those ranging from .615 to .621, appears Vergilian, as does the use of the Fibonacci series. Furthermore, the *Culex* displays, in addition to its tripartite structure, an alternating pattern that is typically Vergilian, in this case an alternation between main theme and descriptive passages; just as the didactic portions and the descriptive passages in the *Georgics* combine to form the exact Golden Section, so the descriptive passages as major and the story itself as minor produce an almost perfect ratio of .619 (No. A1). This surprising similarity between the *Culex* and the *Georgics* is to me the most convincing argument for the authenticity of the *Culex*. I find it far more difficult to ascribe the subtle technique of the poem to a later forger or imitator than to the youthful Vergil composing by the same Golden Mean ratios and the same structural devices which appear so frequently in the *Bucolics*, the *Georgics*, and the *Aeneid*.

# Notes to Chapter 5

1. See above, pp. 61-63.
2. See Chapter 4, "The Main Divisions of the Books."
3. In making the tripartite subdivisions I have in general followed the paragraphing of Hirtzel's edition, but not always; my divisions, where they differ from those of Hirtzel, are favored by other editors as follows:

| New paragraph after | Printed by |
|---|---|
| I, 747 | Sabbadini |
| II, 227 | Jahn; cf. Cartault, *L'Art de Virgile*, pp. 181, 184; Büchner, *P. Vergilius Maro*, col. 327 |
| II, 729 | Jahn, Sabbadini, Mackail |
| VI, 807 | Sabbadini, Mackail |
| VIII, 625 | Deuticke, Janell, Sabbadini, Mackail, Durand; cf. Cartault, p. 620; Büchner, col. 383 |
| X, 255 | Deuticke, Mackail; cf. Cartault, pp. 725, 727 |
| XII, 288 | Mackail. This is the end, not only of a subdivision, but also of the first main division of Book XII. See Chapter 2, note 22. |

This paragraphing is confirmed by the mathematical ratios in many passages, both small and large. My divisions after III, 273, and VIII, 596, are, to the best of my knowledge, my own, but they are supported by both sense and mathematical structure. See above, p. 89.

4. It is possible, of course, that the correlation between the component parts of the ratios in the main divisions and the subdivisions of the narrative is even more exact than I have been able to discover. Or would the correlation have been made still more perfect in the final revision?
5. See above, pp. 60-63.
6. Cf. Courcelle, in *Recherches sur la tradition platonicienne*, p. 107, who describes Anchises' revelation to Aeneas in Book VI as "sorte de somme métaphysique stoïcienne imprégnée de platonisme et de pythagorisme." For Orpheus, cf. *Ecl.* III, 46 *(Orpheaque in medio posuit)*; IV, 55 ff.; VI, 30; VIII, 55 f.; *Georg.* IV, 454 ff.; *Aen.* VI, 645 ff.; see Desport, *L'Incantation Virgilienne*, pp. 136 ff.
7. See Alföldi, *Hermes* 65 (1930), pp. 369 ff.; Nock, *Cambridge Ancient History*, X, pp. 472 f.; Alföldi, *Corvina*, Ser. III, 1 (1952), pp. 34 ff.; *Essays in Roman Coinage*, pp. 88 f.
8. *Virgile et le mystère de la IVe Églogue*; cf. the second edition of 1943, pp. 221 ff., where Carcopino discusses the criticisms directed at the first edition and defends his views.
9. See Boyancé, *RA* 25 (1927), pp. 361-379; Guillemin, *Virgile*, pp. 132 ff.; Saint-Denis, *Virgile, Géorgiques*, pp. 101 f. Scazzoso, *Paideia* 11 (1956), pp. 11-25, takes *felix* as that person who has undergone a complete initiation into the Mysteries, *fortunatus* as one who has only a partial revelation of the truth; the names Spercheus, Taygetus, and Haemus in II, 487 f., are chosen expressly to denote the mystical religions favored by the Neo-Pythagoreans.
10. The old gardener of Tarentum, described in *Georg.* IV, 125-146, has been believed to symbolize Pythagoras; see above, Chapter 1, note 21.
11. *LEC* 17 (1949), pp. 139-235, especially pp. 152 ff., 159 ff., and 222 ff.
12. Cf. Delatte, *Études sur la littérature pythagoricienne*, pp. 259 f.
13. *Lettres d'Humanité* 3 (1944), pp. 71-147; cf. especially pp. 111 ff.
14. Cf. Stégen, *Commentaire sur cinq Bucoliques*, pp. 131-154, who modifies Maury's conclusions and shows additional numerical symmetries based upon multiples of 10; for references to writings for and against Maury's theories, see Duckworth, *CW* 51 (1957-58), p. 123.
15. For additional ratios in Catullus LXIV, many of them likewise in the framework pattern, see Appendix A and Table XVII.
16. See above, Chapter 1, "The Alternating Rhythm"; cf. also Duckworth, *AJPh* 80 (1959), pp. 232 f., 237.
17. Cf. above, Chapter 4, note 40.
18. For additional ratios in Lucretius, both in the main divisions and in Book I, see Appendix B and Tables XIX and XX.
19. On this division, cf. Rolfe, *Sermones et Epistulae*, pp. 144 ff. (1901 ed.); Wili, *Horaz*, pp. 316 (and note 2), 325; Stégen, *Les Épîtres Littéraires*, pp. 8, 166 ff. Rostagni, *Arte Poetica*, pp. 3, 12, 86, makes the tripartite division as follows: 1-41, 42-294, 295-476.
20. For additional ratios in the poetry of Horace, see Appendix D and Tables XXIV and XXV.
21. The loss of the *Annales* of Ennius is greatly to be regretted. The fragments, as few and as short as they are, reveal several examples of the Golden Mean ratio, and what is even more significant, the proportions are determined by the Fibonacci series. I list the following four fragments as given in Vahlen's edition (pp. 8 f., 14, 35 f., 73).

| Book | Fragm. | m/M: type | Major | Total lines | Minor | Total lines | Ratio: M/(M+m) |
|---|---|---|---|---|---|---|---|
| I | XXVIII | a/b | 39-43 | 5 | 36-38 | 3 | .625 |
| | | b/a | 39-43 | 5 | 44-46 | 3 | .625 |
| | | a/b | 44-51 | 8 | 39-43 | 5 | .615 |
| | | a/b | 47-51 | 5 | 44-46 | 3 | .625 |
| I | XLVII | a/b | 89-96 | 8 | 84-88 | 5 | .615 |
| | | (a+c)/b | 90-94 | 5 | 89, 95-96 | 3 | .625 |
| VI | XII | a/b | 197-201 | 5 | 194-196 | 3 | .625 |
| XV | VI | b/a | 401-405 | 5 | 406-408 | 3 | .625 |

The first proportion in Fragment XXVIII of Book I (36-43) suggests that Vahlen's punctuation, with a full stop after *lumen* in 35, is to be preferred to that of Warmington *(Remains of Old Latin,* I, p. 14), who places a comma after *lumen* and joins the first line of the fragment to the following sentence. The appearance of Golden Mean ratios derived from the Fibonacci series in these few extant fragments is most striking, and Ennius' technique resembles Vergil's procedure in similarly short passages so closely that it is tempting to draw the conclusion that Ennius' use of the Golden Section influenced later Roman poets in general and Vergil in particular.

22. *A History of Science,* p. 443. Cf. Weyl, *Symmetry,* p. 72 (quoted above in Chapter 3, "The Golden Mean Ratio"); Ottaviano, *Sophia* 22 (1954), pp. 3-46.

23. *Vorschule der Aesthetik* (3rd ed.), I, pp. 184-202. For discussion of Fechner's poll, see e.g., Bosanquet, *A History of Aesthetic* (2nd ed.), pp. 41, 382 ff.; Birkhoff, *Aesthetic Measure,* pp. 27 ff.; Thompson, *JEP* 36 (1946), pp. 50-58; Osborne, *Theory of Beauty,* pp. 179 ff.; Borissavlievitch, *The Golden Number,* pp. 34 ff.

24. Borissavlievitch, *The Golden Number,* pp. 37 ff., gives what he describes as the first scientific explanation of the beauty of the Golden Section: "(1) it represents the balance between two unequal asymmetrical parts, which means that the dominant is neither too big nor too small, so that this ratio appears at once clear and 'of just measure.' The perception of such a ratio is easy and rapid because of this clarity. And ease and clearness are essentially aesthetic factors, because the opposite factors would cause a painful feeling, and because they agree with the hedonistic and aesthetic law, the law of least effort, hence the beauty of the Golden Number. (2) The visual field for both eyes represents an oval shape exactly inscribed in a Golden Rectangle . . . (3) . . . the rectangle 3 : 5 represents a surface fitted to our vision, hence also its beauty . . . ." For a recent survey of views on the aesthetic value of the Golden Section, see Wittkower, *Daedalus* (Winter, 1960), pp. 199-215.

25. I include in this total V, 595; XI, 391; XII, 218. Most editors bracket the ends of these verses as interpolations, and the MSS support the rejection in the case of V, 595, and XI, 391; on XII, 218, see Mackail, *ad loc.*

26. *Studien über die Compositionskunst Vergils,* pp. 113 f.; on Belling's view, cf. Sparrow, *Half-Lines and Repetitions in Virgil,* pp. 24 f. For additional discussion and bibliography on the half-lines, see Pease, *Publi Vergili Maronis Aeneidos Liber Quartus,* pp. 123-125 (on IV, 44); Büchner, *P. Vergilius Maro,* cols. 403 f.

27. *P. Vergilius Maro,* col. 404.

28. *Half-Lines and Repetitions in Virgil,* pp. 42-45; see his comment on each hemistich. Cf. Belling, *Studien über die Compositionskunst Vergils,* pp. 115-132, who concludes (p. 134) that the half-lines show careful work on the part of the poet.

29. *Half-Lines and Repetitions in Virgil,* p. 46.

30. *Virgil's "Gathering of the Clans,"* pp. 93 f.

31. *Die Entstehung der Halbverse,* p. 67.

32. Cf. above, pp. 47 f.

33. With II, 767, as a whole line, the ratio of No. 59 moves from .627 to .634; the shorter proportion (No. 499 : .603 ⟶ .615) is to preferred here, especially since the Fibonacci series appears in the totals.

34. The change in the ratio here is in itself inconclusive; I list V, 574, in Table XI because the minor and the major are in the Fibonacci series without fractions if we count 574 as a whole line. See above, p. 80, on V, 595.

35. For the occurrence of these series in Tables I-VII, see above, pp. 61-63.

36. Cf. Mackail, *ad loc.*

37. Janell, Sabbadini, and Durand retain IX, 151; Hirtzel and Mackail follow Wagner and Ribbeck in bracketing the verse; the fact that the ratios are less accurate if we accept 151 supports those who reject the verse; see below, Table XII.

38. I reject also as interpolations the conclusions of three verses: V, 595; XI, 391; and XII, 218; see above, note 25.

39. The rejection of XI, 404, produces a slightly better ratio in Nos. 297 (.635 ⟶ .632) and 905 (.630 ⟶ .628), but again the shorter proportions (Nos. 909 and 676) should be considered as the more decisive. I agree with Mackail *(ad loc.)* that the mention of Diomedes and Achilles adds immensely to the effectiveness of the speech.

# THE VALUE OF THE PROPORTIONS

40. Mackail *(ad loc.)* refers to "the clumsy phrasing, and the needlessness, of ll. 256-258," and considers the passage 245-258 unfinished. Pease *(Publi Vergili Maronis Aeneidos Liber Quartus)* accepts 256-258, as well as 126 and 285-286, and finds the arguments for rejection inadequate in each case; cf. his comments on the three passages.

41. These passages which Mackail italicizes as interpolations are bracketed by Ribbeck, with the exception of IV, 285, and IX, 651. On the other hand, all are accepted by Janell (with the exception of VI, 901), Hirtzel, Sabbadini, and Goelzer-Durand.

42. Sparrow, in his discussion of repetitions, comments on many passages suspected by earlier editors and commentators *(Half-Lines and Repetitions in Virgil*, pp. 140-154). These include IV, 126, 285-286, and VI, 901, all italicized by Mackail. Sparrow believes that in many instances temporary half-lines were filled in by early interpolators, often from similar passages elsewhere in the *Aeneid* (I, 380; II, 360, 775; III, 153, 691; IV, 343, 498; IX, 53, 160), and that sometimes a line and a half were added (VI, 438b-439; X, 870b-871; XI, 27b-28; XII, 439b-440). In other words, Sparrow suggests that Vergil may have written 72 half-lines instead of the 59 usually attributed to him. We have seen above that at least 41, and possibly 43, of the half-lines in the text provide better ratios if they are counted as fractions. Sparrow's suggested half-lines, however, receive almost no support from the mathematical structure of the poem. If we delete the conclusions of the lines as interpolations, the ratios in the shorter proportions change as follows:

| Passage | Fraction deducted | Proportion | Ratio | Ratio, with part of line omitted |
|---|---|---|---|---|
| I, 380 | .6 | No. 478 | .605 → .630 | |
| | | 476 | .610 → .617 | |
| | | 717 | .608 → .612 | |
| II, 360 | .8 | No. 372 | .609 → .595 | |
| II, 775 | .6 | No. 500 | .614 → .607 | |
| | | 379 | .619 → .627 | |
| III, 153 | .6 | No. 502 | .622 → .631 | |
| III, 691 | .6 | No. 522 | .609 → .598 | |
| | | 90 | .632 → .626 | |
| | | 521 | .619 → .625 | |
| IV, 343 | .6 | No. 530 | .621 → .613 | |
| | | 395 | .630 → .634 | |
| | | 529 | .630 → .634 | |
| | | 396 | .637 → .641 | |

| Passage | Fraction deducted | Proportion | Ratio | Ratio, with part of line omitted |
|---|---|---|---|---|
| IV, 498 | .4 | No. 766 | .626 → .634 | |
| | | 534 | .607 → .611 | |
| | | 105 | .626 → .622 | |
| IX, 53 | .6 | No. 842 | .613 → .605 | |
| | | 626 | .608 → .613 | |
| IX, 160 | .3 | No. 845 | .622 → .615 | |
| | | 628 | .623 → .626 | |

For III, 691, IV, 498, and IX, 53, the results are conflicting, but the shortest proportions argue against Sparrow's conclusions; in the case of I, 380, the ratios favor his theory (p. 140) that *et genus ab Iove summo* is possibly an interpolation from VI, 123. The other passages receive no support from the Golden Mean ratios. When Sparrow suggests that a line and a half were added later to certain passages, the results of the ratios are even more decisive:

| Passage | Lines deducted | Proportion | Ratio | Ratio with passage deducted |
|---|---|---|---|---|
| VI, 438b-439 (rejected also by Goelzer) | 1.4 | No. 162 | .609 → .583 | |
| | | 796 | .625 → .602 | |
| X, 870b-871 | 1.3 | No. 670 | .620 → .587 | |
| | | 887 | .623 → .656 | |
| XI, 27b-28 | 1.6 | No. 672 | .60 → .552 | |
| | | 288 | .60 → .573 | |
| | | 286 | .607 → .583 | |
| XII, 439b-440 | 1.8 | No. 927 | .633 → .674 | |
| | | 926 | .621 → .609 | |

Sparrow also suggests (pp. 93 f., 149 f.) that I, 530-534 (530-533 = III, 163-166) and II, 792-793 (= VI, 700-701) should be deleted. The ratios on these two passages change as follows, when the lines are omitted:

| Passage | Lines deducted | Proportion | Ratio | Ratio with passage deducted |
|---|---|---|---|---|
| I, 530-534 | 4.3 | No. 722 | .624 → .555 | |
| | | 22 | .624 → .529 | |
| | | 721 | .613 → .682 | |
| | | 359 | .612 → .585 | |
| | | S1001 | .615 → .602 | |
| II, 792-793 | 2 | No. 60 | .60 → .692 | |
| | | 500 | .614 → .590 | |
| | | 379 | .619 → .639 | |
| | | 737 | .621 → .611 | |

Here again we have mathematical evidence to support the text of the *Aeneid* as it has come down in the MSS, against the suggestions of scholars who wish to delete passages on the basis of repetitions and so-called inconsistencies.

43. For this passage I deduct 6.4 lines only; VIII, 46, is already excluded as an interpolated verse.

44. Cf., however, Mackail on VIII, 41: "It is to be observed, that if the intermediate lines *iamque tibi . . . haud incerta cano* were removed, the *nunc qua ratione quod instat* of l. 49 would follow on naturally, completing the broken line"; Durand (on 42-49) notes that *nunc* in 49 lacks meaning with 42-49a removed. Mackail suggests that both 42-49a and 81-85 were later insertions by Vergil, not fully incorporated into the narrative. The disastrous effect upon several ratios caused by the removal of 42-49a was seen above, and the deletion of 81-85 has a similar result; since 81-85 is the entire minor of No. 207, this proportion disappears, and the ratio of No. 828 changes from .619 to .813. Both passages thus seem a necessary part of the poem's mathematical symmetry.

45. Cf. Leo, *Plautinische Forschungen* (ed. 2), p. 42, note 3; Heinze, *Virgils epische Technik* (ed. 3), pp. 45 ff.; Norden, *Aeneis, Buch VI* (ed. 3), pp. 261 f. Norden says that attempts to defend the authenticity of the passage ("Rettungsversuche") are an exercise-ground ("Tummelplatz") for dilettantes; on this, see Büchner, *P. Vergilius Maro*, col. 333. For a more recent view of the passage as spurious, see Liebing, *Die Aeneasgestalt bei Vergil*, pp. 40, 189f. [on Liebing, cf. Duckworth, *CW* 51 (1957-58), p. 159 and note 42].

46. Fairclough, *CPh* 1 (1906), pp. 221-230; Gerloff, *Vindiciae Vergilianae;* Crump, *The Growth of the Aeneid*, pp. 44 ff.; Shipley, *TAPhA* 66 (1925), pp. 172-184 (an answer to Norden's metrical arguments); Knight, *Vergil's Troy*, pp. 45 ff. (cf. p. 47, where Knight says that the passage "is certainly Vergilian and necessary in its context . . . . If the Helen scene is deleted from the second *Aeneid*, there is a dislocation. The passage is not detachable or inorganic?'); Pease, *Publi Vergili Maronis Aeneidos Liber Quartus*, p. 44, note 323; Palmer, *Mnemosyne*, 3rd Ser. 6 (1938), pp. 368-379; Büchner, *P. Vergilius Maro*, cols. 331 ff.; Hatch, *CPh* 54 (1959), pp. 255 ff.

47. *Mnemosyne*, 3rd Ser. 6 (1938), p. 379.

48. *Virgil's "Gathering of the Clans,"* pp. 46 ff. See also Mackail (on 664-665), who suggests that the six verses do not belong to the catalogue at all, but come from an earlier cancelled draft in which certain chiefs were represented as meeting at the palace of Latinus, such a meeting being suggested or implied in the opening paragraph of Book VIII.

49. This transposition violates the structure of Anchises' speech; the first half (760 ff.) begins with the Alban kings, the second half (808 ff.) with the kings of Rome; cf. Duckworth, *TAPhA* 87 (1956), p. 304 and note 79.

50. *The Growth of the Aeneid*, pp. 74 f., 117. Crump's desire to transfer the description of Nisus and Euryalus to Book V results from a misunderstanding of Vergil's technique. The poet describes his characters in detail not when they first appear but just before they play an active part in the narrative; e.g., the detailed description of Drances comes in XI, 336-341, just before his important speech to Latinus, not when he addresses Aeneas in XI, 122 ff.; similarly, the account of Camilla's youth appears in XI, 535-584, just before her aristeia and death, not when she is first mentioned in VII, 803 ff. Cf. Cartault, *L'Art de Virgile*, pp. 795 f.

51. *The Aeneid*, p. lxiii.

52. This may be seen most readily by referring to the chart-index for each book; e.g., in Book I, No. 17 ends with 409; No. 480 begins with 466; No. 723 ends with 558, etc. In most instances one ratio ends with the verse listed and another begins with the following verse, e.g., in Book II, No. 485 ends with 227 and No. 730 begins with 228.

53. Fairclough in his Loeb edition of Vergil indicates a paragraph after 596 in both text and translation, as also do Guinagh and Humphries in their translations.

54. Knight, in his recent translation of the *Aeneid*, begins a new paragraph after VI, 449, also after VIII, 422, 674, and IX, 280a, 398. Fairclough in his Loeb translation (but not his text) paragraphs after VI, 607, IX, 280a, and X, 17. Paragraphs appear in Guinagh's translation after VI, 449, 607, VIII, 312, 394, IX, 280a, X, 17, XI, 71, and XII, 480.

55. *The Aeneid*, p. 47. Actually the second part of the episode ends at verse 227; see above, note 3 and Table XIII. On the importance of the Laocoon story, see Büchner, *P. Vergilius Maro*, cols. 326 ff.

56. Cf. No. 39: 268-297 (appearance of Hector) = major, 318-335 (Panthus episode) = minor; for these and other noncontiguous ratios, see above, p. 50.

57. *The Aeneid*, p. 130. Mackail also suggests that "the four lines, 248-251, *Atlantis . . . barba*, look like an early draft meant to be struck out and replaced by the single line *Atlantis . . . fulcit*." If we delete 248-251, No. 102 changes

58. Cf. Cartault, *L'Art de Virgile*, pp. 644 f.

59. *The Growth of the Aeneid*, p. 117. On this, see above, "Transpositions," and note 50.

60. *The Aeneid*, p. 463; cf. pp. 517 f.

61. Cf. XII, 598 f. See Duckworth, *CJ* 51 (1955-56), pp. 361 f. and note 32.

62. A change in the ratios of Book XII may have been contemplated, however, see below, note 70.

63. For VI, 93-94, cf. Nos. 563 (.634 → .610), 562 (.612 → .627), 152 (.620 → .628); for VII, 128-129, cf. Nos. 188 (.623 → .60), 814 (.626 → .616), 810 (.626 → .621), 182 (.629 → .635).

64. See above, p. 78.

65. *The Aeneid*, p. 89; cf. Crump, *The Growth of the Aeneid*, p. 115: "Vergil had intended to give three years to the final revision of the Aeneid ... The most important task was the rewriting of III."

66. *AJPh* 78 (1957), pp. 133-151, 382-400.

67. See *The Aeneid*, pp. 1, 129, 165, 255 f., 463; he says of Book VI (p. 207): "Virgil was clearly making material alterations up to the last. While in one sense it was his highest achievement, it is at the same time one of his least complete."

68. See above, p. 58; cf. Chapter 4, note 40.

69. See above, p. 60.

70. If a Fibonacci proportion of 68 verses (or two of 34 each) were added to the major of No. 978 (the first main division of XII) and a similar proportion of 34 verses were added to the minor, the three component parts of No. 978, still in the pattern (a + b)/c, would produce a ratio of 243.6/389.6, or .626. The three main divisions of XII would then total (a) 389.6; (b) 405.2, as at present; (c) 256, as at present, and the following ratio would appear:

|  | M | m | M/(M + m) |
|---|---|---|---|
| b/(a + c) | 645.6 | 405.2 | .614 |

A similar result could be achieved by the judicious omission or compression of portions of the second main division, and this procedure would seem the more probable, since XII is already the longest book of the *Aeneid*; e.g., with 69 lines removed from the second main division, the totals would be (a) 287.6, as at present, (b) 336.2; (c) 256, as at present, and the ratio would appear in this form:

|  | M | m | M/(M + m) |
|---|---|---|---|
| b/(a + c) | 543.6 | 336.2 | .618 |

71. See above, p. 76, and cf. Table XIX.

72. The ratios in the *Aeneid* as a whole would be affected by any changes in XII such as I have suggested above (note 70). It seems significant that the first procedure, the addition of 102 verses to XII, not only brings the book into conformity with the other even-numbered books but improves materially the ratios in Table VIII. For No. 1046 (the second half of the *Aeneid*) we have 3255.1/5219.3 = .624, instead of the present .636. The proportions in the poem as a whole are similarly improved; the ratio in No. 1047 moves from .632 to .625 (6223.4/9954.2) and that in No. 1048 changes from .637 to .631 (6276.9/9954.2). Vergil could, of course, have achieved this same result by additions or deletions in the other books, but then the anomalous feature of XII would not have been remedied.

73. The manuscripts of the Donatus-Suetonius Life vary to a slight degree; Σ of the fifteenth century adds the *Moretum* (which is included in the manuscripts of the *Appendix*, dating from the ninth century on), and G of the tenth (or ninth) century, one of the most important codices, omits the phrase *de qua ambigitur* concerning the *Aetna*. Rand, *HSCPh* 30 (1919), pp. 107 f., considers the expression of doubt a late addition and thinks that Donatus as well as Servius accepted the *Aetna* as genuine; against this, cf. Büchner, *P. Vergilius Maro*, col. 43.

74. Nonius in the fourth century A.D. knew the extant poem and considered it Vergil's work; he cites (211) *Culex* 53 as *Vergilius in Culice*.

75. *Latin Literature of the Empire*, II, p. 1.

76. Rand, *HSCPh* 30 (1919), pp. 103-185; Frank, *Vergil, A Biography*; DeWitt, *Virgil's Biographia Litteraria* (DeWitt reads too much autobiography into the poems, seeing in most of the *Catalepton* a consistent attack on Mark Antony). See also Rostagni, *Virgilio Minore*, for a discussion of the *Appendix* as the work of Vergil.

77. *The Development of Virgil's Art*, pp. 19, 21; cf. also Prescott's useful article on the status of the Vergilian *Appendix*, *CJ* 26 (1930-31), pp. 49-62.

78. Bickel, *RhM* 93 (1949-50), pp. 289-324; Büchner, *P. Vergilius Maro*, cols. 42-160; for other recent studies, see Duckworth, *CW* 51 (1957-58), pp. 92, 116 f.

79. In addition to Bickel and Büchner (cf. above, note 78), see Fraenkel, *JRS* 42 (1952), pp. 1-9; according to Fraenkel, the forger vouches in lines 1 and 24 for the early friendship of Vergil

80. See above, p. 41.

81. The corrupt state of the text, especially of the *Culex, Ciris,* and *Aetna,* makes the examination of the mathematical structure all the more difficult, but in my analysis of the proportions I have attempted to avoid an arbitrary handling of the text. I have profited much from the outlines of the *Culex* and the *Ciris,* as given by Richardson, *Poetical Theory,* pp. 72 ff., and of the same poems, plus the *Aetna,* in Büchner, *P. Vergilius Maro,* cols. 71 ff.; Büchner's divisions for the *Moretum* (1-50, 51-84, 85-116; cf. col. 151) are inaccurate. The paragraphing and punctuation in the text and translation of Fairclough's edition of the *Appendix* have also been helpful. I exclude from my calculations *Moretum* 36 and 75, and *Aetna* 186, 188, and 236, which are bracketed as interpolations.

82. For additional information concerning Catullus, Lucretius, and *Georgics* IV, see Appendices A, B, and C, and Tables XVII-XXIII.

83. For the *Aeneid,* see Table IX: the component parts of the final ratio (or ratios) correspond *in every case* to the main divisions of each book, every main division contains a ratio, and the component parts of the ratios in the main divisions are identical with the narrative subdivisions in 74 per cent of the instances.

84. *The Dirae,* p. 133.

# CONCLUSION

SOCRATES SAYS in the *Philebus* (64e): "If measure and symmetry are absent from any composition in any degree, ruin awaits both the ingredients and the composition . . . . Measure and symmetry are beauty and virtue all the world over." We now see that measure and symmetry play an even greater part in Latin poetry, and especially in the works of Vergil, than has hitherto been realized.

The purpose of this book has been threefold: (1) to analyze the architecture of the *Aeneid* as a whole and the structure of the individual books, with special attention to alternation, parallels and contrasts, and tripartite divisions; (2) to show how mathematical symmetry produced by Golden Mean ratios contributes to the composition of the epic; and (3) to use these ratios as a guide to a more accurate text of the poem.

I make no attempt to summarize the preceding chapters in any detail. My study of the structure of the *Aeneid* and particularly of the tripartite nature of each book led me to the surprising discovery that the main divisions of the books reveal proportions close to the famous Golden Section and are made up of dozens and hundreds of smaller units also containing Golden Mean ratios, with an almost exact correspondence everywhere between the narrative units and the mathematical structure.[1] This agreement between narrative and proportions, the large number of framework and alternating patterns which emphasize significant speeches and episodes, the hundreds of almost perfect Golden Mean ratios, the use of the Fibonacci series more than three hundred times to produce the ratios, and other important features of the structure all combine to provide convincing proof that such mathematical symmetry was deliberate on Vergil's part; neither chance nor intuition can explain so many related phenomena.

Many lovers of Vergil will find all this very disturbing and will be reluctant to admit that Vergil counted lines and arranged his material on the basis of mathematical proportion. On the other hand, such a procedure explains the statements in the Donatus-Suetonius Life that Vergil devoted especial attention to the study of mathematics and that he composed the *Aeneid* in short and separate units working from a prose outline.

Numerous problems connected with the text of the *Aeneid* receive new light from the mathematical structure of the poem; about three-quarters of the half-lines seem intentional and play their part in producing more accurate ratios, whereas the remainder would evidently have been replaced by whole lines; passages bracketed as interpolations should be omitted when missing from the better manuscripts, but not otherwise; the transposition of passages from one context to another is usually impossible; certain passages considered spurious should not be rejected, and the evidence is particularly strong for the Vergilian authorship of the famous Helen passage (*Aen.* II, 567-588); furthermore, the ratios provide helpful information for the proper paragraphing of the text and support in this respect the editions of the *Aeneid* by Mackail and Sabbadini. Knight's metrical patterns based on the alternation of heterodyned and homodyned lines also receive added confirmation since the major and minor parts of many proportions reveal strikingly different metrical combinations.[2]

Also, the presence of similar Golden Mean ratios in the longer poems of the *Appendix Vergiliana* raises again the question of their authenticity and provides special arguments to support the genuineness of the *Culex*. The problem of the minor poems would receive a more definite answer if Vergil were the only poet to compose in this fashion, but Golden Mean ratios and a mathematical symmetry with a similar correlation of content and structure had appeared earlier in Catullus and Lucretius, a fact which likewise poses new queries and suggests more work to be done.

This book, devoted primarily to the *Aeneid*, is in a sense a pioneer work, for it has been necessary to develop new methods and techniques and to devise tables and charts for the presentation of the material. I am confident that the final word has not yet been said on the structure of the *Aeneid*; new parallels and contrasts between the books in each half may be detected; on the mathematical side, some ratios may be rejected and others, more accurate, may be discovered. Additional investigation may reveal an even closer correlation between the units of the narrative and the passages containing proportions. New material may provide still more information concerning the half-lines and the interpolations. All this could lead to a new edition of the *Aeneid*, in which the text, the paragraphing, and even the punctuation could reproduce more accurately the units of thought as the poem left Vergil's hand, and possibly even as it might have been revised.

My preliminary work on Catullus and Lucretius should be expanded, and particularly in the case of Lucretius, where the mathematical structure of the poem may help to solve disputed matters of the text. The question arises inevitably concerning the use of Golden Mean ratios by other poets of Virgil's day and later. A tentative examination of Horace's hexameter poems reveals clear evidence of his use of proportions and a striking fondness for the Fibonacci series. I have found several ratios in Ovid's *Metamorphoses*, but these may be the result of chance; I have not yet discovered the close correlation between the narrative units and a mathematical structure which is so characteristic of Vergil's procedure. Another fruitful area for future investigation might be the Silver Latin epic poets who are in other respects so indebted to Vergil. If Golden Mean ratios and a similar mathematical symmetry appear in their works,[3] it would be interesting to determine, if possible, whether they deliberately copied Vergil's practice or merely displayed a subconscious feeling for Vergil's symmetry; the latter seems to be true of Maphaeus Vegius when in the Renaissance he composed a Thirteenth Book of the *Aeneid*.[4]

The basic fact, that mathematical ratios and structural proportion are as important in literature as in art and architecture, seems now clearly established. Latin poetry of the first century B.C., and especially the poetry of Vergil, reveals a rigid adherence to mathematical principles and displays the Golden Section, reached most frequently by the Fibonacci series. Whether the knowledge of such numerical ratios is to be attributed to Pythagorean doctrine or to Greek mathematical theory such as that of Eratosthenes, whether it was introduced into Latin poetry by Ennius, as I suggested above,[5] or by Parthenius, who had close contacts both with the *novi poetae* and with Vergil, may never be determined definitely.[6] But the presence of mathematical ratios coinciding with narrative units, both large and small, should not detract from our enjoyment of the poetic artistry of Vergil and his contemporaries; on the contrary, it adds to our understanding of their learning and of their appreciative attention to basic principles of aesthetics.[7]

## Notes to Conclusion

1. This correlation is best seen in Table IX, the chart-index of the books of the *Aeneid*.

2. See Appendix F.

3. Cf. Getty, *TAPhA* 91 (1960), pp. 317 ff., who has recently discovered several Golden Mean ratios in Lucan, Book I; the most perfect of these, based on the Fibonacci series 21, 34, 55, 89, appear in a passage concerning the Neo-Pythagorean Nigidius Figulus.

4. See Appendix E and Tables XXVI and XXVII.

5. See Chapter 5, note 21.

6. Edwin L. Brown in his dissertation, *Studies in the Eclogues and Georgics*, pp. 128-141, argues plausibly for the scientist Eratosthenes as an early exponent of numerical proportion in verse and also favors him as the person to whom Vergil alludes in *Ecl.* III, 40 *(quis fuit alter)*; Brown has discovered (pp. 150-153) the Golden Section in Aratus' *Phaenomena*, and suggests (pp. 205 f.) that it was Parthenius to whom Vergil and others were directly indebted for their knowledge and use of the Golden Mean ratio. But cf. Getty, *TAPhA* 91 (1960), p. 323: "Indeed, from Ennius onward, the influence of Pythagoreanism or Neopythagoreanism upon Roman poetry may be imponderable, but it is far from being negligible."

7. Cf. Whaler, *Counterpoint and Symbol*, p. 8, on his discovery of symbolic numerical progressions in Milton: "When we find a poet calculating numerical progressions and stringing them through the lines of his masterpiece, we are forced to redefine 'inspiration.'"

# APPENDICES

## APPENDIX A

## CATULLUS LXIV

In establishing the proportions in Catullus LXIV I have used the new Oxford text of Catullus by Mynors, adding one line for the lacuna after 253, and deleting 378. I do not include 23b, following in this respect Ellis' earlier Oxford text.

The structure of the poem is tripartite, with the sections on Peleus and Thetis enclosing the story of Ariadne in the central portion:

        1-51     Peleus and Thetis. Wedding and drapery
        52-250   The story of Ariadne
        251-408  Peleus and Thetis. Drapery and wedding

For these main divisions I follow Murley, *TAPhA* 68 (1937), p. 308. Richardson, *Poetical Theory*, p. 71, and Mendell, *YClS* 12 (1951), p. 212, give the Ariadne story as 52-264, but Murley seems correct: 251-266 balance 47-51 in that both passages give details about the drapery, with Bacchus providing an appropriate transition from the story of Ariadne to the other figures on the tapestry. The three main divisions produce a ratio of the type $c/(a + b)$; $M/(M + m) = 250/408 = .613$ (listed in Table XVII as No. C1). This is the same $c/(a + b)$ pattern which appears in the *De Rerum Natura* as a whole, in the *Eclogues* as a group, and in the *Aeneid* (both as a whole and in each half). This ratio in the main divisions of Catullus LXIV supports Murley's analysis of the poem.

Table XVII contains the ratios which I have found in Catullus LXIV, those for the three main divisions (cited above in Chapter 5, "Vergil and Pythagoreanism") being numbered C4, C11, and C27. The most accurate ratios, those ranging from .615 to .621, comprise 29 per cent of the total; this is very similar to the percentage for this range in Vergil's *Aeneid* (28.8 per cent); cf. Table XV.

If with Ellis we indicate a lacuna after 354 and ignore the lacuna after 253, the changes in the ratios are as follows: No. C26, .616 ⟶ .625; No. C27, .620 ⟶ .614; and No. C29, .618 ⟶ .60. The mathematical symmetry of the poem thus favors the lacuna after 253 (with Mynors), not after 354. The rejection of 378 (with Mynors) is supported by No. C30; if we retain the verse, the ratio changes from .636 to .666. Ellis, however, seems correct in refusing to print 23b in his text; if the verse is added, the ratios in Nos. C3 and C5 move from .633 to .645 and from .632 to .650 respectively.

Table XVIII presents an outline of the poem and the chart-index showing the correspondence of the narrative units and the mathematical structure.


## APPENDIX B

## LUCRETIUS

For my calculations concerning the Golden Mean ratios in Lucretius, I have used Bailey's 1947 edition of the *De Rerum Natura*, deducting the bracketed passages and adding one line for each lacuna indicated. The total lines for each book are thus as follows: I, 1121; II, 1180; III, 1094; IV, 1280; V, 1458; VI, 1289. The books are in proportion in the pattern $c/(a + b)$, with $a = I + II = 2301$; $b = III + IV = 2374$; $c = V + VI = 2747$; $M/(M + m) = 4675/7422 = .630$. The $c/(a + b)$ pattern appears also in Catullus LXIV, the *Eclogues* as a group, and in the *Aeneid* (both as a whole and in each half).

The main divisions of each book of Lucretius' poem are likewise in proportion, usually with two or more main divisions providing additional ratios. I have used the main divisions as given by Bailey (1947 edition) at the beginning of his commentary on each book, with the following exceptions: in Book I, I agree with Leonard and Smith who regard 921-950 (digression on Lucretius' mission) as the introduction to the final division [Bailey himself in his translation of Lucretius (Oxford, 1910), p. 24, begins the final division with 921]; in Book V, I treat 772-924 (the beginnings of life) and 925-1427 (primitive man and the rise of civilization) as two main divisions; in Book VI, I end the first division with verse 42 and, following Leonard and Smith, I begin the final division at 1090, rather than at 1138 with Bailey; the sections on pestilences (1090-1137) and the plague at Athens (1138-1286) appropriately go together; this last change affects in no way the ratios; see Table XIX, Nos. L20 and L21.

Table XIX gives the ratios in the main divisions. Since each book of the *De Rerum Natura* has four or five main parts, the patterns here resemble those in Table IV (four or five parts, usually interlocked) rather than those in Table VI (main divisions of the *Aeneid* in proportion); no book of the *Aeneid* has more than three main divisions. In six books of the *Aeneid* only two of the three main divisions are in proportion, but in each book of the *De Rerum Natura* all the main divisions combine to form a Golden Mean ratio, and a ratio which is surprisingly accurate, considering the uncertainty of the text; the range is from .608 to .620, and in four of the books the ratios are in the area from .617 to .620. I denote the main divisions by the letters a, b, c, d, and e. The ratios derived from all the main divisions in each book are Nos. L1, L3, L6, L9, L15, L20, and these are cited above in Chapter 5, "Vergil and Pythagoreanism."

Bailey in his Oxford text of Lucretius (2nd ed., 1921) bracketed three passages (IV, 858-876; V, 509-533; and VI, 608-638) which in his 1947 edition he now accepts. Leonard and Smith likewise print these passages without brackets. It seems significant that the ratios are more accurate with the passages retained; e.g., the ratio in No. L9 moves from .617 to .627 with IV, 858-876, deleted, and that in No. L15 from .608 to .602 with V, 509-533, removed (in No. L18 the change is from .614 to .637); if we delete VI, 608-638, the ratio in No. L20 moves from .619 to .610. As in the case of Vergil's *Aeneid*, the mathematical structure of the *De Rerum Natura* supports the reading of the MSS and indicates that the passages are not to be rejected as interpolations.

The mathematical symmetry to be found in Lucretius' poem is a separate field of investigation, and I shall not attempt to discuss in detail the Golden Mean ratios which appear in the work. It is important, however, to realize that numerous proportions occur in the shorter passages and that these combine to form the ratios in the main divisions. I have examined Book I, using the subdivisions listed in the editions of Bailey and of Leonard and Smith and following the smaller units as given in Bailey's commentary. The ratios which I have found in Book I appear in Table XX, with Nos. L1 and L2 being repeated from Table XIX. The most accurate ratios, ranging from .615 to .621, constitute 21 per cent of the total; this is lower than in either Catullus LXIV (29 per cent) or the *Aeneid* (28.8 per cent). On the other hand, the ratios in this range in Table XIX (main divisions of the *De Rerum Natura* in proportion) occur six times, or 28.6 per cent; this is very close to the ratios in the same range in Table VI (main divisions of the *Aeneid* in proportion): 5/16, or 31.3 per cent.

Table XXI presents the outline of Book I and the chart-index of the proportions. Again the correspondence between the mathematical structure and Lucretius' argument is amazing; e.g., the four component parts of the ratio in the book as a whole (No. L1) are also the four main divisions of the book (cf. Nos. L24, L32, L59, and L75; these are underlined in the chart).

Additional investigation will undoubtedly reveal similar Golden Mean ratios in the shorter passages of Books II-VI, and these should likewise be helpful for the structure and the text of the poem. It is important to know that Vergil's use of Golden Mean ratios is not unique and that both Catullus (in LXIV) and Lucretius composed their poetry on the basis of the Golden Section.

# APPENDIX C

## THE ARISTAEUS STORY (*GEORG.* IV, 281-558)

In Chapter 3, "Proportions in the *Georgics*," I described the ratios in the Aristaeus story as a whole and in each of its three parts. As in the case of the books of the *Aeneid* (also Catullus LXIV and Lucretius, Book I), the smaller units of the narrative contain ratios, with the passages combining to create the proportions in the main sections. Table XXII gives the ratios which I have found in the Aristaeus story (including those listed in Chapter 3). The ratios in the range from .615 to .621 comprise 22.7 per cent of the total; in this respect the Aristaeus episode is closer to Lucretius I (21 per cent) than to Catullus LXIV (29 per cent) and the *Aeneid* (28.8 per cent). The outline and chart-index in Table XXIII reveal the exact correspondence between the story and the mathematical structure of the episode; this correlation closely resembles Vergil's procedure in the *Aeneid*.

# APPENDIX D

## HORACE

In Chapter 5, "Vergil and Pythagoreanism," I pointed out that in Horace's *Ars Poetica* the division of the poem into Poetry (1-294) and the Poet and Critic (295-476) produces the exact Golden Section: $M/(M + m) = 294/476 = .618$, and that these totals are multiples (by 14) of 13, 21, 34, the Fibonacci series which we find so often in Vergil's *Aeneid*. The existence of this ratio was discovered also by K. Gantar, *ZAnt* 4 (1954), p. 277, whose article came to my attention only after the completion of Chapter 5. That this proportion in the *Ars Poetica* as a whole can hardly be accidental is supported by the presence of many approximate Golden Mean ratios both in the *Ars Poetica* and elsewhere in Horace's poetry, especially in the *Satires* and the other *Epistles*.

The first book of the *Satires* is divided into two halves (cf. Port, *Philologus* 81 [1925-26], pp. 288 ff.; Knoche, *Die römische Satire*, p. 48) as follows: 1-5 (644 verses) and 6-10 (386 verses, or 394, if we retain the disputed eight lines at the beginning of 10); the halves produce the following ratio: $M/(M + m) = 644/1030 = .625$, or (including the extra eight verses of 10) $644/1038 = .620$. The eight verses at the beginning of 10 are lacking in the best MSS and are considered spurious by most editors and commentators. If we delete the eight verses, the two parts of 10 (1-35, criticisms of Lucilius; 36-92, Horace's own ideal; cf. Rolfe's analysis) are in proportion: $M/(M + m) = 57/92 = .620$. The presence of this ratio in the tenth satire strengthens the view that the eight verses are not by Horace; if the lines are included, the ratio changes to $57/100 = .570$.

An interesting illustration of the four-part interlocked pattern, $(b + d)/(a + c)$, appears in *Sat.* II, 3, where four vices are described: avarice (82-157), ambition (158-223), self-indulgence including that of lovers (224-280), and superstition (281-295). It is appropriate that avarice and self-indulgence, condemned by Horace so frequently elsewhere in his poetry, form the major: $M/(M + m) = 133/214 = .621$. The ratio in this satire as a whole is achieved by the framework pattern $b/(a + c)$, with 1-157 (introduction; avarice) and 281-326 (superstition; conclusion) combining to form the major and 158-280 (ambition; self-indulgence) providing the minor; $M/(M + m) = 203/326 = .623$.

I list below in Table XXIV the instances of the Golden Section which I have found in the *Satires* and the *Epistles* (each poem considered as a whole): that in *Epist.* I, 6, was noted also by Gantar, *ZAnt* 3 (1953), pp. 79-81.

Horace's conversational style in the hexameter poems is such that it is often difficult to determine the correct divisions; for most of the ratios listed in Table XXIV, I have followed the paragraphing and the analyses of the poems as given in Rolfe's first edition, but for *Epistles* II, 1, and II, 2, see Stégen, *Les Épîtres Littéraires,* pp. 181 ff., 207 ff. Further investigation of this aspect of Horace's poetry will undoubtedly reveal many additional ratios, indicating that he, like Vergil, devoted great attention to mathematical symmetry. One surprising feature of the hexameter poems is this: in numerous passages Horace's thought breaks naturally into units of 3 and 5, or 5 and 8, or 8 and 13 verses, and we have the Fibonacci series producing Golden Mean ratios: e.g., *Sat*. I, 5, 34-38 and 39-46; I, 8, 1-5 and 6-13; *Epist*. I, 14, 1-5 and 6-13 (8/13 = .615); and *Sat*. I, 7, 1-3 and 4-8 (5/8 = .625); 1-8 and 9-21 (13/21 = .619). Such Fibonacci series are very frequent in the *Ars Poetica,* especially in the early part of the poem. I list those which I have noted in Table XXV.

The Roman Odes of Horace (III, 1-6) are often divided into two halves; the first three are Roman and general, the second three more concerned with Augustus (4) and his military (5), religious, and social policies (6). The two poems concluding each half (3 and 6) deal with Rome and its greatness, the tone of 3 being positive but containing a note of warning, that of 6 negative with the note of warning sounded loudly; see above, pp. 14 f., and cf. Duckworth, *TAPhA* 87 (1956), p. 301 and note 69, p. 304 and note 79. The parallelism of the two halves may be shown as follows (and I list in parentheses the line totals for each poem):

1. Simplicity of living (48)
2. *Virtus* (32)
3. Augustus and Rome (72)
   = climax of first half

4. *Consilium* and *vis temperata* (80)
5. *Virtus* (56)
6. Augustan program (48)
   = climax both of second half and of the cycle

The approximate Golden Mean ratio appears in each pair of balancing odes, as follows:

1 and 4: $M/(M + m) = 80/128 = .625$
2 and 5: $M/(M + m) = 56/88 = .636$
3 and 6: $M/(M + m) = 72/120 = .60$

In the first and third proportions the totals are multiples of the Fibonacci series (16 times 3, 5, 8, and 24 times 2, 3, 5, respectively), and the second proportion contains multiples of the series 1, 3, 4, 7, 11, etc. (8 times 4, 7, 11). The appearance of these ratios in the two halves of the Roman Odes supports the analysis of the cycle as given above; as in Vergil, structure and content go hand in hand. Furthermore, the Golden Section appears in the Roman Odes as a whole; if we take the first ode of each half as the minor (1 = 48, 4 = 80; m = 128) and the remaining four odes as the major (32 + 72 + 56 + 48 = 208), we have the following ratio: $M/(M + m) = 208/336 = .619$, in the Fibonacci series 8, 13, 21 (times 16).

In *TAPhA* 87 (1956), pp. 308 ff., I favored a structural analysis of the *Carmen Saeculare* which stresses the two groups of prayers of six stanzas each (9-32, 37-60); these are framed by stanzas to Apollo and Diana. If we view the prayers as the major (48 verses) and the framing passages as the minor (28 verses), we have the alternating pattern (a + c + e)/(b + d) and the ratio $M/(M + m) = 48/76 = .632$.

# APPENDIX E

# MAPHAEUS VEGIUS

In 1428 Maphaeus Vegius, at the age of twenty-one, composed *A Supplement to the Twelfth Book of the Aeneid,* and this was linked to the *Aeneid* in the 1471 edition, only two years after the *editio princeps* of 1469; for a century and a half this "Thirteenth Book of the *Aeneid*" was regularly published with Vergil's epic and was read and admired. The author was known as "alter Maro" and "alter Parthenias" (Brinton, *Maphaeus Vegius and His Thirteenth Book of the Aeneid*, pp. 5, 30). Since he knew his beloved Vergil almost by heart, he echoed the language, style, and meter of the epic to an amazing degree. The greatest flaw in the composition of the work (as Brinton points out, p. 4) is the excess of speeches—thirteen ranging from four to forty-four lines.

An examination of the structure of the "Thirteenth Book" reveals the presence of a number of approximate Golden Mean ratios and a mathematical symmetry which at first sight appears almost Vergilian—but the differences are as striking as the similarities. The tripartite divisions and subdivisions so characteristic of the books of the *Aeneid* are lacking. The "Thirteenth Book" contains 630 verses and is thus 75 lines shorter than Book IV, the shortest book of the *Aeneid*. I have found in it 34 Golden Mean ratios ranging from .60 to .636, an average of one in every 18.5 verses; the average for the *Aeneid* is one in every 9.4 verses, and for Book VII, which contains the smallest number of proportions (74), the average is one in 11 verses. The distribution of ratios in the range from .615 to .621 is not Vergilian: 14.7 per cent as opposed to 28.8 in the *Aeneid*. When the Fibonacci series produces ratios in the "Thirteenth Book," the units are 2, 3, 5, 8, and do not continue to 13, 21, 34, the numbers used so frequently by Vergil. The four main divisions contain ratios, but the component parts of these ratios fail to correspond with Vergilian accuracy to the smaller narrative units; also, the main divisions do not combine into a proportion for the work as a whole. When we chart the ratios, many gaps appear among the shorter passages; an attempt to find ratios here produces proportions from .529 to .580 and from .641 to .655, and these bear no relation to the Golden Section.

I give in Tables XXVI and XXVII a list of ratios and a chart-index to clarify the preceding summary. The conclusion to be drawn in the case of Maphaeus Vegius seems inevitable. From his close knowledge of the *Aeneid* and his feeling for Vergil's style and structure, he reproduced to a degree the mathematical symmetry of Vergil's epic—but only to a degree. He was not consciously or deliberately introducing Golden Mean ratios into his work or imitating the structure of the individual books of the *Aeneid*; rather, a subconscious feeling for Vergilian proportions guided him, and this not only accounts for the presence of the ratios and the appearance of the Fibonacci series in his poem but explains also the striking dissimilarities between his structural procedure and that of his revered master.

## APPENDIX F

## METRICAL PATTERNS AND GOLDEN MEAN RATIOS

Another possible area of investigation is the relation of Vergil's metrical patterns to the mathematical symmetries appearing throughout the *Aeneid*.

Knight in his book *Accentual Symmetry in Vergil*, stressing the importance for the hexameter of the texture of the fourth foot, revealed the different effects achieved by the use in the fourth foot of heterodyne (i.e., clash of accent and ictus, as in *Aen.* I, 1: *arma virumque cano, Troiae qui primus ab oris*) and homodyne (coincidence of accent and ictus, as in I, 7: *Albanique patres atque altae moenia Romae*). Whereas Catullus (in LXIV) and Lucretius (in Book I) had used fourth-foot homodyne 61.25 per cent and 51.49 per cent respectively, Vergil in his authentic works reduced the amount of fourth-foot homodyne to the following percentages: *Eclogues,* 37.27; *Georgics,* 33.45; *Aeneid,* 35.95. These figures are given by Knight in the "Table of Fourth-Foot Texture" (*op. cit.,* p. 108), and it is interesting to note that in this respect also the *Culex* is Vergilian, with 35.95 per cent fourth-foot homodyne. (On the *Culex* and the *Ciris,* see Knight, pp. 42, 53, and cf. above, Chapter 5, "The Authenticity of the Minor Poems.")

Straight sequences of both heterodyne and homodyne occur; e.g., the simile in II, 626-631, is all homodyne except the final verse (cf. Knight, p. 47); if we denote heterodyne by the letter a and homodyne by the letter b, we have the pattern b b b b b a. Vergil's increased use of heterodyne makes possible the "released movement," defined by Knight (pp. 48 f.) as "a movement in which the rhythm is steadily maintained by fourth-foot heterodyne until the last verse, which is homodyned in the fourth foot . . . . The effect of this pattern is easily recognized; a kind of rhythmic force is generated and pent, to be released in the last verse." A typical released movement appears in IV, 431-436: a  a  a  a . a  b . (p. 50). Such a sequence is sometimes introduced by an initial homodyne. Often released movements of equal length are combined, as in IV, 165-172, where we find a  a  a  b . a  a  a  b . (p. 57).

Knight considers the released movement one of two primary patterns, the other being alternation, with five or more verses alternately homodyned and heterodyned, as in I, 748-756:

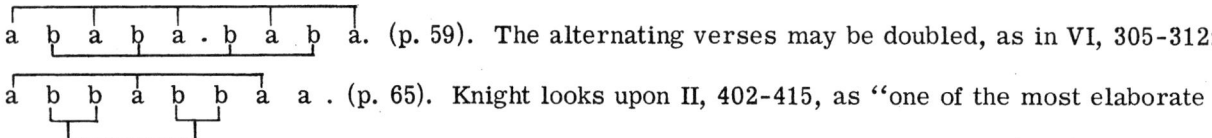

a  b  a  b  a . b  a  b  a. (p. 59). The alternating verses may be doubled, as in VI, 305-312:

a  b  b  a  b  b  a  a . (p. 65). Knight looks upon II, 402-415, as "one of the most elaborate symmetries of pure expanded alternation" (p. 66); the pattern is as follows:

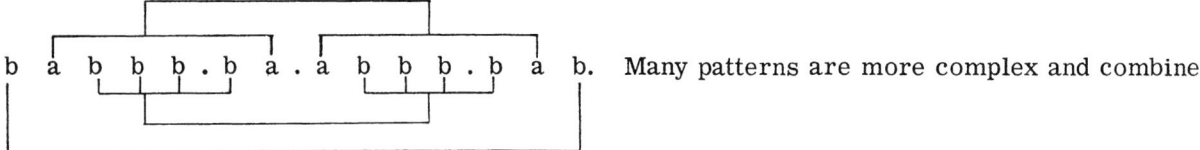

b  a  b  b  b . b  a . a  b  b  b . b  a  b.   Many patterns are more complex and combine released movements and alternations; e.g., in XII, 938-952, we have "a released movement added to an alternation, with additional verses, giving the effect of a released movement, but strengthened" (p. 68):

a  a  b  a  b  a  b . a  a  a  a  b . a  b  b.   For more intricate patterns see Knight, pp. 69 ff.; he praises highly IV, 584-705 (pp. 80 ff.); this is the final subdivision of Book IV and ends in 693-705 with the neat and effective scheme: a  a  b . a  b  b  a . b  b  b  b . a  b.

Knight's patterns may thus be summarized as follows: sequences and the released movement, alternations often expanded with intricate interlocking, and complex patterns combining in various ways the released movements and alternations. My purpose in adding this appendix and in describing Knight's conclusions is to reveal the correlation which exists between his metrical patterns and many of the shorter mathematical proportions in the *Aeneid*. This is a large field and one which deserves much fuller treatment. I shall cite merely a few instances where the majors and minors which determine the Golden Mean ratios reveal strikingly different metrical structures.

First of all, I present in outline form Golden Mean ratios which incorporate some of Knight's patterns as described above.

No. 496. (Ratio = .60). II, 624-633. Major = simile in homodyned sequence, ending with a heterodyne; minor = framing passages in heterodynes.
    M = 626-631: b  b  b  b  b  a.
    m = 624-625: a  a;
        632-633: a  a.

No. 97. (Ratio = .615). IV, 160-172. Major = two released movements; minor = homodynes followed by heterodynes.
    M = 165-172: a  a  a  b . a  a  a  b.
    m = 160-164: b  b  b  a  a.

APPENDIX F

No. 567. (Ratio = .611). VI, 295-312. Major = homodyned sequence and an expanded alternation; minor = heterodyned sequence.

M = 295-297:  b  b  b .

      305-312:  a  b  b  a  b  b  a  a .

m = 298-304:  a  a  a  a . a  b  a .

No. 44. (Ratio = .629). II, 402-436. Major = expanded alternations (on 402-415, see above); minor = simple alternations with a released movement.

M = 402-423:

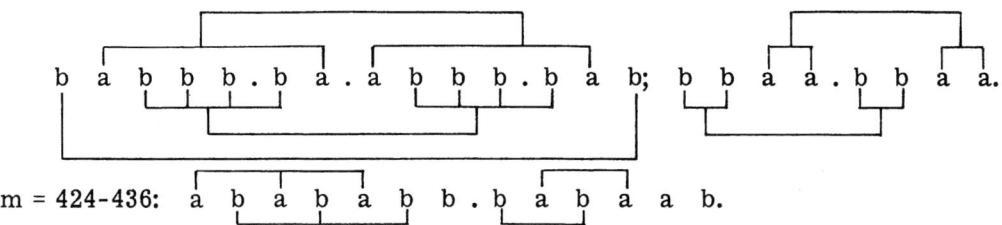

m = 424-436:  a  b  a  b  a  b  b . b  a  b  a  a  b .

The conclusion of the minor with a released movement (434-436) indicates that the punctuation with a full stop after 436, as in Mackail's edition, is correct; see below, Table XIII.

No. 710. (Ratio = .615). XII, 940-952. Major = alternation and a homodyned ending; minor = released movement.

M = 940-944:  b  a  b  a  b .

      950-952:  a  b  b .

m = 945-949:  a  a  a  a  b .

(See also No. 352:  M = 945-949, m = 950-952.)

No. 538. (Ratio = .615). IV, 693-705. Major = "framework" pattern and homodyned sequence; minor = released movement, and a heterodyne followed by a homodyne.

M = 696-703:  a  b  b  a . b  b  b  b .

m = 693-695:  a  a  b .

      704-705:  a  b.

One type of alternation occurs so frequently that it deserves a descriptive term of its own; above in IV, 696-699, we find  a  b  b  a ;  a  b  a  is likewise common. Vergil's tripartite structure in the books of the *Aeneid* (see Chapter 2) and his many framework patterns which produce Golden Mean ratios (see Table III) suggest that this type of metrical alternation should also be called a framework pattern. For example:

No. 404. (Ratio = .625). IV, 648-671. Major = a released movement followed by a lengthy alternation; m = two frameworks, with heterodynes enclosed by homodynes.

M = 648-650:  a  a  b:

      651-662:  b  b . a  b . a  b  a  b . a  b . a  a .

m = 663-671:

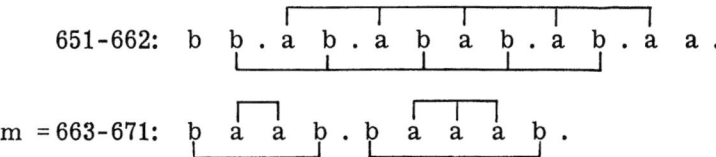

I now add a few more of the ratios in which majors and minors display a surprising variation in the metrical patterns. This correlation between the Golden Mean ratios and Knight's "accentual symmetry" is most significant and should be examined in much fuller detail.

No. 7. (Ratio = .611). I, 124-141. Major = framework and alternation; minor = heterodyned sequence.

M = 131-141:  a  a  a  b  a . a  b  a  b  a  b .

m = 124-130:  b  a  a  a . a  a  a .

No. 64. (Ratio = .60). III, 69-83. Major = two released movements; minor = alternation (see Knight, p. 72).

M = 69-77:  a  a  a . b . a  a  a  a  b .

m = 78-83:  a  b . a  b  a . a .

No. 103. (Ratio = .618). IV, 416-449. Major = two simple framework patterns enclosing a heterodyned sequence and followed by a released movement; minor = expanded frameworks.

M = 416-436:  b  a  b . a  a  a  a  a : a  b  b  a  a . a  a . a  a  a  a . a  b .

m = 437-449:  b  b  a  a . a  b  a  b  b  b  b  b  a .

No. 113. (Ratio = .618). IV, 672-705. Major = two released movements enclosed by alternations; minor described above, see No. 538.

M = 672-692:  b  a  b  a . a . a  a  b . a  a . a  a  b  a  b  b . b  a . b  b  b .

m = 693-705:  a  a  b . a  b  b  a . b  b  b  b . a  b .

No. 149. (Ratio = .636). V, 843-853. Major = alternation plus two homodynes; minor = released movement (see Knight, p. 71).

M = 847-853:  a  b  a  b  a . a  a .

m = 843-846:  a  a  a . b .

No. 162. (Ratio = .609). VI, 417-439. Major = released movement followed by an expanded alternation; minor = expanded alternation almost identical with that in 430-439.

M = 426-439:  a  a  a  b . b . a  a  b . a  a  b  a . a  b .

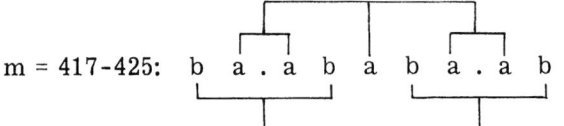

m = 417-425:  b  a . a  b  a  b  a . a  b .

APPENDIX F

No. 164. (Ratio = .630). VI, 450-476. Major = expanded alternations; minor = framework and simple alternation.

M = 450-466:  a  a  b  a  a  b  a  b . a  a . b  a  a  b . a  a .

m = 467-476:  b  a . a  a  b . b  a  b . a  b .

No. 193. (Ratio = .611). VII, 323-340. Major = released movement followed by an expanded alternation; minor = simple alternation.

M = 330-340:  a  a  a  a  b . b  a  a  b  a  a .

m = 323-329:  a  b  a  b . a  b  b .

Cf. Knight, p. 71: "The development of one or more released movements into one or more alternations, so that alternation gives a mitigated finality to the passage, is sufficiently frequent to be called characteristic of Vergil."

No. 351. (Ratio = .618). XII, 919-952. Major = elaborate alternating and interlocking pattern; minor described above, cf. No. 710. This passage as a whole provides a most effective ending to the *Aeneid*.

M = 919-939:  a  a  b  a  b  a  a  a  b . a  b . a  a . a  a  b  b  a  a  a .

m = 940-952:  b  a  b  a  b . a  a  a  a  b . a  b  b .

No. 356. (Ratio = .607). I, 297-324. Major = alternations; minor = heterodyned sequence ending in a framework pattern.

M = 297-304:  a  a  b  b  a . b  a  a .

305-313:  b  a  b  a  b . a  b  a  a .

m = 314-324:  a  a  a  a . a  a  a . a  b  b  a .

No. 364: (Ratio = .618). I, 723-756. Major = heterodyned sequence followed by a simple alternation (on 748-756, see above); minor = two framework patterns and two homodynes.

M = 736-747:  a  a  a  a  a  a . a  a  a  a  a  a .

748-756:  a  b  a  b  a . b  a  b  a .

m = 723-735:  a  a . b  a  a . a  a  b  b  a  a . a  a .

No. 381. (Ratio = .618). III, 1-68. Major = sequences of heterodynes and homodynes with occasional framework and alternating patterns; minor = a series of framework and alternating patterns.

M = 27-48:  a  b  a  a . a  a  a . a  a  a . b  b  a  b  a  a  b . a  b  a . a  b .

49-68:  a  a  b  b . b  b  b  b  a  a  b . b  b . b  a  a  a  b  a  b .

m = 1-26:

b  a  a  b  a  b  a  b  b  a  a  b . a  a  a  b  a  a . a  b . a  b . a  b  a .

No. 390. (Ratio = .618). III, 472-505. Major = heterodyned sequence and framework pattern followed by an alternation enclosing a framework; minor = released movement blending with an alternation.

M = 472-481:  a  a . a  a  a  a . b  a  a  b .

482-492:  b  a  a  a  b  a  b  b . a  a . a .

m = 493-505:  b  a . a  a  a  a  a . b  a  a  b  a  a .

No. 663. (Ratio = .611). X, 653-688. Major = alternation followed by sequences; minor = framework patterns closing with a released movement.

M = 653-665:  b  a  b . a  b  b . a  b . b  b  a  a . a .

680-688:  b  b  b  a  a . a  a . b  a .

m = 666-679:  b  a . a  b . b  a  b . b  a  b  a  a  a  b .

No. 769. (Ratio = .609). IV, 607-629. Major = alternation followed by sequences; minor = framework pattern plus an alternation. See Knight, p. 81.

M = 607-620:  a  b  a  b  a  a  a  a  a  b  b  b  b  b .

m = 621-629:  a . b  b  a . a  b  a  b  a .

No. 794. (Ratio = .612). VI, 384-416. Major = expanded framework and alternating patterns, followed by a heterodyned sequence; minor = framework and two released movements.

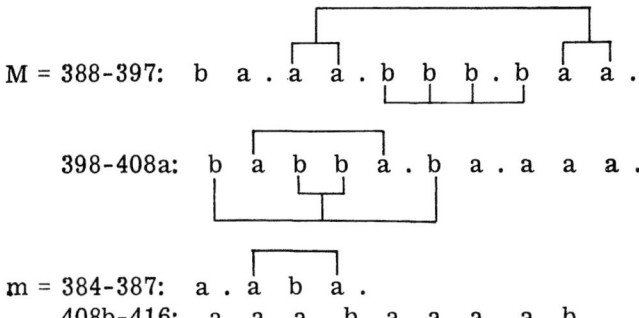

```
M = 388-397:    b  a . a  a . b  b  b . b  a  a .
    398-408a:   b  a  b  b  a . b  a . a  a  a .
```

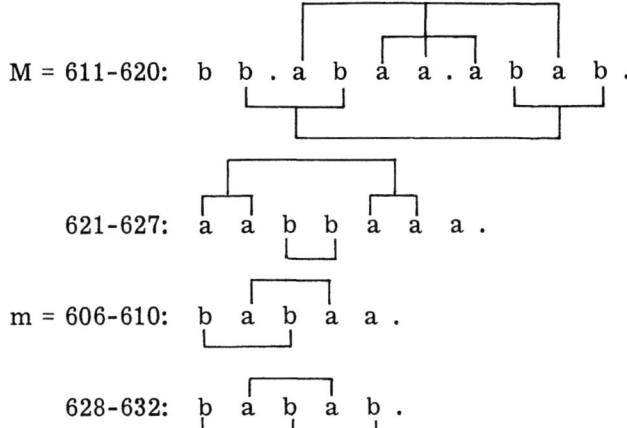

```
m = 384-387:    a . a  b  a .
    408b-416:   a  a  a . b  a  a  a . a  b .
```

No. 879. (Ratio = .630). X, 606-632. Major = expanded alternation and expanded framework; minor = simple alternations.

```
M = 611-620:    b  b . a  b  a  a . a  b  a  b .
    621-627:    a  a  b  b  a  a  a .
m = 606-610:    b  a  b  a  a .
    628-632:    b  a  b  a  b .
```

I have listed above proportions from Tables I-IV, with ratios ranging from .60 to .636, and I have included six perfect .618 ratios (Nos. 103, 113, 351, 364, 381, 390). It is interesting, but perhaps not surprising, to discover that the passages with the most exact proportions reveal elaborate metrical symmetries and perhaps the most striking variations between the metrical rhythms in the major and the minor parts. These are the passages where Vergil seems to have devoted the greatest attention to mathematical symmetry, metrical effects, structure, and content.

The important fact is that in so many instances Vergil's units of narrative, his rhythmical effects (or "accentual symmetry," to use Knight's phrase), and the mathematical ratios go hand in hand. The existence of this triple correlation implies that Vergil has aimed deliberately at the metrical patterns and in many cases has consciously combined them with the mathematical structure.

# APPENDIX G

## ADDENDUM ON *AEN.* II, 76

In Chapter 5, "Interpolations," I pointed out that the verses which are bracketed by modern editors because they are missing in the better MSS produce inferior ratios if retained in the text. We thus have additional evidence to confirm the soundness of the textual tradition.

One of the bracketed verses, II, 76, is unique in that, unlike the other interpolations, it is also involved in the matter of paragraphing. Most modern editors who bracket the line add it to the preceding paragraph, 57-75; Goelzer likewise brackets it, but indicates that it belongs with the speech beginning with verse 77. The ratios in Table XII (No. 33: .615 ⟶ .583, and No. 34: .606 ⟶ .593) show that II, 76, should be rejected *if* we consider it a part of the preceding paragraph. *But if* we join the verse to Sinon's speech in 77-104, the results are very different; cf. the following ratios:

    No. 34.    .606 ⟶ .614
         35.    .607 ⟶ .621
        368.    .613 ⟶ .617
        367.    .615 ⟶ .618

These surprisingly more accurate ratios suggest that in this instance the verse should be retained as the beginning of a new paragraph, to introduce the first of the three speeches of Sinon. The larger proportions containing II, 76, are not materially affected by the inclusion of the verse; cf. No. 729: .628 ⟶ .630; No. 483: .619 ⟶ .620; No. 945: .621 ⟶ .623.

    Fairclough, who brackets or omits from his text most of the interpolated lines, retains II, 76. For a defense of the verse, see Heyne-Wagner, Vol. II, *ad loc.* (4th ed.; Leipzig, 1832), and cf. Forbiger, *ad loc.* Conington-Nettleship, Vol. II, *ad loc.* (4th ed.; London, 1884), likewise bracket II, 76, but Conington's note implies that he is not quite convinced that the verse should be rejected.

# TABLES

## TABLE I. SHORT PASSAGES: BIPARTITE PATTERN

| No. | m/M: type | Book | Major | Total lines | Minor | Total lines | Ratio: M/(M+m) |
|---|---|---|---|---|---|---|---|
| 1. | a/b | I | 34-75 | 42 | 8-33 | 26 | .618 |
| 2. | b/a | I | 34-75 | 42 | 76-101 | 26 | .618 |
| 3. | b/a | I | 76-80 | 5 | 81-83 | 3 | .625 |
| 4. | a/b | I | 106-112 | 7 | 102-105 | 4 | .636 |
| 5. | a/b | I | 124-156 | 33 | 102-123 | 22 | .60 |
| 6. | b/a | I | 113-119 | 7 | 120-123 | 4 | .636 |
| 7. | a/b | I | 131-141 | 11 | 124-130 | 7 | .611 |
| 8. | a/b | I | 148-156 | 9 | 142-147 | 6 | .60 |
| 9. | a/b | I | 162-169 | 8 | 157-161 | 5 | .615 |
| 10. | a/b | I | 174-179 | 6 | 170-173 | 4 | .60 |
| 11. | a/b | I | 187-197 | 11 | 180-186 | 7 | .611 |
| 12. | b/a | I | 198-203 | 6 | 204-207 | 4 | .60 |
| 13. | a/b | I | 214-222 | 9 | 208-213 | 6 | .60 |
| 14. | a/b | I | 254-296 | 43 | 227-253 | 27 | .614 |
| 15. | b/a | I | 297-371 | 75 | 372-417 | 46 | .620 |
| 16. | a/b | I | 393-401 | 9 | 387-392 | 6 | .60 |
| 17-18. | | | | | | | |
| 17. | b/a | I | 402-406 | 5 | 407-409 | 3 | .625 |
| 18. | a/c | | 410-417 | 8 | 402-406 | 5 | .615 |
| 19. | b/a | I | 453-465 | 13 | 466-473 | 8 | .619 |
| 20. | b/a | I | 453-493 | 41 | 494-519 | 26 | .612 |
| 21. | b/a | I | 522-526 | 5 | 527-529 | 3 | .625 |
| 22. | a/b | I | 530-543 | 13.3 | 522-529 | 8 | .624 |
| 23. | b/a | I | 561-571 | 11 | 572-578 | 7 | .611 |
| 24. | a/b | I | 595b-610a | 15.2 | 586-595a | 9.4 | .618 |
| 25. | b/a | I | 613-630 | 18 | 631-642 | 11.6 | .608 |
| 26. | a/b | I | 673-688 | 16 | 663-672 | 10 | .615 |
| 27. | b/a | I | 723-730 | 8 | 731-735 | 5 | .615 |
| 28. | a/b | I | 740b-747 | 7.6 | 736-740a | 4.4 | .633 |

| No. | m/M: type | Book | Major | Total lines | Minor | Total lines | Ratio: M/(M+m) |
|---|---|---|---|---|---|---|---|
| 29. | b/a | II | 1-8a | 7.4 | 8b-13a | 4.8 | .607 |
| 30. | b/a | II | 1-13a | 12.2 | 13b-20 | 7.8 | .610 |
| 31. | b/a | II | 13b-17 | 4.8 | 18-20 | 3 | .615 |
| 32. | a/b | II | 199-227 | 29 | 40-56 | 17 | .630 |
| 33. | b/a | II | 57-68 | 11.2 | 69-75 | 7 | .615 |
| 34. | a/b | II | 77-104 | 28 | 57-75 | 18.2 | .606 |
| 35. | b/a | II | 77-93 | 17 | 94-104 | 11 | .607 |
| 36. | a/b | II | 122-144 | 23 | 108-121 | 14 | .622 |
| 37. | a/b | II | 268-297 | 30 | 250-267 | 18 | .625 |
| 38. | b/a | II | 250-369 | 119.2 | 370-441 | 72 | .623 |
| 39. | b/a | II | 268-297 | 30 | 318-335 | 18 | .625 |
| 40. | b/a | II | 298-310a | 12.2 | 310b-317 | 7.8 | .610 |
| 41. | a/b | II | 376-385 | 10 | 370-375 | 6 | .625 |
| 42. | a/b | II | 402-452 | 51 | 370-401 | 32 | .614 |
| 43. | b/a | II | 386-395 | 10 | 396-401 | 6 | .625 |
| 44. | b/a | II | 402-423 | 22 | 424-436 | 13 | .629 |
| 45-46. | | | | | | | |
| 45. | b/a | II | 437-441 | 5 | 442-444 | 3 | .625 |
| 46. | a/c | | 445-452 | 8 | 437-441 | 5 | .615 |
| 47. | b/a | II | 506-517 | 12 | 518-525 | 8 | .60 |
| 48. | a/b | II | 526-558 | 33 | 506-525 | 20 | .623 |
| 49. | b/a | II | 559-563 | 5 | 564-566 | 3 | .625 |
| 50. | a/b | II | 588-633 | 44.8 | 559-587 | 29 | .607 |
| 51. | a/b | II | 575-587 | 13 | 567-574 | 8 | .619 |
| 52. | a/b | II | 588-623 | 34.8 | 567-587 | 21 | .624 |
| 53. | a/b | II | 604-620 | 16.4 | 594-603 | 10 | .621 |
| 54. | a/b | II | 634-649 | 15.4 | 624-633 | 10 | .606 |
| 55. | a/b | II | 679-691 | 13 | 671-678 | 8 | .619 |
| 56. | b/a | II | 679-686 | 8 | 687-691 | 5 | .615 |
| 57. | a/b | II | 707-720 | 13.2 | 699-706 | 8 | .623 |
| 58. | a/b | II | 712-720 | 8.2 | 707-711 | 5 | .621 |
| 59. | a/b | II | 745-770 | 25.2 | 730-744 | 15 | .627 |
| 60. | a/b | II | 796-804 | 9 | 790-795 | 6 | .60 |

TABLE I. SHORT PASSAGES: BIPARTITE PATTERN

| No. | m/M: type | Book | Major | Total lines | Minor | Total lines | Ratio: M/(M+m) |
|---|---|---|---|---|---|---|---|
| 61. | b/a | III | 1-8a | 7.4 | 8b-12 | 4.6 | .617 |
| 62. | a/b | III | 22-26 | 5 | 19-21 | 3 | .625 |
| 63. | b/a | III | 27-40 | 14 | 41-48 | 8 | .636 |
| 64. | b/a | III | 69-77 | 9 | 78-83 | 6 | .60 |
| 65. | b/a | III | 69-89 | 21 | 90-101 | 12 | .636 |
| 66. | a/b | III | 102-120 | 19 | 90-101 | 12 | .613 |
| 67. | b/a | III | 94-98 | 5 | 99-101 | 3 | .625 |
| 68. | b/a | III | 102-113 | 12 | 114-120 | 7 | .632 |
| 69. | b/a | III | 147-191 | 45 | 192-218 | 26.2 | .632 |
| 70. | a/b | III | 161-171 | 11 | 154-160 | 7 | .611 |
| 71. | b/a | III | 192-199 | 8 | 200-204 | 5 | .615 |
| 72. | a/b | III | 210b-218 | 8 | 205-210a | 5.2 | .606 |
| 73. | a/b | III | 229-244 | 15 | 219-228 | 10 | .60 |
| 74. | b/a | III | 245-252 | 8 | 253-257 | 5 | .615 |
| 75. | b/a | III | 258-267 | 10 | 268-273 | 6 | .625 |
| 76. | b/a | III | 344-348 | 5 | 349-351 | 3 | .625 |
| 77. | a/b | III | 356-373 | 18 | 344-355 | 12 | .60 |
| 78. | b/a | III | 374-395 | 22 | 396-409 | 14 | .611 |
| 79. | a/b | III | 410-432 | 23 | 396-409 | 14 | .622 |
| 80. | b/a | III | 410-423 | 14 | 424-432 | 9 | .609 |
| 81. | b/a | III | 410-440 | 31 | 441-460 | 20 | .608 |
| 82. | b/a | III | 420-432 | 13 | 433-440 | 8 | .619 |
| 83. | b/a | III | 441-452 | 12 | 453-460 | 8 | .60 |
| 84. | a/b | III | 532-547 | 16 | 521-531 | 10.4 | .606 |
| 85. | b/a | III | 570-577 | 8 | 578-582 | 5 | .615 |
| 86. | b/a | III | 570-595 | 26 | 596-611 | 16 | .619 |
| 87. | a/b | III | 588-595 | 8 | 583-587 | 5 | .615 |
| 88. | a/b | III | 613-654 | 41.2 | 588-612 | 25 | .622 |
| 89. | b/a | III | 613-638 | 26 | 639-654 | 15.2 | .631 |
| 90. | a/b | III | 669-691 | 23 | 655-668 | 13.4 | .632 |
| 91. | b/a | III | 692-706 | 15 | 707-715 | 9 | .625 |
| 92. | b/a | IV | 30-44 | 14.4 | 45-53 | 9 | .615 |
| 93. | a/b | IV | 68-89 | 22 | 54-67 | 14 | .611 |
| 94. | b/a | IV | 90-104 | 15 | 105-114a | 9.2 | .620 |

| No. | m/M: type | Book | Major | Total lines | Minor | Total lines | Ratio: M/(M+m) |
|---|---|---|---|---|---|---|---|
| 95. | a/b | IV | 114b-128 | 14.8 | 105-114a | 9.2 | .617 |
| 96. | b/a | IV | 136-150 | 15 | 151-159 | 9 | .625 |
| 97. | a/b | IV | 165-172 | 8 | 160-164 | 5 | .615 |
| 98. | a/b | IV | 198-237 | 40 | 173-197 | 25 | .615 |
| 99. | a/b | IV | 206-218 | 13 | 198-205 | 8 | .619 |
| 100. | a/b | IV | 211-218 | 8 | 206-210 | 5 | .615 |
| 101. | b/a | IV | 223-231 | 9 | 232-237 | 6 | .60 |
| 102. | a/b | IV | 246b-258 | 12.8 | 238-246a | 8.2 | .610 |
| 103. | b/a | IV | 416-436 | 21 | 437-449 | 13 | .618 |
| 104. | a/b | IV | 450-503 | 53.4 | 416-449 | 34 | .611 |
| 105. | b/a | IV | 474-503 | 29.4 | 504-521 | 17.6 | .626 |
| 106. | b/a | IV | 504-514 | 11 | 515-521 | 6.6 | .625 |
| 107. | a/b | IV | 522-552 | 30 | 504-521 | 17.6 | .630 |
| 108. | a/b | IV | 534-552 | 19 | 522-533 | 11 | .633 |
| 109. | b/a | IV | 522-552 | 30 | 553-570 | 18 | .625 |
| 110. | a/b | IV | 560-570 | 11 | 553-559 | 7 | .611 |
| 111. | b/a | IV | 571-579a | 8.2 | 579b-583 | 4.8 | .631 |
| 112. | b/a | IV | 672-685a | 13.2 | 685b-692 | 7.8 | .629 |
| 113. | b/a | IV | 672-692 | 21 | 693-705 | 13 | .618 |
| 114. | b/a | V | 1-7 | 7 | 8-11 | 4 | .636 |
| 115. | b/a | V | 1-113 | 113 | 114-182 | 69 | .621 |
| 116. | a/b | V | 114-285 | 172 | 1-113 | 113 | .604 |
| 117. | b/a | V | 35-39a | 4.2 | 39b-41 | 2.8 | .60 |
| 118. | a/b | V | 84-103 | 20 | 72-83 | 12 | .625 |
| 119. | b/a | V | 94-99 | 6 | 100-103 | 4 | .60 |
| 120. | a/b | V | 159-171 | 13 | 151-158 | 8 | .619 |
| 121. | b/a | V | 286-303 | 17.6 | 304-314 | 11 | .615 |
| 122. | a/b | V | 315-361 | 46.4 | 286-314 | 28.6 | .619 |
| 123. | a/b | V | 362-484 | 123 | 286-361 | 75 | .621 |
| 124. | a/b | V | 327-339 | 13 | 318-326 | 8.4 | .607 |
| 125. | b/a | V | 340-347 | 8 | 348-352 | 5 | .615 |
| 126. | b/a | V | 353-358a | 5.4 | 358b-361 | 3.6 | .60 |
| 127. | b/a | V | 362-436 | 75 | 437-484 | 48 | .610 |
| 128. | b/a | V | 387-400a | 13.6 | 400b-408 | 8.4 | .618 |

TABLE I. SHORT PASSAGES: BIPARTITE PATTERN

| No. | m/M: type | Book | Major | Total lines | Minor | Total lines | Ratio: M/(M+m) |
|---|---|---|---|---|---|---|---|
| 129. | b/a | V | 410-416 | 7 | 417-420 | 4 | .636 |
| 130. | a/b | V | 437-460 | 24 | 421-436 | 16 | .60 |
| 131. | b/a | V | 485-521 | 37 | 522-544 | 23 | .617 |
| 132. | a/b | V | 529-544 | 16 | 519-528 | 10 | .615 |
| 133. | b/a | V | 545-576 | 31.2 | 577-595 | 18.6 | .627 |
| 134. | b/a | V | 563-587 | 24.2 | 588-603 | 15.6 | .608 |
| 135. | a/b | V | 618-640 | 23 | 604-617 | 14 | .622 |
| 136. | b/a | V | 604-640 | 37 | 641-663 | 22.2 | .625 |
| 137. | b/a | V | 604-663 | 59.2 | 664-699 | 36 | .622 |
| 138. | a/b | V | 670-679 | 10 | 664-669 | 6 | .625 |
| 139. | a/b | V | 670-674 | 5 | 667-669 | 3 | .625 |
| 140. | a/b | V | 685-692 | 8 | 680-684 | 5 | .615 |
| 141. | b/a | V | 700-778 | 79 | 779-826 | 47.3 | .625 |
| 142. | b/a | V | 719-745 | 27 | 746-761 | 16 | .628 |
| 143. | b/a | V | 746-758 | 13 | 759-766 | 8 | .619 |
| 144. | b/a | V | 746-766 | 21 | 767-778 | 12 | .636 |
| 145. | b/a | V | 762-766 | 5 | 767-769 | 3 | .625 |
| 146. | b/a | V | 799-815 | 16.7 | 816-826 | 11 | .603 |
| 147. | a/b | V | 833-842 | 10 | 827-832 | 6 | .625 |
| 148. | b/a | V | 827-853 | 27 | 854-871 | 18 | .60 |
| 149. | a/b | V | 847-853 | 7 | 843-846 | 4 | .636 |
| 150. | b/a | VI | 1-8 | 8 | 9-13 | 5 | .615 |
| 151. | a/b | VI | 14-33a | 19.6 | 1-13 | 13 | .601 |
| 152. | b/a | VI | 1-76 | 76 | 77-123 | 46.6 | .620 |
| 153. | b/a | VI | 1-55 | 55 | 124-155 | 32 | .632 |
| 154. | a/b | VI | 42-55 | 14 | 33b-41 | 8.4 | .625 |
| 155. | b/a | VI | 56-68 | 13 | 69-76 | 8 | .619 |
| 156. | a/b | VI | 156-211 | 56 | 124-155 | 32 | .636 |
| 157. | a/b | VI | 194-211 | 18 | 183-193 | 11 | .621 |
| 158. | b/a | VI | 282-289 | 8 | 290-294 | 5 | .615 |
| 159. | b/a | VI | 295-316 | 22 | 317-330 | 14 | .611 |
| 160. | b/a | VI | 347-361 | 15 | 362-371 | 10 | .60 |
| 161. | a/b | VI | 376-381 | 6 | 372-375 | 4 | .60 |
| 162. | a/b | VI | 426-439 | 14 | 417-425 | 9 | .609 |

| No. | m/M: type | Book | Major | Total lines | Minor | Total lines | Ratio: M/(M+m) |
|---|---|---|---|---|---|---|---|
| 163. | a/b | VI | 445-449 | 5 | 442-444 | 3 | .625 |
| 164. | b/a | VI | 450-466 | 17 | 467-476 | 10 | .630 |
| 165. | b/a | VI | 450-476 | 27 | 477-493 | 17 | .614 |
| 166. | b/a | VI | 456-468 | 13 | 469-476 | 8 | .619 |
| 167. | a/b | VI | 500-534 | 35 | 477-499 | 23 | .603 |
| 168. | a/b | VI | 500-508 | 9 | 494-499 | 6 | .60 |
| 169. | a/b | VI | 509-534 | 26 | 494-508 | 15 | .634 |
| 170. | a/b | VI | 548-636 | 89 | 494-547 | 54 | .622 |
| 171. | b/a | VI | 562-572 | 11 | 573-579 | 7 | .611 |
| 172. | a/b | VI | 580-607 | 28 | 562-579 | 18 | .609 |
| 173. | b/a | VI | 608-615 | 8 | 616-620 | 5 | .615 |
| 174. | a/b | VI | 616-627 | 12 | 608-615 | 8 | .60 |
| 175. | b/a | VI | 703-712 | 10 | 713-718 | 6 | .625 |
| 176. | a/b | VI | 735-751 | 17 | 724-734 | 11 | .607 |
| 177. | a/b | VI | 756-807 | 52 | 724-755 | 32 | .619 |
| 178. | a/b | VI | 760-766 | 7 | 756-759 | 4 | .636 |
| 179. | a/b | VI | 771-776 | 6 | 767-770 | 4 | .60 |
| 180. | a/b | VI | 781-787 | 7 | 777-780 | 4 | .636 |
| 181. | b/a | VI | 789b-800 | 11.6 | 801-807 | 7 | .624 |
| 182. | b/a | VII | 1-106 | 106 | 107-169 | 62.6 | .629 |
| 183. | b/a | VII | 5-24 | 20 | 25-36 | 12 | .625 |
| 184. | a/b | VII | 45b-58 | 13.6 | 37-45a | 8.4 | .618 |
| 185. | b/a | VII | 37-80 | 44 | 81-106 | 26 | .629 |
| 186. | b/a | VII | 45b-58 | 13.6 | 59-67 | 9 | .602 |
| 187. | a/b | VII | 59-80 | 22 | 45b-58 | 13.6 | .618 |
| 188. | a/b | VII | 117b-134 | 17.2 | 107-117a | 10.4 | .623 |
| 189. | b/a | VII | 135-147 | 13 | 148-155 | 8 | .619 |
| 190. | a/b | VII | 148-169 | 22 | 135-147 | 13 | .629 |
| 191. | a/b | VII | 192-285 | 93.6 | 135-191 | 57 | .622 |
| 192. | a/b | VII | 259-273 | 15 | 249-258 | 10 | .60 |
| 193. | a/b | VII | 330-340 | 11 | 323-329 | 7 | .611 |
| 194. | a/b | VII | 406-539 | 133 | 323-405 | 83 | .616 |
| 195. | a/b | VII | 346-353 | 8 | 341-345 | 5 | .615 |
| 196. | b/a | VII | 354-372 | 19 | 373-384 | 12 | .613 |

TABLE I. SHORT PASSAGES: BIPARTITE PATTERN

| No. | m/M: type | Book | Major | Total lines | Minor | Total lines | Ratio: M/(M+m) |
|---|---|---|---|---|---|---|---|
| 197. | a/b | VII | 552-571 | 20 | 540-551 | 12 | .625 |
| 198. | b/a | VII | 641-654 | 14 | 655-663 | 9 | .609 |
| 199. | a/b | VII | 678-690 | 13 | 670-677 | 8 | .619 |
| 200. | b/a | VII | 706-722 | 17 | 723-732 | 10 | .630 |
| 201. | b/a | VII | 723-782 | 59.6 | 783-817 | 35 | .630 |
| 202. | b/a | VIII | 1-101 | 99.4 | 306-369 | 64 | .608 |
| 203. | b/a | VIII | 18-25 | 8 | 26-30 | 5 | .615 |
| 204. | a/b | VIII | 36-65 | 28.4 | 18-35 | 18 | .612 |
| 205. | b/a | VIII | 36-54 | 17.4 | 55-65 | 11 | .613 |
| 206. | a/b | VIII | 71-78 | 8 | 66-70 | 5 | .615 |
| 207. | a/b | VIII | 86-93 | 8 | 81-85 | 5 | .615 |
| 208. | a/b | VIII | 97-101 | 5 | 94-96 | 3 | .625 |
| 209. | a/b | VIII | 107-114 | 8 | 102-106 | 5 | .615 |
| 210. | b/a | VIII | 126-141 | 16 | 142-151 | 10 | .615 |
| 211. | a/b | VIII | 190-199 | 10 | 184-189 | 6 | .625 |
| 212. | a/b | VIII | 219-267 | 49 | 190-218 | 29 | .628 |
| 213. | a/b | VIII | 205-212 | 8 | 200-204 | 5 | .615 |
| 214. | b/a | VIII | 200-240 | 41 | 241-267 | 27 | .603 |
| 215. | a/b | VIII | 247-255 | 9 | 241-246 | 6 | .60 |
| 216. | a/b | VIII | 306-369 | 64 | 268-305 | 38 | .627 |
| 217. | b/a | VIII | 306-369 | 64 | 370-406 | 37 | .634 |
| 218. | b/a | VIII | 313-327 | 15 | 328-336 | 9 | .625 |
| 219. | b/a | VIII | 337-350 | 14 | 351-358 | 8 | .636 |
| 220. | b/a | VIII | 374-386 | 13 | 387-394 | 8 | .619 |
| 221. | b/a | VIII | 454-540 | 85.6 | 541-596 | 56 | .605 |
| 222. | b/a | VIII | 481-488 | 8 | 489-493 | 5 | .615 |
| 223. | a/b | VIII | 503b-519 | 16.4 | 494-503a | 9.6 | .631 |
| 224. | a/b | VIII | 528-540 | 12.4 | 520-527 | 8 | .608 |
| 225. | a/b | VIII | 546-553 | 8 | 541-545 | 5 | .615 |
| 226. | a/b | VIII | 558-584 | 27 | 541-557 | 17 | .614 |
| 227. | b/a | VIII | 597-607 | 11 | 608-614 | 7 | .611 |
| 228. | a/b | VIII | 608-625 | 18 | 597-607 | 11 | .621 |
| 229. | a/b | VIII | 615-625 | 11 | 608-614 | 7 | .611 |
| 230. | b/a | VIII | 635-641 | 7 | 642-645 | 4 | .636 |

| No. | m/M: type | Book | Major | Total lines | Minor | Total lines | Ratio: M/(M+m) |
|---|---|---|---|---|---|---|---|
| 231. | b/a | VIII | 689-703 | 15 | 704-713 | 10 | .60 |
| 232. | b/a | VIII | 689-713 | 25 | 714-728 | 15 | .625 |
| 233. | a/b | VIII | 720-728 | 9 | 714-719 | 6 | .60 |
| 234. | b/a | VIII | 724-728 | 5 | 729-731 | 3 | .625 |
| 235. | b/a | IX | 1-15 | 15 | 16-24 | 9 | .625 |
| 236. | b/a | IX | 1-46 | 45 | 47-76 | 30 | .60 |
| 237. | b/a | IX | 25-76 | 51 | 77-106 | 30 | .630 |
| 238. | a/b | IX | 77-122 | 45 | 47-76 | 30 | .60 |
| 239. | b/a | IX | 126-145 | 20 | 146-158 | 12 | .625 |
| 240. | a/b | IX | 184-196 | 13 | 176-183 | 8 | .619 |
| 241. | b/a | IX | 184-191 | 8 | 192-196 | 5 | .615 |
| 242. | b/a | IX | 342-356 | 15 | 357-366 | 10 | .60 |
| 243. | a/b | IX | 379-398 | 20 | 367-378 | 12 | .625 |
| 244. | a/b | IX | 399-449 | 51 | 367-398 | 32 | .614 |
| 245. | b/a | IX | 399-430 | 32 | 431-449 | 19 | .627 |
| 246. | a/b | IX | 503-589 | 85.4 | 450-502 | 52.4 | .620 |
| 247. | b/a | IX | 450-589 | 137.8 | 590-671 | 82 | .627 |
| 248. | b/a | IX | 459-467 | 8.4 | 468-472 | 5 | .627 |
| 249. | b/a | IX | 477-489 | 13 | 490-497 | 8 | .619 |
| 250. | b/a | IX | 490-497 | 8 | 498-502 | 5 | .615 |
| 251. | b/a | IX | 569-576a | 7.4 | 576b-580 | 4.6 | .617 |
| 252. | b/a | IX | 590-594 | 5 | 595-597 | 3 | .625 |
| 253. | a/b | IX | 621-671 | 51 | 590-620 | 31 | .622 |
| 254. | a/b | IX | 646b-658 | 12.6 | 638-646a | 8.4 | .60 |
| 255. | b/a | IX | 638-658 | 21 | 659-671 | 13 | .618 |
| 256. | a/b | IX | 664-671 | 8 | 659-663 | 5 | .615 |
| 257. | b/a | IX | 683-687 | 5 | 688-690 | 3 | .625 |
| 258. | a/b | IX | 756-818 | 62.4 | 717-755 | 38.7 | .617 |
| 259. | b/a | IX | 756-777 | 21.4 | 778-790 | 13 | .622 |
| 260. | a/b | IX | 768-777 | 10 | 762-767 | 6 | .625 |
| 261. | b/a | IX | 762-777 | 16 | 778-787 | 10 | .615 |

[cf. No. 642 in Table III]

TABLE I. SHORT PASSAGES: BIPARTITE PATTERN

| No. | m/M: type | Book | Major | Total lines | Minor | Total lines | Ratio: M/(M+m) |
|---|---|---|---|---|---|---|---|
| 262. | b/a | X | 118-145 | 28 | 146-162 | 17 | .622 |
| 263. | b/a | X | 146-156a | 10.6 | 156b-162 | 6.4 | .624 |
| 264. | a/b | X | 204-212 | 9 | 198-203 | 6 | .60 |
| 265. | b/a | X | 215-307 | 91.6 | 308-361 | 54 | .629 |
| 266. | b/a | X | 260-275 | 16 | 276-286 | 9.6 | .625 |
| 267. | b/a | X | 287-293 | 7 | 294-298a | 4.6 | .603 |
| 268. | a/b | X | 332b-344 | 12.8 | 324-332a | 8.2 | .610 |
| 269. | a/b | X | 369-379 | 11 | 362-368 | 7 | .611 |
| 270. | b/a | X | 399-438 | 40 | 439-463 | 25 | .615 |
| 271. | a/b | X | 431-438 | 8 | 426-430 | 5 | .615 |
| 272. | b/a | X | 510-530a | 20.2 | 530b-542 | 12.8 | .612 |
| 273. | b/a | X | 575-580 | 5.2 | 581-584a | 3.4 | .605 |
| 274. | b/a | X | 633-644 | 12 | 645-652 | 8 | .60 |
| 275. | b/a | X | 633-652 | 20 | 653-665 | 13 | .606 |
| 276. | b/a | X | 653-660 | 8 | 661-665 | 5 | .615 |
| 277. | a/b | X | 689-746 | 57.2 | 653-688 | 36 | .614 |
| 278. | b/a | X | 666-679 | 14 | 680-688 | 9 | .609 |
| 279. | b/a | X | 689-706 | 18 | 707-718 | 12 | .60 |
| 280. | a/b | X | 789-832 | 44 | 762-788 | 27 | .620 |
| 281. | a/b | X | 833-908 | 74.6 | 789-832 | 44 | .629 |
| 282. | b/a | X | 791-816 | 26 | 817-832 | 16 | .619 |
| 283. | b/a | X | 833-840 | 8 | 841-845 | 5 | .615 |
| 284. | b/a | X | 873-894 | 21.6 | 895-908 | 14 | .607 |
| 285. | a/b | X | 896-908 | 13 | 888-895 | 8 | .619 |
| 286. | a/b | XI | 12-28 | 17 | 1-11 | 11 | .607 |
| 287. | a/b | XI | 39-99 | 61 | 1-38 | 38 | .616 |
| 288. | b/a | XI | 14-28 | 15 | 29-38 | 10 | .60 |
| 289. | a/b | XI | 148-181 | 34 | 39-58 | 20 | .630 |
| 290. | a/b | XI | 64-71 | 8 | 59-63 | 5 | .615 |
| 291. | a/b | XI | 108-121 | 14 | 100-107 | 8 | .636 |
| 292. | b/a | XI | 106-121 | 16 | 122-131 | 10 | .615 |
| 293. | b/a | XI | 148-181 | 34 | 182-202 | 21 | .618 |
| 294. | a/b | XI | 231-241a | 10.4 | 225-230 | 6 | .634 |
| 295. | a/b | XI | 252-295 | 44 | 225-251 | 27 | .620 |

| No. | m/M: type | Book | Major | Total lines | Minor | Total lines | Ratio: M/(M+m) |
|---|---|---|---|---|---|---|---|
| 296. | b/a | XI | 336-359 | 24 | 360-375 | 15.2 | .612 |
| 297. | a/b | XI | 376-444 | 68.2 | 336-375 | 39.2 | .635 |
| 298. | b/a | XI | 445-458 | 14 | 459-467 | 9 | .609 |
| 299. | a/b | XI | 473-480 | 8 | 468-472 | 5 | .615 |
| 300. | b/a | XI | 468-485 | 18 | 486-497 | 12 | .60 |
| 301. | b/a | XI | 473-480 | 8 | 481-485 | 5 | .615 |
| 302. | a/b | XI | 507-521 | 15 | 498-506 | 9 | .625 |
| 303. | a/b | XI | 535b-594 | 59.8 | 498-535a | 37.2 | .616 |
| 304. | a/b | XI | 526-531 | 6 | 522-525 | 4 | .60 |
| 305. | a/b | XI | 629-647 | 19 | 618-628 | 11 | .633 |
| 306. | a/b | XI | 664-689 | 26 | 648-663 | 16 | .619 |
| 307. | a/b | XI | 741-767 | 27 | 725-740 | 16 | .628 |
| 308. | b/a | XI | 725-767 | 43 | 768-793 | 26 | .623 |
| 309. | a/b | XI | 778-793 | 16 | 768-777 | 10 | .615 |
| 310. | a/b | XI | 794-835 | 42 | 768-793 | 26 | .618 |
| 311. | b/a | XI | 816-828a | 12.6 | 828b-835 | 7.4 | .630 |
| 312. | a/b | XI | 836-867 | 32 | 816-835 | 20 | .615 |
| 313. | a/b | XI | 841-849a | 8.4 | 836-840 | 5 | .627 |
| 314. | b/a | XI | 836-849a | 13.4 | 849b-857 | 8.6 | .609 |
| 315. | a/b | XI | 868-915 | 48 | 836-867 | 32 | .60 |
| 316. | b/a | XI | 849b-857 | 8.6 | 858-862 | 5 | .632 |
| 317. | a/b | XII | 29-45a | 16.2 | 19-28 | 10 | .618 |
| 318. | b/a | XII | 81-100 | 20 | 101-112 | 12 | .625 |
| 319. | a/b | XII | 116-120 | 5 | 113-115 | 3 | .625 |
| 320. | a/b | XII | 121-133 | 13 | 113-120 | 8 | .619 |
| [cf. No. 691 in Table III] | | | | | | | |
| 321. | a/b | XII | 161-237 | 76.6 | 113-160 | 48 | .615 |
| 322. | b/a | XII | 121-128 | 8 | 129-133 | 5 | .615 |
| 323. | a/b | XII | 142-153 | 12 | 134-141 | 8 | .60 |
| 324. | b/a | XII | 142-153 | 12 | 154-160 | 7 | .632 |
| 325. | b/a | XII | 161-237 | 76.6 | 238-288 | 51 | .60 |
| 326. | a/b | XII | 183-194 | 12 | 175-182 | 8 | .60 |
| 327. | a/b | XII | 257-288 | 32 | 238-256 | 19 | .627 |
| 328. | b/a | XII | 257-276 | 20 | 277-288 | 12 | .625 |

TABLE I. SHORT PASSAGES: BIPARTITE PATTERN

| No. | m/M: type | Book | Major | Total lines | Minor | Total lines | Ratio: M/(M+m) |
|---|---|---|---|---|---|---|---|
| 329. | a/b | XII | 297-310 | 14 | 289-296 | 8 | .636 |
| 330. | b/a | XII | 289-323 | 35 | 324-345 | 22 | .614 |
| 331. | a/b | XII | 383-440 | 58 | 346-382 | 37 | .611 |
| 332. | a/b | XII | 473-480 | 8 | 468-472 | 5 | .615 |
| 333. | a/b | XII | 500-553 | 54 | 468-499 | 32 | .628 |
| 334. | a/b | XII | 488-499 | 12 | 481-487 | 7 | .632 |
| 335. | a/b | XII | 521-528 | 8 | 500-504 | 5 | .615 |
| 336. | a/b | XII | 554-592 | 39 | 529-553 | 25 | .609 |
| 337. | b/a | XII | 631-642 | 11.2 | 643-649 | 7 | .615 |
| 338. | b/a | XII | 631-671 | 40.2 | 672-696 | 25 | .617 |
| 339. | b/a | XII | 650-696 | 47 | 697-724 | 28 | .627 |
| 340. | a/b | XII | 707b-724 | 17.4 | 697-707a | 10.6 | .621 |
| 341. | b/a | XII | 735-741 | 7 | 742-745 | 4 | .636 |
| 342 | b/a | XII | 746-757 | 12 | 758-765 | 8 | .60 |
| 343. | b/a | XII | 766-780 | 15 | 781-790 | 10 | .60 |
| 344. | a/b | XII | 808-828 | 21 | 793-806a | 13.5 | .609 |
| 345. | b/a | XII | 808-828 | 21 | 830-842 | 13 | .618 |
| 346. | b/a | XII | 830-837 | 8 | 838-842 | 5 | .615 |
| 347. | b/a | XII | 843-848 | 6 | 849-852 | 4 | .60 |
| 348. | a/b | XII | 853-868 | 16 | 843-852 | 10 | .615 |
| 349. | b/a | XII | 869-880a | 11.2 | 880b-886 | 6.8 | .622 |
| 350. | a/b | XII | 926b-939 | 13.3 | 919-926a | 7.7 | .633 |
| 351. | b/a | XII | 919-939 | 21 | 940-952 | 13 | .618 |
| 352. | b/a | XII | 945-949 | 5 | 950-952 | 3 | .625 |

## TABLE II. SHORT PASSAGES: TRIPARTITE (NONFRAMEWORK) PATTERN

| No. | m/M: type | Book | Major | Total lines | Minor | Total lines | Ratio: M/(M+m) |
|---|---|---|---|---|---|---|---|
| 353. | a/(b+c) | I | 50-64<br>65-75 | 26 | 34-49 | 16 | .619 |
| 354. | c/(a+b) | I | 76-80<br>81-91 | 16 | 92-101 | 10 | .615 |
| 355. | (b+c)/a | I | 157-197 | 41 | 198-207<br>208-222 | 25 | .621 |
| 356. | c/(a+b) | I | 297-304<br>305-313 | 17 | 314-324 | 11 | .607 |
| 357. | c/(a+b) | I | 402-417<br>418-440 | 39 | 441-465 | 25 | .609 |
| 358. | a/(b+c) | I | 430-440<br>441-449 | 20 | 418-429 | 12 | .625 |
| 359. | (b+c)/a | I | 520-560 | 39.5 | 561-578<br>579-585 | 25 | .612 |
| 360. | c/(a+b) | I | 520-578<br>579-642 | 121.1 | 643-722 | 80 | .602 |
| 361. | a/(b+c) | I | 613-630<br>631-656 | 43.6 | 586-612 | 27 | .618 |
| 362. | a/(b+c) | I | 695-722<br>723-756 | 62 | 657-694 | 38 | .620 |
| 363. | c/(a+b) | I | 695-706<br>707-711 | 17 | 712-722 | 11 | .607 |
| 364. | a/(b+c) | I | 736-747<br>748-756 | 21 | 723-735 | 13 | .618 |
| 365. | c/(a+b) | II | 1-20<br>21-24 | 24 | 25-39 | 15 | .615 |
| 366. | c/(a+b) | II | 13b-20<br>21-39 | 26.8 | 40-56 | 17 | .612 |
| 367. | c/(a+b) | II | 57-104<br>105-144 | 86.2 | 145-198 | 54 | .615 |
| 368. | c/(a+b) | II | 77-104<br>108-144 | 65 | 154-194 | 41 | .613 |
| 369. | c/(a+b) | II | 154-161<br>162-179 | 26 | 180-194 | 15 | .634 |
| 370. | c/(a+b) | II | 195-198<br>199-227 | 33 | 228-249 | 21.4 | .607 |

TABLE II. SHORT PASSAGES: TRIPARTITE (NONFRAMEWORK) PATTERN

| No. | m/M: type | Book | Major | Total lines | Minor | Total lines | Ratio: M/(M+m) |
|---|---|---|---|---|---|---|---|
| 371. | a/(b+c) | II | 279b-286<br>287-297 | 18.6 | 268-279a | 11.4 | .620 |
| 372. | c/(a+b) | II | 347-354<br>355-360 | 14 | 361-369 | 9 | .609 |
| 373. | c/(a+b) | II | 453-468<br>469-485 | 32.7 | 486-505 | 20 | .620 |
| 374. | a/(b+c) | II | 650-670<br>671-691 | 42 | 624-649 | 25.4 | .623 |
| 375. | a/(b+c) | II | 671-698<br>699-729 | 58.2 | 634-670 | 36.4 | .615 |
| 376. | a/(b+c) | II | 699-729<br>730-804 | 104.1 | 634-698 | 64.4 | .618 |
| 377. | (b+c)/a | II | 657-670 | 14 | 671-674<br>675-678 | 8 | .636 |
| 378. | a/(b+c) | II | 687-691<br>692-698 | 12 | 679-686 | 8 | .60 |
| 379. | c/(a+b) | II | 730-744<br>745-770 | 40.2 | 771-795 | 24.7 | .619 |
| 380. | a/(b+c) | III | 19-26<br>27-48 | 30 | 1-18 | 18 | .625 |
| 381. | a/(b+c) | III | 27-48<br>49-68 | 42 | 1-26 | 26 | .618 |
| 382. | a/(b+c) | III | 37-48<br>49-52 | 16 | 27-36 | 10 | .615 |
| 383. | c/(a+b) | III | 69-89<br>90-101 | 33 | 102-120 | 19 | .635 |
| 384. | (a+b)/c | III | 147-191 | 45 | 121-134<br>135-146 | 26 | .634 |
| 385. | c/(a+b) | III | 192-273<br>274-293 | 100.2 | 294-355 | 61 | .622 |
| 386. | c/(a+b) | III | 274-278<br>279-293 | 20 | 294-305 | 12 | .625 |
| 387. | a/(b+c) | III | 374-462<br>463-505 | 131.6 | 294-373 | 79 | .625 |
| 388. | (a+b)/c | III | 320-343 | 23.4 | 306-313a<br>313b-319 | 13.6 | .632 |
| 389. | c/(a+b) | III | 374-380<br>381-387 | 14 | 388-395 | 8 | .636 |
| 390. | c/(a+b) | III | 472-481<br>482-492 | 21 | 493-505 | 13 | .618 |

| No. | m/M: type | Book | Major | Total lines | Minor | Total lines | Ratio: M/(M+m) |
|---|---|---|---|---|---|---|---|
| 391. | c/(a+b) | III | 506-520<br>521-531 | 25.4 | 532-547 | 16 | .614 |
| 392. | c/(a+b) | IV | 68-89<br>90-104 | 37 | 105-128 | 24 | .607 |
| 393. | a/(b+c) | IV | 105-114a<br>114b-128 | 24 | 90-104 | 15 | .615 |
| 394. | a/(b+c) | IV | 141b-150<br>151-159 | 18.8 | 129-141a | 12.2 | .606 |
| 395. | c/(a+b) | IV | 279-304<br>305-330 | 52 | 331-361 | 30.6 | .630 |
| 396. | a/(b+c) | IV | 362-387<br>388-415 | 53.6 | 331-361 | 30.6 | .637 |
| 397. | a/(b+c) | IV | 402-407<br>408-415 | 14 | 393-401 | 8.6 | .619 |
| 398-400. | | | | | | | |
| 398. | a/(b+c) | IV | 424-428<br>429-436 | 13 | 416-423 | 8 | .619 |
| 399. | b/a | | 416-423 | 8 | 424-428 | 5 | .615 |
| 400. | b/c | | 429-436 | 8 | 424-428 | 5 | .615 |
| 401. | a/(b+c) | IV | 478-503<br>504-521 | 43 | 450-477 | 28 | .606 |
| 402. | a/(b+c) | IV | 504-521<br>522-552 | 47.6 | 474-503 | 29.4 | .618 |
| 403. | a/(b+c) | IV | 630-692<br>693-705 | 76 | 584-629 | 46 | .623 |
| 404. | c/(a+b) | IV | 648-650<br>651-662 | 15 | 663-671 | 9 | .625 |
| 405. | a/(b+c) | V | 55-63<br>64-71 | 17 | 45-54 | 10 | .630 |
| 406. | a/(b+c) | V | 779-826<br>827-871 | 92.3 | 545-603 | 57.8 | .615 |
| 407. | a/(b+c) | VI | 249b-263<br>264-267 | 18.6 | 236-249a | 12.4 | .60 |
| 408. | c/(a+b) | VI | 236-267<br>268-294 | 58 | 295-330 | 36 | .617 |

TABLE II. SHORT PASSAGES: TRIPARTITE (NONFRAMEWORK) PATTERN

| No. | m/M: type | Book | Major | Total lines | Minor | Total lines | Ratio: M/(M+m) |
|---|---|---|---|---|---|---|---|
| 409. | c/(a+b) | VI | 331-336<br>337-383 | 53 | 384-416 | 33 | .616 |
| 410. | c/(a+b) | VI | 337-341a<br>341b-343a | 6.2 | 343b-346 | 3.8 | .620 |
| 411. | (a+b)/c | VI | 440-476 | 37 | 417-425<br>426-439 | 23 | .617 |
| 412. | (b+c)/a | VI | 426-476 | 51 | 477-493<br>494-508 | 32 | .614 |
| 413. | c/(a+b) | VI | 426-476<br>477-493 | 68 | 494-534 | 41 | .624 |
| 414. | a/(b+c) | VI | 648-655<br>656-665 | 18 | 637-647 | 11 | .621 |
| 415. | c/(a+b) | VI | 679-702<br>703-723 | 45 | 724-751 | 28 | .616 |
| 416. | c/(a+b) | VI | 756-776<br>777-787 | 32 | 788-807 | 20 | .615 |
| 417. | a/(b+c) | VI | 808-853<br>854-892 | 84.7 | 756-807 | 52 | .620 |
| 418. | c/(a+b) | VI | 819-825<br>826-835 | 16.7 | 836-846 | 11 | .603 |
| 419. | c/(a+b) | VII | 135-147<br>148-169 | 35 | 170-191 | 22 | .614 |
| 420. | c/(a+b) | VII | 540-571<br>572-600 | 61 | 601-640 | 40 | .604 |
| 421. | a/(b+c) | VII | 583-590<br>591-600 | 18 | 572-582 | 11 | .621 |
| 422. | a/(b+c) | VII | 616-622<br>623-640 | 25 | 601-615 | 15 | .625 |
| 423. | c/(a+b) | VII | 733-743<br>744-749 | 17 | 750-760 | 10.6 | .616 |
| 424. | a/(b+c) | VII | 783-802<br>803-817 | 35 | 761-782 | 22 | .614 |
| 425. | c/(a+b) | VIII | 102-125<br>126-151 | 50 | 152-183 | 32 | .610 |
| 426. | (b+c)/a | VIII | 152-171 | 20 | 172-174<br>175-183 | 12 | .625 |
| 427. | c/(a+b) | VIII | 280-305<br>306-336 | 57 | 337-369 | 33 | .633 |

| No. | m/M: type | Book | Major | Total lines | Minor | Total lines | Ratio: M/(M+m) |
|---|---|---|---|---|---|---|---|
| 428. | c/(a+b) | IX | 1-46<br>47-76 | 75 | 77-122 | 45 | .625 |
| 429. | a/(b+c) | IX | 77-122<br>123-158 | 80 | 25-76 | 51 | .611 |
| 430. | a/(b+c) | IX | 207-218<br>219-223 | 17 | 197-206 | 10 | .630 |
| 431. | a/(b+c) | IX | 292b-302<br>303-313 | 21 | 280b-292a | 12 | .636 |
| 432. | c/(a+b) | IX | 292b-302<br>303-305 | 13 | 306-313 | 8 | .619 |
| 433. | a/(b+c) | IX | 438-445<br>446-449 | 12 | 431-437 | 7 | .632 |
| 434. | a/(b+c) | IX | 542-555<br>556-568 | 27 | 525-541 | 16 | .628 |
| 435. | c/(a+b) | IX | 691-716<br>717-730 | 39.7 | 731-755 | 25 | .614 |
| 436. | c/(a+b) | X | 163-165<br>166-184 | 22 | 185-197 | 13 | .629 |
| 437. | a/(b+c) | X | 185-197<br>198-214 | 30 | 166-184 | 19 | .612 |
| 438. | a/(b+c) | X | 267-275<br>276-286 | 18.6 | 256-266 | 11 | .628 |
| 439. | c/(a+b) | X | 294-298a<br>298b-302a | 8.4 | 302b-307 | 5.6 | .60 |
| 440. | c/(a+b) | X | 439-456<br>457-463 | 25 | 464-478 | 15 | .625 |
| 441. | a/(b+c) | X | 479-509<br>510-542 | 63.6 | 439-478 | 40 | .614 |
| 442. | a/(b+c) | X | 491-500<br>501-509 | 19 | 479-490 | 11.6 | .621 |
| 443. | c/(a+b) | X | 543-605<br>606-632 | 89.2 | 633-688 | 56 | .614 |
| 444. | a/(b+c) | X | 755-761<br>762-768 | 14 | 747-754 | 8 | .636 |
| 445. | c/(a+b) | X | 873-877<br>878-882a | 8.8 | 882b-887 | 5.8 | .603 |
| 446. | c/(a+b) | XI | 39-58<br>59-99 | 61 | 100-138 | 39 | .610 |
| 447. | c/(a+b) | XI | 336-342<br>343-345 | 10 | 346-351 | 6 | .625 |

TABLE II. SHORT PASSAGES: TRIPARTITE (NONFRAMEWORK) PATTERN

| No. | m/M: type | Book | Major | Total lines | Minor | Total lines | Ratio: M/(M+m) |
|---|---|---|---|---|---|---|---|
| 448. | c/(a+b) | XI | 597-617<br>618-628 | 32 | 629-647 | 19 | .627 |
| 449. | a/(b+c) | XI | 759b-867<br>868-915 | 156.6 | 664-759a | 95.4 | .621 |
| 450. | c/(a+b) | XI | 690-724<br>725-767 | 78 | 768-815 | 48 | .619 |
| 451. | c/(a+b) | XI | 725-767<br>768-815 | 91 | 816-867 | 52 | .636 |
| 452. | c/(a+b) | XI | 794-835<br>836-867 | 74 | 868-915 | 48 | .607 |
| 453. | c/(a+b) | XI | 841-849a<br>849b-857 | 17 | 858-867 | 10 | .630 |
| 454. | a/(b+c) | XI | 887-895<br>896-915 | 29 | 868-886 | 19 | .604 |
| 455. | a/(b+c) | XII | 64-69<br>70-80 | 17 | 54-63 | 10 | .630 |
| 456. | c/(a+b) | XII | 81-106<br>107-112 | 32 | 113-133 | 21 | .604 |
| 457. | a/(b+c) | XII | 142-153<br>154-155 | 14 | 134-141 | 8 | .636 |
| 458.<br>[cf. No. 693 in Table III] | c/(a+b) | XII | 161-174<br>175-194 | 34 | 195-215 | 21 | .618 |
| 459. | c/(a+b) | XII | 289-296<br>297-310 | 22 | 311-323 | 13 | .629 |
| 460.<br>[cf. No. 699 in Table III] | c/(a+b) | XII | 346-382<br>383-440 | 95 | 441-499 | 59 | .617 |
| 461. | a/(b+c) | XII | 451-461<br>462-467 | 17 | 441-450 | 10 | .630 |
| 462. | a/(b+c) | XII | 500-553<br>554-592 | 93 | 441-499 | 59 | .612 |
| 463. | a/(b+c) | XII | 521-528<br>529-553 | 33 | 500-520 | 21 | .611 |
| 464. | a/(b+c) | XII | 521-528<br>529-547 | 27 | 505-520 | 16 | .628 |
| 465. | c/(a+b) | XII | 554-564<br>565-573 | 20 | 574-586 | 13 | .606 |
| 466. | c/(a+b) | XII | 614-630<br>631-649 | 35.2 | 650-671 | 22 | .615 |

| No. | m/M: type | Book | Major | Total lines | Minor | Total lines | Ratio: M/(M+m) |
|---|---|---|---|---|---|---|---|
| 467. | c/(a+b) | XII | 725-745<br>746-765 | 41 | 766-790 | 25 | .621 |
| 468. | c/(a+b) | XII | 791-842<br>843-868 | 78 | 869-918 | 50 | .609 |
| 469. | a/(b+c) | XII | 887-918<br>919-952 | 66 | 843-886 | 44 | .60 |

## TABLE III. SHORT PASSAGES: TRIPARTITE FRAMEWORK PATTERN

| No. | m/M: type | Book | Major | Total lines | Minor | Total lines | Ratio: M/(M+m) |
|---|---|---|---|---|---|---|---|
| 470. | b/(a+c) | I | 8-22<br>33 | 16 | 23-32 | 10 | .615 |
| 471. | b/(a+c) | I | 34-75<br>124-156 | 75 | 76-123 | 48 | .610 |
| 472. | b/(a+c) | I | 81-83<br>92-101 | 13 | 84-91 | 8 | .619 |
| 473. | b/(a+c) | I | 81-101<br>157-222 | 87 | 102-156 | 55 | .613 |
| 474. | (a+c)/b | I | 106-119 | 14 | 102-105<br>120-123 | 8 | .636 |
| 475. | b/(a+c) | I | 223-237<br>250-253 | 19 | 238-249 | 12 | .613 |
| 476. | (a+c)/b | I | 335-370a | 35.4 | 326-334<br>372-385a | 22.6 | .610 |
| 477. | (a+c)/b | I | 343-364 | 22 | 335-342<br>365-370a | 13.4 | .621 |
| 478. | b/(a+c) | I | 370b-377<br>384-385a | 9.2 | 378-383 | 6 | .605 |
| 479. | b/(a+c) | I | 430-440<br>450-452 | 14 | 441-449 | 9 | .609 |
| 480. | b/(a+c) | I | 466-473<br>485-493 | 17 | 474-484 | 11 | .607 |
| 481. | (a+c)/b) | I | 657-688 | 32 | 643-656<br>689-694 | 20 | .615 |
| 482. | (a+c)/b | I | 740b-752 | 12.6 | 736-740a<br>753-756 | 8.4 | .60 |
| 483. | (a+c)/b | II | 57-198 | 140.2 | 21-56<br>199-249 | 86.4 | .619 |
| 484. | (a+c)/b | II | 162-194 | 33 | 145-161<br>195-198 | 21 | .611 |
| 485. | b/(a+c) | II | 195-198<br>212-227 | 20 | 199-211 | 13 | .606 |
| 486. | (a+c)/b | II | 254-264 | 11 | 250-253<br>265-267 | 7 | .611 |
| 487. | b/(a+c) | II | 268-273<br>281-286 | 12 | 274-280 | 7 | .632 |

| No. | m/M: type | Book | Major | Total lines | Minor | Total lines | Ratio: M/(M+m) |
|---|---|---|---|---|---|---|---|
| 488. | (a+c)/b | II | 289-295 | 7 | 287-288<br>296-297 | 4 | .636 |
| 489. | b/(a+c) | II | 298-317<br>336-346 | 30.2 | 318-335 | 18 | .627 |
| 490. | b/(a+c) | II | 298-317<br>347-369 | 43 | 318-346 | 28.2 | .604 |
| 491. | (a+c)/b | II | 322-338 | 17 | 318-321<br>339-346 | 11.2 | .603 |
| 492. | b/(a+c) | II | 370-401<br>453-505 | 84.7 | 402-452 | 51 | .624 |
| 493. | (a+c)/b | II | 491-502 | 12 | 486-490<br>503-505 | 8 | .60 |
| 494. | b/(a+c) | II | 526-530<br>544-558 | 20 | 531-543 | 13 | .606 |
| 495. | b/(a+c) | II | 531-534<br>540-543 | 8 | 535-539 | 5 | .615 |
| 496. | (a+c)/b | II | 626-631 | 6 | 624-625<br>632-633 | 4 | .60 |
| 497. | b/(a+c) | II | 634-638a<br>645-649 | 9.4 | 638b-644 | 6 | .610 |
| 498. | b/(a+c) | II | 730-734<br>741-744 | 9 | 735-740 | 6 | .60 |
| 499. | b/(a+c) | II | 745-749<br>760-770 | 15.2 | 750-759 | 10 | .603 |
| 500. | (a+c)/b | II | 775-795 | 20.7 | 771-774<br>796-804 | 13 | .614 |
| 501. | b/(a+c) | III | 121-123<br>128-131 | 7 | 124-127 | 4 | .636 |
| 502. | (a+c)/b | III | 154-181 | 28 | 147-153<br>182-191 | 17 | .622 |
| 503. | (a+c)/b | III | 209-273 | 63.2 | 192-208<br>274-293 | 37 | .630 |
| 504. | (a+c)/b | III | 219-258 | 39 | 209-218<br>259-273 | 24.2 | .617 |
| 505. | b/(a+c) | III | 219-228<br>245-258 | 24 | 229-244 | 15 | .615 |
| 506. | b/(a+c) | III | 274-293<br>320-343 | 43.4 | 294-319 | 25.6 | .629 |
| 507. | b/(a+c) | III | 294-319<br>344-355 | 37.6 | 320-343 | 23.4 | .616 |

TABLE III. SHORT PASSAGES: TRIPARTITE FRAMEWORK PATTERN

| No. | m/M: type | Book | Major | Total lines | Minor | Total lines | Ratio: M/(M+m) |
|---|---|---|---|---|---|---|---|
| 508. | (a+c)/b | III | 306-355 | 49 | 294-305<br>356-373 | 30 | .620 |
| 509. | (a+c)/b | III | 358-368 | 11 | 356-357<br>369-373 | 7 | .611 |
| 510. | b/(a+c) | III | 374-409<br>461-505 | 80.6 | 410-460 | 51 | .612 |
| 511. | (a+c)/b | III | 414-432 | 19 | 410-413<br>433-440 | 12 | .613 |
| 512. | b/(a+c) | III | 410-440<br>461-462 | 33 | 441-460 | 20 | .623 |
| 513. | (a+c)/b | III | 463-469a | 6.4 | 461-462<br>469b-471 | 4.2 | .604 |
| 514. | b/(a+c) | III | 493-494<br>500-505 | 8 | 495-499 | 5 | .615 |
| 515. | b/(a+c) | III | 506-511<br>518-520 | 9 | 512-517 | 6 | .60 |
| 516. | b/(a+c) | III | 548-557<br>566-569 | 14 | 558-565 | 8 | .636 |
| 517. | b/(a+c) | III | 548-569<br>588-595 | 30 | 570-587 | 18 | .625 |
| 518. | b/(a+c) | III | 570-577<br>588-595 | 16 | 578-587 | 10 | .615 |
| 519. | b/(a+c) | III | 570-595<br>655-718 | 89.4 | 596-654 | 58.2 | .606 |
| 520. | (a+c)/b | III | 612-691 | 78.6 | 588-611<br>692-718 | 51 | .606 |
| 521. | b/(a+c) | III | 655-668<br>692-715 | 37.4 | 669-691 | 23 | .619 |
| 522. | b/(a+c) | III | 669-674<br>684-691 | 14 | 675-683 | 9 | .609 |
| 523. | b/(a+c) | IV | 1-8<br>20-29 | 18 | 9-19 | 11 | .621 |
| 524. | b/(a+c) | IV | 54-89<br>129-159 | 67 | 90-128 | 39 | .632 |
| 525. | (a+c)/b | IV | 115-123 | 9 | 114b<br>124-128 | 5.8 | .608 |
| 526. | b/(a+c) | IV | 173-197<br>238-278 | 65 | 198-237 | 40 | .619 |
| 527. | b/(a+c) | IV | 219-222<br>238-258 | 25 | 223-237 | 15 | .625 |

| No. | m/M: type | Book | Major | Total lines | Minor | Total lines | Ratio: M/(M+m) |
|---|---|---|---|---|---|---|---|
| 528. | (a+c)/b | IV | 309-324 | 16 | 305-308<br>325-330 | 10 | .615 |
| 529. | b/(a+c) | IV | 305-330<br>362-387 | 52 | 331-361 | 30.6 | .630 |
| 530. | (a+c)/b | IV | 337-355 | 19 | 331-336<br>356-361 | 11.6 | .621 |
| 531. | (a+c)/b | IV | 365-380 | 16 | 362-364<br>381-387 | 10 | .615 |
| 532. | (a+c)/b | IV | 441-448 | 8 | 437-440<br>449 | 5 | .615 |
| 533. | b/(a+c) | IV | 450-455<br>465b-473 | 14.4 | 456-465a | 9.6 | .60 |
| 534. | b/(a+c) | IV | 450-477<br>499-503 | 32.4 | 478-498 | 21 | .607 |
| 535. | (a+c)/b | IV | 560-579a | 19.2 | 553-559<br>579b-583 | 11.8 | .619 |
| 536. | b/(a+c) | IV | 630-633<br>641-647 | 11 | 634-640 | 7 | .611 |
| 537. | b/(a+c) | IV | 630-647<br>672-692 | 39 | 648-671 | 24 | .619 |
| 538. | (a+c)/b | IV | 696-703 | 8 | 693-695<br>704-705 | 5 | .615 |
| 539. | b/(a+c) | V | 1-12<br>26-34 | 21 | 13-25 | 13 | .618 |
| 540. | (a+c)/b | V | 35-103 | 69 | 1-34<br>104-113 | 44 | .611 |
| 541. | b/(a+c) | V | 35-44<br>72-103 | 42 | 45-71 | 27 | .609 |
| 542. | b/(a+c) | V | 114-123<br>151-182 | 42 | 124-150 | 27 | .609 |
| 543. | b/(a+c) | V | 124-128<br>139-150 | 17 | 129-138 | 10 | .630 |
| 544. | b/(a+c) | V | 183-200<br>225-243 | 37 | 201-224 | 24 | .607 |
| 545. | b/(a+c) | V | 225-243<br>268-285 | 37 | 244-267 | 24 | .607 |
| 546. | b/(a+c) | V | 315-317<br>323-326 | 7 | 318-322 | 4.4 | .614 |
| 547. | b/(a+c) | V | 318-326<br>340-352 | 21.4 | 327-339 | 13 | .622 |

TABLE III. SHORT PASSAGES: TRIPARTITE FRAMEWORK PATTERN

| No. | m/M: type | Book | Major | Total lines | Minor | Total lines | Ratio: M/(M+m) |
|---|---|---|---|---|---|---|---|
| 548. | b/(a+c) | V | 327-339 353-361 | 22 | 340-352 | 13 | .629 |
| 549. | b/(a+c) | V | 362-367 378-386 | 15 | 368-377 | 10 | .60 |
| 550. | b/(a+c) | V | 362-385 409-420 | 36 | 386-408 | 23 | .610 |
| 551. | b/(a+c) | V | 421-436 461-484 | 40 | 437-460 | 24 | .625 |
| 552. | b/(a+c) | V | 461-467 477-484 | 15 | 468-476 | 9 | .625 |
| 553. | b/(a+c) | V | 545-603 664-699 | 93.8 | 604-663 | 59.2 | .613 |
| 554. | b/(a+c) | V | 641-644a 653-663 | 13.4 | 644b-652 | 8.8 | .604 |
| 555. | b/(a+c) | V | 680-684 693-699 | 12 | 685-692 | 8 | .60 |
| 556. | (a+c)/b | V | 719-766 | 48 | 700-718 767-778 | 31 | .608 |
| 557. | (a+c)/b | V | 724-740 | 17 | 719-723 741-745 | 10 | .630 |
| 558. | b/(a+c) | V | 746-778 827-871 | 78 | 779-826 | 47.3 | .623 |
| 559. | b/(a+c) | V | 779-784 793-798 | 12 | 785-792 | 7.6 | .612 |
| 560. | b/(a+c) | V | 854-856 864-871 | 11 | 857-863 | 7 | .611 |
| 561. | b/(a+c) | VI | 14-22 30b-33a | 12 | 23-30a | 7.6 | .612 |
| 562. | b/(a+c) | VI | 56-76 103b-123 | 41.4 | 77-103a | 26.2 | .612 |
| 563. | (a+c)/b | VI | 81-97 | 16.6 | 77-80 98-103a | 9.6 | .634 |
| 564. | b/(a+c) | VI | 124-136a 149-155 | 19.6 | 136b-148 | 12.4 | .613 |
| 565. | (a+c)/b | VI | 160-176a | 16.6 | 156-159 176b-182 | 10.4 | .615 |
| 566. | b/(a+c) | VI | 212-219 229-235 | 15 | 220-228 | 9 | .625 |
| 567. | b/(a+c) | VI | 295-297 305-312 | 11 | 298-304 | 7 | .611 |

| No. | m/M: type | Book | Major | Total lines | Minor | Total lines | Ratio: M/(M+m) |
|---|---|---|---|---|---|---|---|
| 568. | (a+c)/b | VI | 347-371 | 25 | 341b-346<br>372-381 | 15.4 | .619 |
| 569. | b/(a+c) | VI | 426-449<br>477-493 | 41 | 450-476 | 27 | .603 |
| 570. | b/(a+c) | VI | 509-519<br>529b-534 | 16.4 | 520-529a | 9.6 | .631 |
| 571. | b/(a+c) | VI | 535-538<br>544-547 | 8 | 539-543 | 5 | .615 |
| 572. | (a+c)/b | VI | 679-751 | 73 | 637-678<br>752-755 | 46 | .613 |
| or | | | | | | | |
| 573. | b/(a+c) | VI | 637-678<br>724-755 | 74 | 679-723 | 45 | .622 |
| 574. | b/(a+c) | VI | 637-723<br>808-853 | 132.7 | 724-807 | 84 | .612 |
| 575. | (a+c)/b | VI | 669-676 | 8 | 666-668<br>677-678 | 5 | .615 |
| 576. | (a+c)/b | VI | 684-698 | 15 | 679-683<br>699-702 | 9 | .625 |
| 577. | (a+c)/b | VI | 819-846 | 27.7 | 808-818<br>847-853 | 18 | .606 |
| 578. | b/(a+c) | VI | 854-866<br>882-892 | 24 | 867-881 | 15 | .615 |
| 579. | b/(a+c) | VI | 854-867<br>886b-901 | 29.8 | 868-886a | 18.2 | .621 |
| 580. | b/(a+c) | VII | 81-91<br>102-106 | 16 | 92-101 | 10 | .615 |
| 581. | b/(a+c) | VII | 135-140<br>146-147 | 8 | 141-145 | 5 | .615 |
| 582. | b/(a+c) | VII | 148-151<br>160-169 | 14 | 152-159 | 8 | .636 |
| 583. | (a+c)/b | VII | 173-186 | 14 | 170-172<br>187-191 | 8 | .636 |
| 584. | b/(a+c) | VII | 192-211<br>234-248 | 34.6 | 212-233 | 22 | .611 |
| 585. | b/(a+c) | VII | 192-211<br>249-285 | 57 | 212-248 | 36.6 | .609 |
| 586. | b/(a+c) | VII | 249-259a<br>274-285 | 22.4 | 259b-273 | 14.6 | .605 |
| 587. | b/(a+c) | VII | 286-298<br>313-322 | 23 | 299-312 | 14 | .622 |

TABLE III. SHORT PASSAGES: TRIPARTITE FRAMEWORK PATTERN

| No. | m/M: type | Book | Major | Total lines | Minor | Total lines | Ratio: M/(M+m) |
|---|---|---|---|---|---|---|---|
| 588. | b/(a+c) | VII | 385-391 400b-405 | 12.9 | 392-400a | 8.1 | .614 |
| 589. | b/(a+c) | VII | 406-420 435-444 | 24.4 | 421-434 | 14 | .635 |
| 590. | b/(a+c) | VII | 445-455 467-474 | 18.6 | 456-466 | 11 | .628 |
| 591. | b/(a+c) | VII | 511-523a 535-539 | 17.4 | 523b-534 | 11.6 | .60 |
| 592. [cf. No. 824 in Table IV] | (a+c)/b | VII | 647-669 | 23 | 641-646 670-677 | 14 | .622 |
| 593. | (a+c)/b | VII | 647-677 | 31 | 641-646 678-690 | 19 | .620 |
| 594. | b/(a+c) | VII | 647-654 664-669 | 14 | 655-663 | 9 | .609 |
| 595. | (a+c)/b | VII | 655-690 | 36 | 647-654 691-705 | 22.2 | .618 |
| 596. | (a+c)/b | VII | 706-782 | 76.6 | 691-705 783-817 | 49.2 | .609 |
| 597. | b/(a+c) | VII | 706-722 744-760 | 33.6 | 723-743 | 21 | .615 |
| 598. | (a+c)/b | VIII | 18-67 | 48.4 | 1-17 68-80 | 30 | .617 |
| 599. | (a+c)/b | VIII | 18-80 | 61.4 | 1-17 81-101 | 38 | .618 |
| 600. | b/(a+c) | VIII | 102-114 124-125 | 15 | 115-123 | 9 | .625 |
| 601. | b/(a+c) | VIII | 102-125 152-171 | 44 | 126-151 | 26 | .629 |
| 602. | (a+c)/b | VIII | 152-279 | 128 | 102-151 280-305 | 76 | .627 |
| 603. | (a+c)/b | VIII | 117-123 | 7 | 115-116 124-125 | 4 | .636 |
| 604. | b/(a+c) | VIII | 126-151 175-183 | 35 | 152-174 | 23 | .603 |
| 605. | (a+c)/b | VIII | 154b-168 | 14.6 | 152-154a 169-174 | 8.4 | .635 |
| 606. | (a+c)/b | VIII | 184-279 | 96 | 152-183 280-305 | 58 | .623 |
| 607. | (a+c)/b | VIII | 184-267 | 84 | 172-183 268-305 | 50 | .627 |

| No. | m/M: type | Book | Major | Total lines | Minor | Total lines | Ratio: M/(M+m) |
|---|---|---|---|---|---|---|---|
| 608. | b/(a+c) | VIII | 200-218<br>241-255 | 34 | 219-240 | 22 | .607 |
| 609. | b/(a+c) | VIII | 280-293a<br>303-305 | 16.2 | 293b-302 | 9.8 | .623 |
| 610. | b/(a+c) | VIII | 306-312<br>337-369 | 40 | 313-336 | 24 | .625 |
| 611. | b/(a+c) | VIII | 313-336<br>359-369 | 35 | 337-358 | 22 | .614 |
| 612. | b/(a+c) | VIII | 359-361<br>366-369 | 7 | 362-365 | 4 | .636 |
| 613. | b/(a+c) | VIII | 370-386<br>407-422 | 33 | 387-406 | 20 | .623 |
| 614. | b/(a+c) | VIII | 394-399<br>404b-406 | 7.4 | 400-404a | 4.6 | .617 |
| 615. | b/(a+c) | VIII | 454-458<br>465-468 | 9 | 459-464 | 6 | .60 |
| 616-617. | | | | | | | |
| 616. | b/(a+c) | VIII | 454-468<br>494-519 | 41 | 469-493 | 24.2 | .629 |
| 617. | a/b | | 469-493 | 24.2 | 454-468 | 15 | .617 |
| 618. | b/(a+c) | VIII | 469-493<br>520-540 | 44.6 | 494-519 | 26 | .632 |
| 619. | b/(a+c) | VIII | 554-559<br>572-584 | 19 | 560-571 | 12 | .613 |
| 620. | (a+c)/b | VIII | 558-584 | 27 | 554-557<br>585-596 | 16 | .628 |
| 621. | b/(a+c) | VIII | 626-651<br>671-674 | 30 | 652-670 | 19 | .612 |
| 622. | b/(a+c) | VIII | 626-674<br>714-728 | 64 | 675-713 | 39 | .621 |
| 623. | b/(a+c) | VIII | 646-651<br>663-674 | 18 | 652-662 | 11 | .621 |
| 624. | b/(a+c) | VIII | 675-688<br>704-713 | 24 | 689-703 | 15 | .615 |
| 625. | b/(a+c) | IX | 1-15<br>35-46 | 27 | 16-34 | 18 | .60 |
| 626. | b/(a+c) | IX | 25-46<br>67-76 | 31 | 47-66 | 20 | .608 |

TABLE III. SHORT PASSAGES: TRIPARTITE FRAMEWORK PATTERN

| No. | m/M: type | Book | Major | Total lines | Minor | Total lines | Ratio: M/(M+m) |
|---|---|---|---|---|---|---|---|
| 627. | b/(a+c) | IX | 77-92<br>104-106 | 19 | 93-103 | 11 | .633 |
| 628. | (a+c)/b | IX | 126-158 | 32 | 123-125<br>159-175 | 19.4 | .623 |
| 629. | b/(a+c) | IX | 176-178<br>182-183 | 5 | 179-181 | 3 | .625 |
| 630. | b/(a+c) | IX | 224-233<br>246-256 | 21 | 234-245 | 12 | .636 |
| 631. | b/(a+c) | IX | 224-245<br>280b-313 | 55 | 246-280a | 34.4 | .615 |
| 632. | b/(a+c) | IX | 257-262<br>272-280a | 14.4 | 263-271 | 9 | .615 |
| 633. | (a+c)/b | IX | 324-356 | 33 | 314-323<br>357-366 | 20 | .623 |
| 634. | b/(a+c) | IX | 399-409<br>422-430 | 20 | 410-421 | 12 | .625 |
| 635. | b/(a+c) | IX | 503-511<br>521-524 | 13 | 512-520 | 8.4 | .607 |
| 636. | b/(a+c) | IX | 576b-580<br>586-589 | 8.6 | 581-585 | 5 | .632 |
| 637. | b/(a+c) | IX | 672-676<br>683-687 | 10 | 677-682 | 6 | .625 |
| 638. | (a+c)/b | IX | 691-755 | 64.7 | 672-690<br>756-777 | 40.4 | .616 |
| 639. | (a+c)/b | IX | 694-709 | 16 | 691-693<br>710-716 | 10 | .615 |
| 640. | b/(a+c) | IX | 717-721<br>731-740 | 14.7 | 722-730 | 9 | .620 |
| 641. | b/(a+c) | IX | 740-742<br>749-755 | 10 | 743-748 | 6 | .625 |
| 642.<br>[cf. No. 261 in Table I] | b/(a+c) | IX | 762-767<br>778-787 | 16 | 768-777 | 10 | .615 |
| 643. | (a+c)/b | IX | 781-789a | 8.2 | 778-780<br>789b-790 | 4.8 | .631 |
| 644. | b/(a+c) | IX | 778-790<br>806-818 | 26 | 791-805 | 15 | .634 |
| 645. | (a+c)/b | IX | 808b-816a | 8.2 | 806-808a<br>816b-818 | 4.8 | .631 |

| No. | m/M: type | Book | Major | Total lines | Minor | Total lines | Ratio: M/(M+m) |
|---|---|---|---|---|---|---|---|
| 646. | (a+c)/b | X | 6-15 | 10 | 1-5<br>16-17 | 6.2 | .617 |
| 647. | b/(a+c) | X | 1-17<br>62b-117 | 71.6 | 18-62a | 44.6 | .616 |
| 648. | b/(a+c) | X | 18-30<br>48-62a | 27.6 | 31-47 | 17 | .619 |
| 649. | b/(a+c) | X | 62b-71<br>85-95 | 20.4 | 72-84 | 13 | .611 |
| 650. | (a+c)/b | X | 100-113a | 13.4 | 96-99<br>113b-117 | 8.6 | .609 |
| 651. | b/(a+c) | X | 118-131<br>143-145 | 17 | 132-142 | 11 | .607 |
| 652. | b/(a+c) | X | 118-162<br>215-255 | 86 | 163-214 | 52 | .623 |
| 653. | (a+c)/b<br>[cf. No. 865 in Table IV] | X | 166-197 | 32 | 163-165<br>198-214 | 20 | .615 |
| 654. | b/(a+c) | X | 185-188<br>194-197 | 8 | 189-193 | 5 | .615 |
| 655. | b/(a+c) | X | 215-228a<br>246-259 | 27.6 | 228b-245 | 17.4 | .613 |
| 656. | b/(a+c) | X | 308-323<br>345-361 | 33 | 324-344 | 21 | .611 |
| 657. | b/(a+c) | X | 345-349<br>356b-361 | 10.8 | 350-356a | 6.2 | .635 |
| 658. | b/(a+c) | X | 362-379<br>394-398 | 23 | 380-393 | 14 | .622 |
| 659. | (a+c)/b | X | 379-425 | 47 | 362-378<br>426-438 | 30 | .610 |
| 660. | b/(a+c) | X | 399-411a<br>426-438 | 25.4 | 411b-425 | 14.6 | .635 |
| 661. | b/(a+c) | X | 457-465<br>474-478 | 14 | 466-473 | 8 | .636 |
| 662. | b/(a+c) | X | 606-620<br>653-688 | 51 | 621-652 | 32 | .614 |
| 663. | b/(a+c) | X | 653-665<br>680-688 | 22 | 666-679 | 14 | .611 |
| 664. | b/(a+c) | X | 689-718<br>747-761 | 45 | 719-746 | 27.2 | .623 |
| 665. | b/(a+c) | X | 719-731<br>742-746 | 17.2 | 732-741 | 10 | .632 |

TABLE III. SHORT PASSAGES: TRIPARTITE FRAMEWORK PATTERN

| No. | m/M: type | Book | Major | Total lines | Minor | Total lines | Ratio: M/(M+m) |
|---|---|---|---|---|---|---|---|
| 666. | b/(a+c) | X | 747-754<br>761 | 9 | 755-760 | 6 | .60 |
| 667. | b/(a+c) | X | 747-788<br>821-832 | 54 | 789-820 | 32 | .628 |
| 668. | (a+c)/b | X | 821-830a | 9.7 | 817-820<br>830b-832 | 6.3 | .606 |
| 669. | b/(a+c) | X | 846-850<br>855-856a | 6.2 | 851-854 | 4 | .608 |
| 670. | b/(a+c) | X | 856b-860<br>867-871 | 9.8 | 861-866 | 6 | .620 |
| 671. | b/(a+c) | XI | 1-13<br>29-38 | 23 | 14-28 | 15 | .605 |
| 672. | b/(a+c) | XI | 14-17<br>24-28 | 9 | 18-23 | 6 | .60 |
| 673. | (a+c)/b | XI | 59-147 | 89 | 39-58<br>148-181 | 54 | .622 |
| 674. | (a+c)/b | XI | 108-131 | 24 | 100-107<br>132-138 | 15 | .615 |
| 675. | b/(a+c) | XI | 139-147<br>182-224 | 52 | 148-181 | 34 | .605 |
| 676. | b/(a+c) | XI | 376-409<br>445-467 | 56.2 | 410-444 | 35 | .616 |
| 677. | b/(a+c) | XI | 468-535a<br>584b-596 | 79.8 | 535b-584a | 49.2 | .619 |
| 678. | b/(a+c) | XI | 498-506<br>520-531 | 21 | 507-519 | 13 | .618 |
| 679. | b/(a+c) | XI | 498-531<br>595-663 | 103 | 532-594 | 63 | .620 |
| 680. | (a+c)/b | XI | 544-584a | 40.4 | 532-543<br>584b-596 | 24.6 | .622 |
| 681. | b/(a+c) | XI | 597-599a<br>608-617 | 12.6 | 599b-607 | 8.4 | .60 |
| 682. | b/(a+c) | XI | 597-647<br>725-798 | 125 | 648-724 | 77 | .619 |
| 683. | b/(a+c) | XI | 648-652<br>659-663 | 10 | 653-658 | 6 | .625 |
| 684.<br>[cf. No. 912 in Table IV] | b/(a+c) | XI | 648-767<br>868-915 | 168 | 768-867 | 100 | .627 |
| 685. | (a+c)/b | XI | 670-685 | 16 | 664-669<br>686-689 | 10 | .615 |

| No. | m/M: type | Book | Major | Total lines | Minor | Total lines | Ratio: M/(M+m) |
|---|---|---|---|---|---|---|---|
| 686. | b/(a+c) | XI | 690-698<br>712-724 | 22 | 699-711 | 13 | .629 |
| 687. | (a+c)/b | XI | 768-835 | 68 | 759a-767<br>836-867 | 40.6 | .626 |
| 688. | b/(a+c) | XI | 868-875<br>887-895 | 17 | 876-886 | 11 | .607 |
| 689. | b/(a+c) | XII | 1-4a<br>11-17 | 10.4 | 4b-10 | 6.6 | .612 |
| 690. | (a+c)/b | XII | 18-45a | 27.2 | 10-17<br>45b-53 | 16.8 | .618 |
| 691. [cf. No. 320 in Table I] | b/(a+c) | XII | 113-120<br>129-133 | 13 | 121-128 | 8 | .619 |
| 692. | b/(a+c) | XII | 134-174<br>216-237 | 62.6 | 175-215 | 41 | .604 |
| 693. [cf. No. 458 in Table II] | b/(a+c) | XII | 161-174<br>195-215 | 35 | 175-194 | 20 | .636 |
| 694. | b/(a+c) | XII | 216-228<br>257-288 | 44.6 | 229-256 | 28 | .614 |
| 695. | b/(a+c) | XII | 238-243<br>251-256 | 12 | 244-250 | 7 | .632 |
| 696. | b/(a+c) | XII | 289-323<br>383-440 | 93 | 324-382 | 59 | .612 |
| 697. | b/(a+c) | XII | 311-312<br>318-323 | 8 | 313-317 | 5 | .615 |
| 698. | b/(a+c) | XII | 346-352<br>359-361 | 10 | 353-358 | 6 | .625 |
| 699. [cf. No. 460 in Table II] | b/(a+c) | XII | 346-382<br>441-499 | 96 | 383-440 | 58 | .623 |
| 700. | b/(a+c) | XII | 441-467<br>500-528 | 56 | 468-499 | 32 | .636 |
| 701. | b/(a+c) | XII | 650-656a<br>665-671 | 13.6 | 656b-664 | 8.4 | .618 |
| 702. | b/(a+c) | XII | 725-733a<br>742-745 | 12.6 | 733b-741 | 8.4 | .60 |
| 703. | b/(a+c) | XII | 728-734<br>742-745 | 11 | 735-741 | 7 | .611 |
| 704. | b/(a+c) | XII | 791-796<br>803b-807 | 10.6 | 797-803a | 6.4 | .624 |

TABLE III. SHORT PASSAGES: TRIPARTITE FRAMEWORK PATTERN

| No. | m/M: type | Book | Major | Total lines | Minor | Total lines | Ratio: M/(M+m) |
|---|---|---|---|---|---|---|---|
| 705. | (a+c)/b | XII | 808-840 | 33 | 791-807<br>841-842 | 19 | .635 |
| 706. | b/(a+c) | XII | 808-817<br>826-828 | 13 | 818-825 | 8 | .619 |
| 707. | b/(a+c) | XII | 853-860<br>867-868 | 10 | 861-866 | 6 | .625 |
| 708. | b/(a+c) | XII | 869-886<br>906-918 | 31 | 887-905 | 19 | .620 |
| 709. | b/(a+c) | XII | 887-893<br>906-918 | 20 | 894-905 | 12 | .625 |
| 710. | b/(a+c) | XII | 940-944<br>950-952 | 8 | 945-949 | 5 | .615 |

## TABLE IV. SHORT PASSAGES: FOUR OR MORE PARTS, USUALLY INTERLOCKED

| No. | m/M: type | Book | Major | Total lines | Minor | Total lines | Ratio: M/(M+m) |
|---|---|---|---|---|---|---|---|
| 711. | (b+d)/(a+c) | I | 1a-7<br>12-33 | 33 | 8-11<br>34-49 | 20 | .623 |
| 712. | (a+c)/(b+d) | I | 12-33<br>50-80 | 53 | 1a-11<br>34-49 | 31 | .631 |
| 713. | (b+d)/(a+c) | I | 157-169<br>180-197 | 31 | 170-179<br>198-207 | 20 | .608 |
| 714-716. | | | | | | | |
| 714. | (a+d)/(b+c) | I | 261-271<br>272-285 | 25 | 257-260<br>286-296 | 15 | .625 |
| 715. | b/(a+c) | | 257-260<br>272-285 | 18 | 261-271 | 11 | .621 |
| 716. | c/(b+d) | | 261-271<br>286-296 | 22 | 272-285 | 14 | .611 |
| 717. | (c+d)/(a+b) | I | 326-334<br>335-370a | 44.4 | 372-385a<br>387-401 | 28.6 | .608 |
| 718. | (b+d)/(a+c) | I | 418-429<br>441-449 | 21 | 430-440<br>450-452 | 14 | .60 |
| 719. | (b+d)/(a+c) | I | 418-440<br>453-493 | 64 | 441-452<br>494-519 | 38 | .627 |
| 720. | (a+c)/(b+d) | I | 498-506<br>513-519 | 16 | 494-497<br>507-512 | 10 | .615 |
| 721-722. | | | | | | | |
| 721. | (b+d)/(a+c+e) | I | 520-529<br>539-543<br>551-560 | 24.2 | 530-538<br>544-550 | 15.3 | .613 |
| 722. | c/b | | 530-538 | 8.3 | 539-543 | 5 | .624 |
| 723. | (a+c)/(b+d) | I | 546-550<br>555-558 | 9 | 544-545<br>551-554 | 6 | .60 |
| 724. | (b+c)/(a+d) | I | 561-571<br>582-585 | 15 | 572-578<br>579-581 | 10 | .60 |
| 725. | (a+d)/(b+c) | I | 595b-612<br>613-630 | 35.6 | 586-595a<br>631-642 | 21 | .629 |
| 726. | (a+c)/(b+d) | I | 647-652<br>657-662 | 12 | 643-646<br>653-656 | 8 | .60 |

TABLE IV. SHORT PASSAGES: FOUR OR MORE PARTS, USUALLY INTERLOCKED

| No. | m/M: type | Book | Major | Total lines | Minor | Total lines | Ratio: M/(M+m) |
|---|---|---|---|---|---|---|---|
| 727. | (a+c)/(b+d) | I | 657-694<br>712-722 | 49 | 643-656<br>695-711 | 31 | .613 |
| 728. | (a+c)/(b+d) | II | 25-39<br>50-56 | 22 | 21-24<br>40-49 | 14 | .611 |
| 729. | (a+c)/(b+d) | II | 57-104<br>145-194 | 96.2 | 40-56<br>105-144 | 57 | .628 |
| 730. | (a+c)/(b+d) | II | 234-238a<br>241-249 | 13.2 | 228-233<br>238b-240 | 8.2 | .617 |
| 731. | (a+c)/(b+d) | II | 268-317<br>347-369 | 73 | 250-267<br>318-346 | 46.2 | .612 |
| 732. | (b+d)/(a+c) | II | 370-378<br>383-393 | 20 | 379-382<br>394-401 | 12 | .625 |
| 733. | (b+d)/(a+c+e) | II | 402-409<br>424-430<br>437-452 | 31 | 410-423<br>431-436 | 20 | .608 |
| 734. | (a+c)/(b+d) | II | 458-468<br>476-485 | 20.7 | 453-457<br>469-475 | 12 | .633 |
| 735. | (b+d)/(a+c) | II | 588-603<br>615-620 | 22 | 604-614<br>621-623 | 12.8 | .632 |
| 736. | (a+c)/(b+d) | II | 657-670<br>675-678 | 18 | 650-656<br>671-674 | 11 | .621 |
| 737. | (a+c+e)/(b+d) | II | 745-770<br>775-795 | 45.9 | 730-744<br>771-774<br>796-804 | 28 | .621 |
| 738. | (b+d)/(a+c+e) | III | 1-8a<br>11b-12<br>16b-18 | 11.4 | 8b-11a<br>13-16a | 6.6 | .633 |
| 739. | (a+d)/(b+c+e) | III | 8b-12<br>13-16a<br>19-26 | 16 | 1-8a<br>16b-18 | 10 | .615 |
| 740. | (b+c)/(a+d) | III | 27-36<br>53-68 | 26 | 37-48<br>49-52 | 16 | .619 |
| 741. | (a+c)/(b+d) | III | 57b-59<br>62-68 | 9.6 | 53-57a<br>60-61 | 6.4 | .60 |
| 742. | (b+d)/(a+c) | III | 84-89<br>94-98 | 11 | 90-93<br>99-101 | 7 | .611 |
| 743-744. | | | | | | | |
| 743. | (b+d)/(a+c) | III | 121-131<br>135-139 | 16 | 132-134<br>140-146 | 10 | .615 |
| 744. | b/c | | 135-139 | 5 | 132-134 | 3 | .625 |

| No. | m/M: type | Book | Major | Total lines | Minor | Total lines | Ratio: M/(M+m) |
|---|---|---|---|---|---|---|---|
| 745. | (b+c)/(a+d) | III | 192-199<br>210b-218 | 16 | 200-204<br>205-210a | 10.2 | .611 |
| 746. | (b+d)/(a+c) | III | 306-313a<br>320-343 | 31 | 313b-319<br>344-355 | 18 | .633 |
| 747. | (b+d)/(a+c+e) | III | 356-373<br>396-440<br>463-471 | 71.6 | 374-395<br>441-462 | 44 | .619 |
| <u>748-749</u>. | | | | | | | |
| 748. | (b+d)/(a+c+e) | III | 374-395<br>410-440<br>461-462 | 55 | 396-409<br>441-460 | 34 | .618 |
| 749. | (b+d)/(a+c) | | 374-395<br>410-440 | 53 | 396-409<br>441-460 | 34 | .609 |
| 750. | (a+c+e)/(b+d) | III | 475-481<br>496-491 | 13 | 472-474<br>482-485<br>492 | 8 | .619 |
| 751. | (b+d)/(a+c) | III | 548-560<br>566-577 | 25 | 561-565<br>578-587 | 15 | .625 |
| 752. | (b+d)/(a+c) | III | 588-595<br>599b-606 | 15.4 | 596-599a<br>607-612 | 9.6 | .616 |
| 753. | (b+e)/(a+c+d) | IV | 1-8<br>20-29<br>30-44 | 32.4 | 9-19<br>45-53 | 20 | .618 |
| 754. | (a+c)/(b+d) | IV | 68-89<br>105-128 | 46 | 54-67<br>90-104 | 29 | .613 |
| 755. | (a+c)/(b+d) | IV | 136-150<br>160-172 | 28 | 129-135<br>151-159 | 16 | .636 |
| 756. | (a+c)/(b+d) | IV | 178-183<br>189-197 | 15 | 173-177<br>184-188 | 10 | .60 |
| 757. | (b+d)/(a+c) | IV | 173-183<br>189-197 | 20 | 184-188<br>198-205 | 13 | .606 |
| 758. | (b+c)/(a+d) | IV | 173-188<br>206-218 | 29 | 189-197<br>198-205 | 17 | .630 |
| 759. | (b+d)/(a+c) | IV | 198-218<br>238-264 | 48 | 219-237<br>265-278 | 32 | .60 |
| 760. | (b+d)/(a+c) | IV | 238-246a<br>259-276a | 24.4 | 246b-258<br>276b-278 | 15.6 | .610 |
| 761. | (b+d)/(a+c+e) | IV | 279-282<br>285-287<br>296-304 | 16 | 283-284<br>288-295 | 10 | .615 |

TABLE IV. SHORT PASSAGES: FOUR OR MORE PARTS, USUALLY INTERLOCKED

| No. | m/M: type | Book | Major | Total lines | Minor | Total lines | Ratio: M/(M+m) |
|---|---|---|---|---|---|---|---|
| 762. | (b+d)/(a+c+e) | IV | 279-304<br>331-361<br>388-415 | 84.2 | 305-330<br>362-387 | 52 | .618 |
| or |  |  |  |  |  |  |  |
| 763. | (a+e)/(b+c+d) | IV | 305-330<br>331-361<br>362-387 | 82.6 | 279-304<br>388-415 | 53.6 | .606 |
| 764. | (a+c)/(b+d) | IV | 393-401<br>408-415 | 16.6 | 388-392<br>402-407 | 11 | .601 |
| 765. | (a+c)/(b+d) | IV | 478-521<br>534-552 | 62 | 450-477<br>522-533 | 39 | .614 |
| 766. | (a+c)/(b+d) | IV | 478-491<br>499-503 | 18.4 | 474-477<br>492-498 | 11 | .626 |
| 767-769. |  |  |  |  |  |  |  |
| 767. | (a+d)/(b+c) | IV | 597-606<br>607-620 | 24 | 590b-596<br>621-629 | 15.4 | .609 |
| 768. | a/b |  | 597-606 | 10 | 590b-596 | 6.4 | .610 |
| 769. | d/c |  | 607-620 | 14 | 621-629 | 9 | .609 |
| 770. | (a+c)/(b+d) | V | 45-54<br>64-71 | 18 | 42-44<br>55-63 | 12 | .60 |
| 771. | (b+d)/(a+c) | V | 72-79<br>84-89 | 14 | 80-83<br>90-93 | 8 | .636 |
| 772. | (a+c)/(b+d) | V | 124-150<br>183-224 | 69 | 114-123<br>151-182 | 42 | .622 |
| 773. | (a+c)/(b+d) | V | 139-182<br>225-285 | 105 | 114-138<br>183-224 | 67 | .610 |
| 774. | (a+c)/(b+d) | V | 183-224<br>244-285 | 84 | 151-182<br>225-243 | 51 | .622 |
| 775. | (b+d)/(a+c) | V | 172-175<br>178-180 | 7 | 176-177<br>181-182 | 4 | .636 |
| 776. | (a+c)/(b+d) | V | 258-265<br>268-285 | 26 | 244-257<br>266-267 | 16 | .619 |
| 777. | (b+d)/(a+c) | V | 315-317<br>327-352 | 29 | 318-326<br>353-361 | 17.4 | .625 |
| 778. | (b+d)/(a+c) | V | 485-499<br>507-512 | 21 | 500-506<br>513-518 | 13 | .618 |
| 779. | (b+d)/(a+c) | V | 545-552<br>563-574 | 19.2 | 553-562<br>575-576 | 12 | .615 |
| 780. | (b+c)/(a+d) | V | 604-663<br>719-745 | 86.2 | 664-679<br>680-718 | 55 | .610 |

| No. | m/M: type | Book | Major | Total lines | Minor | Total lines | Ratio: M/(M+m) |
|---|---|---|---|---|---|---|---|
| 781. | (b+d)/(a+c) | V | 664-679<br>687-692 | 22 | 680-686<br>693-699 | 14 | .611 |
| 782. | (b+d)/(a+c+e) | V | 664-699<br>746-761<br>779-826 | 99.3 | 700-745<br>762-778 | 63 | .612 |
| 783. | (b+c)/(a+d) | V | 700-703<br>711-718 | 12 | 704-707<br>708-710 | 7 | .632 |
| 784. | (a+c+e)/(b+d) | V | 708-718<br>724-740 | 28 | 700-707<br>719-723<br>741-745 | 18 | .609 |
| 785. | (a+c)/(b+d) | V | 781-798<br>816-826 | 28.6 | 779-780<br>799-815 | 18.7 | .605 |
| 786. | (a+c)/(b+d) | V | 833-842<br>847-853 | 17 | 827-832<br>843-846 | 10 | .630 |
| 787. | (a+c)/(b+d) | VI | 14-33a<br>42-55 | 33.6 | 1-13<br>33b-41 | 21.4 | .611 |
| 788. | (b+d)/(a+c) | VI | 103b-109<br>116b-122a | 12.4 | 110-116a<br>122b-123 | 8 | .608 |
| 789. | (b+d)/(a+c) | VI | 156-182<br>190-211 | 49 | 183-189<br>212-235 | 31 | .613 |
| 790. | (b+d)/(a+c+e) | VI | 183-200<br>212-219<br>229-235 | 33 | 201-211<br>220-228 | 20 | .623 |
| 791. | (b+d)/(a+c) | VI | 236-294<br>331-383 | 111 | 295-330<br>384-416 | 69 | .617 |
| 792. | (a+d)/(b+c) | VI | 273-281<br>282-289 | 17 | 268-272<br>290-294 | 10 | .630 |
| 793. | (a+c)/(b+d) | VI | 317-321<br>325-330 | 11 | 313-316<br>322-324 | 7 | .611 |
| 794. | (a+d)/(b+c) | VI | 388-397<br>398-408a | 20.2 | 384-387<br>408b-416 | 12.8 | .612 |
| 795. | (a+c)/(b+d) | VI | 445-476<br>494-534 | 73 | 417-444<br>477-493 | 45 | .619 |
| 796. | (a+c)/(b+d) | VI | 430-439<br>445-449 | 15 | 426-429<br>440-444 | 9 | .625 |
| 797. | (a+c+e)/(b+d) | VI | 430-437<br>440-449 | 18 | 426-429<br>438-439<br>450-455 | 12 | .60 |
| 798. | (a+c)/(b+d) | VI | 440-449<br>456-476 | 31 | 426-439<br>450-455 | 20 | .608 |
| 799. | (b+d)/(a+c) | VI | 477-493<br>509-534 | 43 | 494-508<br>535-547 | 28 | .606 |

TABLE IV. SHORT PASSAGES: FOUR OR MORE PARTS, USUALLY INTERLOCKED

| No. | m/M: type | Book | Major | Total lines | Minor | Total lines | Ratio: M/(M+m) |
|---|---|---|---|---|---|---|---|
| 800. | (a+c)/(b+d) | VI | 562-607<br>628-636 | 55 | 548-561<br>608-627 | 34 | .618 |
| 801. | (a+c)/(b+d) | VI | 585-594<br>601-607 | 17 | 580-584<br>595-600 | 11 | .607 |
| 802. | (b+d)/(a+c) | VI | 608-620<br>628-632 | 18 | 621-627<br>633-636 | 11 | .621 |
| 803. | (b+d)/(a+c) | VI | 637-755<br>808-853 | 164.7 | 756-807<br>854-901 | 100 | .622 |
| 804. | (b+d)/(a+c) | VI | 703-712<br>719-721 | 13 | 713-718<br>722-723 | 8 | .619 |
| 805. | (a+c)/(b+d) | VI | 760-766<br>771-776 | 13 | 756-759<br>767-770 | 8 | .619 |
| 806. | (b+c)/(a+d) | VI | 808-812a<br>817-818 | 6.6 | 812b-815a<br>815b-816 | 4.4 | .60 |
| 807. | (b+d)/(a+c) | VII | 1-4<br>10-20 | 15 | 5-9<br>21-24 | 9 | .625 |
| 808. | (a+c)/(b+d) | VII | 10-24<br>30b-36 | 21.8 | 1-9<br>25-30a | 14.2 | .606 |
| 809. | (a+c+e)/(b+d) | VII | 5-24<br>37-80 | 64 | 1-4<br>25-36<br>81-106 | 42 | .604 |
| 810. | (a+c)/(b+d) | VII | 25-80<br>107-134 | 83.6 | 1-24<br>81-106 | 50 | .626 |
| 811. | (b+d)/(a+c) | VII | 5-24<br>37-80 | 64 | 25-36<br>81-106 | 38 | .627 |
| 812. | (b+d)/(a+c+e) | VII | 37-45a<br>59-67<br>71-80 | 27.4 | 45b-58<br>68-70 | 16.6 | .623 |
| 813. | (a+c)/(b+d) | VII | 64-67<br>71-80 | 14 | 59-63<br>68-70 | 8 | .636 |
| 814. | (a+c)/(b+d) | VII | 117b-134<br>148-169 | 39.2 | 107-117a<br>135-147 | 23.4 | .626 |
| 815. | (a+b)/(c+d) | VII | 212-248<br>249-285 | 73.6 | 170-191<br>192-211 | 42 | .637 |
| 816. | (b+d)/(a+c+e) | VII | 192-194<br>199-204<br>209-211 | 12 | 195-198<br>205-208 | 8 | .60 |
| 817-819. | | | | | | | |
| 817. | (c+d)/(a+b) | VII | 212-221<br>222-233 | 22 | 234-242<br>243-248 | 14.6 | .601 |

| No. | m/M: type | Book | Major | Total lines | Minor | Total lines | Ratio: M/(M+m) |
|---|---|---|---|---|---|---|---|
| 818. | b/(a+c) | | 212-221<br>234-242 | 19 | 222-233 | 12 | .613 |
| 819. | d/c | | 234-242 | 9 | 243-248 | 5.6 | .616 |
| 820. | (a+c)/(b+d) | VII | 354-372<br>385-405 | 40 | 341-353<br>373-384 | 25 | .615 |
| 821. | (b+d)/(a+c) | VII | 406-434<br>445-457 | 41.6 | 435-444<br>458-474 | 26.4 | .612 |
| 822. | (b+d)/(a+c+e) | VII | 475-482<br>487-492<br>503-510 | 22 | 483-486<br>493-502 | 14 | .611 |
| 823. | (a+d)/(b+c) | VII | 483-510<br>511-522 | 40 | 475-482<br>523-539 | 25 | .615 |
| 824. [cf. No. 592 in Table III] | (a+b)/(c+d) | VII | 655-669<br>670-677 | 23 | 641-646<br>647-654 | 14 | .622 |
| 825. | (a+c)/(b+d) | VII | 655-669<br>678-690 | 28 | 647-654<br>670-677 | 16 | .636 |
| 826. | (b+d)/(a+c+e) | VII | 706-722<br>744-760<br>783-817 | 68.6 | 723-743<br>761-782 | 43 | .615 |
| 827. | (b+d)/(a+c) | VIII | 1-17<br>26-30 | 22 | 18-25<br>31-35 | 13 | .629 |
| 828. | (a+c)/(b+d) | VIII | 86-93<br>97-101 | 13 | 81-85<br>94-96 | 8 | .619 |
| 829. | (a+b)/(c+d) | VIII | 219-267<br>268-279 | 61 | 184-189<br>190-218 | 35 | .635 |
| 830. | (a+c)/(b+d) | VIII | 219-267<br>280-305 | 75 | 184-218<br>268-279 | 47 | .615 |
| 831. | (b+d)/(a+c) | VIII | 213-218<br>222-232 | 17 | 219-221<br>233-240 | 11 | .607 |
| 832. | (b+d)/(a+c) | VIII | 256-267<br>273-275 | 15 | 268-272<br>276-279 | 9 | .625 |

833-834.

| 833. | (a+c+e)/(b+d) | VIII | 374-386<br>395-404a | 22.6 | 370-373<br>387-394<br>404b-406 | 14.4 | .611 |
|---|---|---|---|---|---|---|---|
| 834. | (a+d)/(b+c) | | 374-386<br>387-394 | 21 | 370-373<br>395-404a | 13.6 | .607 |

TABLE IV. SHORT PASSAGES: FOUR OR MORE PARTS, USUALLY INTERLOCKED

| No. | m/M: type | Book | Major | Total lines | Minor | Total lines | Ratio: M/(M+m) |
|---|---|---|---|---|---|---|---|
| 835. | (b+d)/(a+c) | VIII | 407-415<br>423-443a | 29.4 | 416-422<br>443b-453 | 17.6 | .626 |
| 836. | (a+c)/(b+d) | VIII | 475-493<br>508-519 | 31 | 469-474<br>494-507 | 19.2 | .618 |
| 837. | (a+c)/(b+d) | VIII | 630-641<br>652-662 | 23 | 626-629<br>642-651 | 14 | .622 |
| 838. | (b+d)/(a+c+e) | VIII | 630-634<br>642-645<br>652-662 | 20 | 635-641<br>646-651 | 13 | .606 |
| 839. | (b+d)/(a+c+e) | VIII | 635-641<br>646-651<br>663-674 | 25 | 642-645<br>652-662 | 15 | .625 |
| 840. | (b+d)/(a+c) | VIII | 714-719<br>724-728 | 11 | 720-723<br>729-731 | 7 | .611 |
| 841. | (b+d)/(a+c) | IX | 25-34<br>40-43 | 13 | 35-39<br>44-46 | 8 | .619 |
| 842. | (b+d)/(a+c) | IX | 47-53<br>57b-68 | 18.4 | 54-57a<br>69-76 | 11.6 | .613 |
| 843. | (a+c)/(b+d) | IX | 80-92<br>107-122 | 28 | 77-79<br>93-106 | 17 | .622 |
| 844. | (b+d)/(a+c) | IX | 107-109<br>114-120a | 9.2 | 110-113<br>120b-122 | 5.8 | .613 |
| 845. | (b+d)/(a+c+e) | IX | 159-163<br>168-171a<br>174-175 | 10.2 | 164-167<br>171b-173 | 6.2 | .622 |
| 846. | (b+e)/(a+c+d) | IX | 176-183<br>197-206<br>207-218 | 30 | 184-196<br>219-223 | 18 | .625 |
| 847. | (a+d)/(b+c) | IX | 197-223<br>224-280a | 83.4 | 176-196<br>280b-313 | 54 | .607 |
| 848. | (b+d)/(a+c) | IX | 184-196<br>207-218 | 25 | 197-206<br>219-223 | 15 | .625 |
| 849. | (a+c)/(b+d) | IX | 234-245<br>257-280a | 35.4 | 224-233<br>246-256 | 21 | .628 |
| 850. | (b+d)/(a+c+e) | IX | 280b-283a<br>287-289<br>291-292a | 7.4 | 283b-286<br>290 | 4.6 | .617 |
| 851. | (a+c)/(b+d) | IX | 320-323<br>329-341 | 17 | 314-319<br>324-328 | 11 | .607 |
| 852. | (a+c)/(b+d) | IX | 477-489<br>498-502 | 18 | 473-476<br>490-497 | 12 | .60 |

| No. | m/M: type | Book | Major | Total lines | Minor | Total lines | Ratio: M/(M+m) |
|---|---|---|---|---|---|---|---|
| 853. | (a+c)/(b+d) | IX | 512-524<br>542-568 | 39.4 | 503-511<br>525-541 | 25 | .612 |
| 854-855. | | | | | | | |
| 854. | (a+b)/(c+d) | IX | 602-613<br>614-620 | 19 | 590-597<br>598-601 | 12 | .613 |
| 855. | d/c | | 602-613 | 12 | 614-620 | 7 | .632 |
| 856. | (b+d)/(a+c) | IX | 621-624<br>630-636a | 10.4 | 625-629<br>636b-637 | 6.6 | .612 |
| 857. | (a+d)/(b+c) | IX | 683-690<br>691-709 | 27 | 672-682<br>710-716 | 18 | .60 |
| 858. | (b+d)/(a+c+e) | IX | 672-690<br>717-755<br>788-818 | 88.7 | 691-716<br>756-787 | 57.4 | .607 |
| 859. | (a+c)/(b+d) | IX | 722-730<br>741-755 | 24 | 717-721<br>731-740 | 14.7 | .620 |
| 860. | (a+c)/(b+d) | IX | 762-764<br>768-777 | 13 | 756-761<br>765-767 | 8.4 | .607 |
| 861. | (b+d)/(a+c) | IX | 756-761<br>768-777 | 15.4 | 762-767<br>778-790 | 9 | .631 |
| 862-863. | | | | | | | |
| 862. | (a+c)/(b+d) | IX | 791-796<br>806-818 | 19 | 788-790<br>797-805 | 12 | .613 |
| 863. | b/c | | 797-805 | 9 | 791-796 | 6 | .60 |
| 864. | (a+c)/(b+d) | X | 166-169<br>175-184 | 14 | 163-165<br>170-174 | 8 | .636 |
| 865. | (a+d)/(b+c)<br>[cf. No. 653 in Table III] | X | 166-184<br>185-197 | 32 | 163-165<br>198-214 | 20 | .615 |
| 866. | (a+c)/(b+d) | X | 276-286<br>308-361 | 63.6 | 256-275<br>287-307 | 41 | .608 |
| 867. | (a+c)/(b+d) | X | 290b-293<br>298b-307 | 13 | 287-290a<br>294-298a | 8 | .619 |
| 868. | (a+c+e)/(b+d) | X | 310b-314<br>317b-322a | 9.6 | 308-310a<br>315-317a<br>322b-323 | 6.4 | .60 |
| 869. | (a+d)/(b+c) | X | 345-361<br>362-425 | 81 | 308-344<br>426-438 | 50 | .618 |
| 870. | (b+d)/(a+c+e) | X | 362-378<br>399-438<br>464-478 | 72 | 379-398<br>439-463 | 45 | .615 |

TABLE IV. SHORT PASSAGES: FOUR OR MORE PARTS, USUALLY INTERLOCKED

| No. | m/M: type | Book | Major | Total lines | Minor | Total lines | Ratio: M/(M+m) |
|---|---|---|---|---|---|---|---|
| 871. | (b+d)/(a+c) | X | 380-387<br>390-393 | 12 | 388-389<br>394-398 | 7 | .632 |
| 872. | (a+c)/(b+d) | X | 405-416<br>421-425 | 17 | 399-404<br>417-420 | 10 | .630 |
| 873. | (b+d)/(a+c) | X | 439-440<br>444-452 | 11 | 441-443<br>453-456 | 7 | .611 |
| 874. | (a+d)/(b+c) | X | 491-509<br>510-530a | 39.2 | 479-490<br>530b-542 | 24.4 | .616 |
| 875. | (b+d)/(a+c+e) | X | 479-509<br>537-542<br>565-605 | 77.8 | 510-536<br>543-564 | 48 | .618 |
| 876. | (a+c)/(b+d) | X | 545b-560<br>570b-574 | 19.9 | 543-545a<br>561-570a | 12.1 | .622 |
| 877-878. | | | | | | | |
| 877. | (b+d)/(a+c) | X | 575-584a<br>591-600 | 18.6 | 584b-590<br>601-605 | 11.6 | .616 |
| 878. | b/c | | 591-600 | 10 | 584b-590 | 6.6 | .602 |
| 879. | (a+d)/(b+c) | X | 611-620<br>621-627 | 17 | 606-610<br>628-632 | 10 | .630 |
| 880. | (b+d)/(a+c) | X | 633-652<br>666-679 | 34 | 653-665<br>680-688 | 22 | .607 |
| 881. | (b+d)/(a+c+e) | X | 689-706<br>719-731<br>742-746 | 35.2 | 707-718<br>732-741 | 22 | .615 |
| 882-884. | | | | | | | |
| 882. | (b+d)/(a+c) | X | 762-768<br>773-782 | 17 | 769-772<br>783-788 | 10 | .630 |
| 883. | b/a | | 762-768 | 7 | 769-772 | 4 | .636 |
| 884. | d/c | | 773-782 | 10 | 783-788 | 6 | .625 |
| 885. | (a+c)/(b+d) | X | 794-795<br>803-816 | 16 | 791-793<br>796-802 | 10 | .615 |
| 886. | (a+c)/(b+d) | X | 846-856a<br>873-908 | 45.8 | 833-845<br>856b-871 | 28.8 | .614 |
| 887. | (b+d)/(a+c) | X | 846-856a<br>861-866 | 16.2 | 856b-860<br>867-871 | 9.8 | .623 |
| 888. | (b+d)/(a+c) | XI | 1-4<br>8b-11 | 7.8 | 5-8a<br>12-13 | 5.2 | .60 |

| No. | m/M: type | Book | Major | Total lines | Minor | Total lines | Ratio: M/(M+m) |
|---|---|---|---|---|---|---|---|
| 889. | (a+c)/(b+d) | XI | 42-48<br>54-58 | 12 | 39-41<br>49-53 | 8 | .60 |
| 890. | (b+d)/(a+c) | XI | 59-71<br>81-93 | 26 | 72-80<br>94-99 | 15 | .634 |
| 891. | (b+d)/(a+c) | XI | 72-84<br>88-92a | 17.2 | 85-87<br>92b-99 | 10.8 | .614 |
| 892. | (a+c)/(b+d) | XI | 135b-138<br>142-147 | 9.6 | 132-135a<br>139-141 | 6.4 | .60 |
| 893. | (b+d)/(a+c) | XI | 148-155<br>158b-159 | 9.6 | 156-158a<br>160-163 | 6.4 | .60 |
| 894-896. | | | | | | | |
| 894. | (b+c)/(a+d) | XI | 148-163<br>177-181 | 21 | 164-168<br>169-176 | 13 | .618 |
| 895. | b/c | | 169-176 | 8 | 164-168 | 5 | .615 |
| 896. | d/c | | 169-176 | 8 | 177-181 | 5 | .615 |
| 897. | (a+c)/(b+d) | XI | 188-190<br>193-202 | 13 | 182-187<br>191-192 | 8 | .619 |
| 898. | (b+d)/(a+c) | XI | 203-209<br>213-219 | 14 | 210-212<br>220-224 | 8 | .636 |
| 899. | (b+d)/(a+c) | XI | 241b-251<br>261-268 | 18.6 | 252-260<br>269-270 | 11 | .628 |
| 900-901. | | | | | | | |
| 900. | (a+c)/(b+d) | XI | 252-270<br>281-295 | 34 | 241b-251<br>271-280 | 20.6 | .623 |
| 901. | c/d | | 281-295 | 15 | 271-280 | 10 | .60 |
| 902. | (a+c)/(b+d) | XI | 302-313<br>324-335 | 24 | 296-301<br>314-323 | 16 | .60 |
| 903-904. | | | | | | | |
| 903. | (a+d)/(b+c) | XI | 302-335<br>336-351 | 50 | 296-301<br>352-375 | 29.2 | .631 |
| 904. | (a+c)/b | | 302-335 | 34 | 296-301<br>336-351 | 22 | .607 |
| 905. | (a+d)/(b+c) | XI | 336-375<br>376-444 | 107.4 | 296-335<br>445-467 | 63 | .630 |

TABLE IV. SHORT PASSAGES: FOUR OR MORE PARTS, USUALLY INTERLOCKED

| No. | m/M: type | Book | Major | Total lines | Minor | Total lines | Ratio: M/(M+m) |
|---|---|---|---|---|---|---|---|
| 906-907. | | | | | | | |
| 906. | (a+c)/(b+d) | XI | 305-313<br>324-335 | 21 | 302-304<br>314-323 | 13 | .618 |
| 907. | d/(b+c) | | 305-313<br>314-323 | 19 | 324-335 | 12 | .613 |
| 908. | (a+c)/(b+d) | XI | 357-363<br>368-375 | 14.2 | 352-356<br>364-367 | 9 | .612 |
| 909. | (a+d)/(b+c) | XI | 392-409<br>410-433 | 42 | 376-391<br>434-444 | 26.2 | .616 |
| 910. | (a+c)/(b+d) | XI | 532-594<br>618-647 | 93 | 498-531<br>595-617 | 57 | .620 |
| 911. | (a+c)/(b+d) | XI | 690-724<br>768-867 | 135 | 648-689<br>725-767 | 85 | .614 |
| 912.<br>[cf. No. 684 in Table III] | (b+d)/(a+c) | XI | 648-793<br>816-835 | 166 | 794-815<br>836-915 | 102 | .619 |
| 913. | (a+c)/(b+d) | XI | 794-835<br>868-915 | 90 | 768-793<br>836-867 | 58 | .608 |
| 914. | (a+c)/(b+d) | XI | 799-801a<br>805-815 | 13.6 | 794-798<br>801b-804 | 8.4 | .618 |
| 915. | (b+d)/(a+c) | XI | 794-808<br>816-827a | 26.2 | 809-815<br>827b-835 | 15.8 | .624 |
| 916. | (a+c)/(b+d) | XI | 901-902<br>906-915 | 12 | 896-900<br>903-905 | 8 | .60 |
| 917. | (a+c+e)/(b+d) | XII | 10-53<br>64-69 | 50 | 1-9<br>54-63<br>70-80 | 30 | .625 |
| 918. | (b+d)/(a+c+e) | XII | 113-133<br>161-215<br>257-288 | 108 | 134-160<br>216-256 | 67.6 | .615 |
| 919. | (a+c)/(b+d) | XII | 142-153<br>156-160 | 17 | 134-141<br>154-155 | 10 | .630 |
| 920-921. | | | | | | | |
| 920. | (b+d)/(a+c) | XII | 175-182<br>195-211 | 25 | 183-194<br>212-215 | 16 | .610 |
| 921. | b/(c+d) | | 195-211<br>212-215 | 21 | 183-194 | 12 | .636 |

| No. | m/M: type | Book | Major | Total lines | Minor | Total lines | Ratio: M/(M+m) |
|---|---|---|---|---|---|---|---|
| 922. | (b+d)/(a+c+e) | XII | 216-218 222-228 234-237 | 13.6 | 219-221 229-233 | 8 | .630 |
| 923. | (b+d)/(a+c) | XII | 324-327 331-340 | 14 | 328-330 341-345 | 8 | .636 |
| 924-925. | | | | | | | |
| 924. | (a+c)/(b+d) | XII | 387-397 405-410 | 17 | 383-386 398-404 | 11 | .607 |
| 925. | c/b | | 387-397 | 11 | 398-404 | 7 | .611 |
| 926. | (b+c)/(a+d) | XII | 383-397 420-440 | 36 | 398-410 411-419 | 22 | .621 |
| 927-928. | | | | | | | |
| 927. | (c+d)/(a+b) | XII | 411-419 420-429 | 19 | 430-434 435-440 | 11 | .633 |
| 928. | a/(b+c) | | 420-429 430-434 | 15 | 411-419 | 9 | .625 |
| 929. | (a+c)/(b+d) | XII | 468-553 593-611 | 105 | 441-467 554-592 | 66 | .614 |
| 930. | (a+c)/(b+d) | XII | 473-480 488-499 | 20 | 468-472 481-487 | 12 | .625 |
| 931. | (b+d)/(a+c) | XII | 500-528 554-592 | 68 | 529-553 593-611 | 44 | .607 |
| 932. | (a+c+e)/(b+d+f) | XII | 509-512 516-520 535-547 | 22 | 505-508 513-515 529-534 | 13 | .629 |
| 933. | (b+d)/(a+c) | XII | 554-564 574-586 | 24 | 565-573 587-592 | 15 | .615 |
| 934. | (a+c)/(b+d) | XII | 595-603 609-611 | 12 | 593-594 604-608 | 7 | .632 |
| 935. | (b+d)/(a+c) | XII | 614-621 623-625a | 10.6 | 622 625b-630 | 6.4 | .624 |
| 936. | (b+d)/(a+c+e) | XII | 614-630 650-671 684-696 | 52 | 631-649 672-683 | 30.2 | .633 |
| 937. | (b+d)/(a+c+e) | XII | 650-668 672-680 696 | 29 | 669-671 681-695 | 18 | .617 |

TABLE IV. SHORT PASSAGES: FOUR OR MORE PARTS, USUALLY INTERLOCKED

| No. | m/M: type | Book | Major | Total lines | Minor | Total lines | Ratio: M/(M+m) |
|---|---|---|---|---|---|---|---|
| 938. | (b+d)/(a+c) | XII | 672-683<br>693-695 | 15 | 684-692<br>696 | 10 | .60 |
| 939. | (b+d)/(a+c+e) | XII | 697-724<br>746-765<br>781-790 | 58 | 725-745<br>766-780 | 36 | .617 |
| 940. | (b+c)/(a+d) | XII | 725-741<br>758-765 | 25 | 742-745<br>746-757 | 16 | .610 |
| 941. | (b+d)/(a+c) | XII | 746-757<br>766-780 | 27 | 758-765<br>781-790 | 18 | .60 |

## TABLE V. PROPORTIONS IN THE MAIN DIVISIONS

| No. | m/M: type | Book | Major | Total lines | Minor | Total lines | Ratio: M/(M+m) |
|---|---|---|---|---|---|---|---|
| 942. | a/(b+c) | I | 81-156<br>157-222 | 142 | 1a-80 | 84 | .628 |
| 943. | a/b | I | 297-417 | 121 | 223-296 | 74 | .621 |
| 944. | (a+c)/b | I | 520-722 | 201.1 | 418-519<br>723-747 | 127 | .613 |
| 945. | (a+c)/b | II | 40-194 | 153.2 | 1-39<br>195-249 | 93.4 | .621 |
| 946. | a/(b+c) | II | 370-505<br>506-558 | 188.7 | 250-369 | 119.2 | .613 |
| 947. | b/(a+c) | II | 559-633<br>730-804 | 147.7 | 634-729 | 94.6 | .610 |
| 948. | c/(a+b) | III | 1-68<br>69-120 | 120 | 121-191 | 71 | .628 |
| 949. | b/(a+c) | III | 192-355<br>472-505 | 195.2 | 356-471 | 115.6 | .628 |
| 950. | (a+b)/c | III | 588-718 | 129.6 | 506-547<br>548-587 | 81.4 | .614 |
| 951. | (a+c)/b | IV | 54-159 | 106 | 1-53<br>160-172 | 65.4 | .618 |
| 952. | a/(b+c) | IV | 279-415<br>416-449 | 170.2 | 173-278 | 105 | .618 |
| 953. | a/(b+c) | IV | 553-583<br>584-705 | 153 | 450-552 | 101 | .602 |
| 954. | b/(a+c) | V | 1-285<br>485-544 | 345 | 286-484 | 198 | .635 |
| 955. | (b+d)/(a+c) | V | 545-562<br>577-595 | 36.6 | 563-576<br>596-603 | 21.2 | .633 |
| 956. | (a+c)/b | V | 664-826 | 162.3 | 604-663<br>827-871 | 104.2 | .609 |
| 957. | b/(a+c) | VI | 1-123<br>212-235 | 146.6 | 124-211 | 88 | .625 |

# TABLE V. PROPORTIONS IN THE MAIN DIVISIONS

| No. | m/M: type | Book | Major | Total lines | Minor | Total lines | Ratio: M/(M+m) |
|---|---|---|---|---|---|---|---|
| 958. | b/(a+c) | VI | 236-416<br>535-547 | 191.9 | 417-534 | 118 | .619 |
| 959. | (a+c)/b | VI | 637-853 | 216.7 | 548-636<br>854-901 | 137 | .613 |
| 960. | a/(b+c) | VII | 107-169<br>170-285 | 178.2 | 1-106 | 106 | .627 |
| 961. | (a+c)/b | VII | 323-539 | 216 | 286-322<br>540-640 | 138 | .610 |
| 962. | b/(a+c) | VII | 641-690<br>761-817 | 107 | 691-760 | 68.8 | .609 |
| or | | | | | | | |
| 963. | (a+c+e)/(b+d+f) | VII | 655-690<br>706-743<br>783-817 | 109 | 641-654<br>691-705<br>744-782 | 66.8 | .620 |
| 964. | (b+d)/(a+c) | VIII | 1-101<br>184-305 | 221.4 | 102-183<br>306-369 | 146 | .603 |
| 965. | (a+c)/(b+d) | VIII | 387-406<br>423-453 | 51 | 370-386<br>407-422 | 33 | .607 |
| 966. | c/(a+b) | VIII | 454-596<br>597-625 | 170.6 | 626-731 | 106 | .617 |
| 967. | (c+d)/(a+b) | IX | 1-76<br>77-106 | 105 | 107-122<br>123-175 | 66.4 | .613 |
| 968-969. | | | | | | | |
| 968. | c/a | IX | 176-313 | 137.4 | 367-449 | 83 | .623 |
| 969. | b/c | | 367-449 | 83 | 314-366 | 53 | .610 |
| 970. | a/(b+c) | IX | 590-671<br>672-818 | 228.1 | 450-589 | 137.8 | .623 |
| 971. | b/(a+c) | X | 1-117<br>256-361 | 220.8 | 118-255 | 138 | .615 |
| 972. | b/(a+c) | X | 362-478<br>606-688 | 200 | 479-605 | 125.8 | .614 |
| 973. | b/(a+c) | X | 689-746<br>833-908 | 131.8 | 747-832 | 86 | .605 |
| 974. | c/(a+b) | XI | 1-99<br>100-138 | 138 | 139-224 | 86 | .616 |

| No. | m/M: type | Book | Major | Total lines | Minor | Total lines | Ratio: M/(M+m) |
|---|---|---|---|---|---|---|---|
| 975. | c/(a+b) | XI | 225-295<br>296-375 | 150.2 | 376-467 | 91.2 | .622 |
| 976. | b/(a+c) | XI | 468-497<br>648-867 | 250 | 498-647 | 150 | .625 |
| or | | | | | | | |
| 977. | (a+c+e)/(b+d) | XI | 498-647<br>690-815 | 276 | 468-497<br>648-689<br>816-915 | 172 | .616 |
| 978. | (a+b)/c | XII | 113-288 | 175.6 | 1-80<br>81-112 | 112 | .611 |
| 979. | a/(b+c) | XII | 441-611<br>614-696 | 253.2 | 289-440 | 152 | .625 |
| 980. | a/(b+c) | XII | 791-842<br>843-952 | 162 | 697-790 | 94 | .633 |

## TABLE VI. THE MAIN DIVISIONS IN PROPORTION

| No. | m/M: type | Book | Major | Total lines | Minor | Total lines | Ratio: M/(M+m) |
|---|---|---|---|---|---|---|---|
| 981. | b/c | I | 418-747 | 328.1 | 223-417 | 195 | .627 |
| 982. | b/(a+c) | II | 1-249<br>559-804 | 488.8 | 250-558 | 307.9 | .614 |
| 983. | a/b | III | 192-505 | 310.8 | 1-191 | 191 | .619 |
| 984-985. | | | | | | | |
| 984. | b/(a+c) | IV | 1-172<br>450-705 | 425.4 | 173-449 | 275.2 | .607 |
| 985. | a/b | | 173-449 | 275.2 | 1-172 | 171.4 | .616 |
| 986. | (b+c)/a | V | 1-544 | 543 | 545-603<br>604-871 | 324.3 | .626 |
| 987. | c2/(b+c1) | V | 545-603<br>604-745 | 199 | 746-871 | 125.3 | .614 |
| 988-989. | | | | | | | |
| 988. | c/(a+b) | VI | 1-235<br>236-547 | 545.6 | 548-901 | 353.7 | .607 |
| 989. | a/c | | 548-901 | 353.7 | 1-235 | 234.6 | .601 |
| 990. | c/a | VII | 1-285 | 284.2 | 641-817 | 175.8 | .618 |
| 991. | c/(a+b) | VIII | 1-369<br>370-453 | 451.4 | 454-731 | 276.6 | .620 |
| 992. | a/b | IX | 176-449 | 273.4 | 1-175 | 171.4 | .615 |
| 993-994. | | | | | | | |
| 993. | a/(b+c) | X | 362-688<br>689-908 | 543.6 | 1-361 | 358.8 | .602 |
| 994. | c/a | | 1-361 | 358.8 | 689-908 | 217.8 | .622 |
| 995. | b/c | XI | 468-867 | 400 | 225-467 | 241.4 | .624 |
| 996. | c/b | XII | 289-696 | 405.2 | 697-952 | 256 | .613 |

## TABLE VII. SUPPLEMENTARY LIST OF RATIOS (FOR TABLES I-IV)

| No. | m/M: type | Book | Major | Total lines | Minor | Total lines | Ratio: M/(M+m) |
|---|---|---|---|---|---|---|---|
| S997. | (b+d)/(a+c+e) | I | 81-101<br>113-123<br>142-156 | 47 | 102-112<br>124-141 | 29 | .618 |
| S998. | (a+c)/(b+d) | I | 162-169<br>174-179 | 14 | 157-161<br>170-173 | 9 | .609 |
| S999. | (a+d)/(b+c) | I | 187-197<br>198-203 | 17 | 180-186<br>204-207 | 11 | .607 |
| S1000. | b/(a+c) | I | 305-313<br>325-334 | 19 | 314-324 | 11 | .633 |
| S1001. | (b+c)/(a+d) | I | 520-560<br>613-656 | 83.1 | 561-585<br>586-612 | 52 | .615 |
| S1002. | (a+c)/(b+d) | I | 673-688<br>691-694 | 20 | 663-672<br>689-690 | 12 | .625 |
| S1003. | b/(a+c) | II | 707-711<br>721-729 | 14 | 712-720 | 8.2 | .631 |
| S1004. | (b+d)/(a+c) | II | 721-729<br>735-740 | 15 | 730-734<br>741-744 | 9 | .625 |
| S1005. | (a+c)/(b+d) | III | 135-139<br>143-146 | 9 | 132-134<br>140-142 | 6 | .60 |
| S1006. | (a+c)/b | III | 209-258 | 48.2 | 192-208<br>259-273 | 32 | .601 |
| S1007. | (a+c)/b | III | 278-289 | 12 | 274-277<br>290-293 | 8 | .60 |
| S1008. | b/(a+c) | III | 278-283<br>290-293 | 10 | 284-289 | 6 | .625 |
| S1009. | b/a | IV | 1-5 | 5 | 6-8 | 3 | .625 |
| S1010. | (b+d)/(a+c+e) | IV | 1-8<br>30-53<br>68-89 | 53.4 | 9-29<br>54-67 | 35 | .604 |
| S1011-1012. | | | | | | | |
| S1011. | (b+c)/(a+d) | IV | 9-11<br>20-29 | 13 | 12-14<br>15-19 | 8 | .619 |
| S1012. | b/c | | 15-19 | 5 | 12-14 | 3 | .625 |

TABLE VII. SUPPLEMENTARY LIST OF RATIOS (FOR TABLES I-IV)

| No. | m/M: type | Book | Major | Total lines | Minor | Total lines | Ratio: M/(M+m) |
|---|---|---|---|---|---|---|---|
| S1013. | a/b | IV | 265-276a | 10.2 | 259-264 | 6 | .630 |
| S1014. | b/(a+c) | IV | 584-590a<br>597-599 | 9.6 | 590b-596 | 6.4 | .60 |
| S1015. | b/(a+c) | V | 12-16<br>26-34 | 14 | 17-25 | 9 | .609 |
| S1016. | b/(a+c) | V | 35-41<br>72-113 | 49 | 42-71 | 30 | .620 |
| S1017. | b/a | V | 116-120 | 5 | 121-123 | 3 | .625 |
| S1018. | (b+d)/(a+c) | V | 664-686<br>719-766 | 71 | 687-718<br>767-778 | 44 | .617 |
| S1019. | a/b | VI | 337-346 | 10 | 331-336 | 6 | .625 |
| S1020. | (b+d)/(a+c) | VI | 337-341a<br>347-371 | 29.4 | 341b-346<br>372-383 | 17.6 | .626 |
| S1021. | a/b | VI | 489-493 | 5 | 486-488 | 3 | .625 |
| S1022. | (b+d)/(a+c+e) | VI | 548-551<br>554b-556<br>560-561 | 8.6 | 552-554a<br>557-559 | 5.4 | .614 |
| S1023. | b/a | VI | 637-678 | 42 | 679-702 | 24 | .636 |
| S1024. | a/b | VI | 724-755 | 32 | 703-723 | 21 | .604 |
| S1025. | b/a | VI | 836-846 | 11 | 847-853 | 7 | .611 |
| S1026. | (a+c)/(b+d) | VII | 290-291<br>293-298 | 8 | 286-289<br>292 | 5 | .615 |
| S1027. | (a+c)/(b+d) | VII | 301-304a<br>308-312 | 8.6 | 299-300<br>304b-307 | 5.4 | .614 |
| S1028. | a/b | VII | 317-322 | 6 | 313-316 | 4 | .60 |
| S1029. | b/(a+c) | VII | 670-677<br>691-705 | 22.2 | 678-690 | 13 | .631 |
| S1030. | b/(a+c) | VII | 691-705<br>733-760 | 41.8 | 706-732 | 27 | .608 |
| S1031. | b/(a+c) | VIII | 306-307a<br>310-312 | 4.4 | 307b-309 | 2.6 | .629 |
| S1032. | (a+c)/(b+d) | VIII | 374-380<br>387-394 | 15 | 370-373<br>381-386 | 10 | .60 |
| S1033. | (a+c)/b | VIII | 426-445a | 19.2 | 423-425<br>445b-453 | 11.8 | .619 |
| S1034. | (b+d)/(a+c) | VIII | 454-464<br>469-474 | 16.2 | 465-468<br>475-480 | 10 | .618 |

| No. | m/M: type | Book | Major | Total lines | Minor | Total lines | Ratio: M/(M+m) |
|---|---|---|---|---|---|---|---|
| S1035. | b/(a+c) | IX | 450-458<br>468-472 | 14 | 459-467 | 8.4 | .625 |
| S1036. | (a+c+e)/(b+d+f) | X | 6-15<br>18-30<br>48-62a | 37.6 | 1-5<br>16-17<br>31-47 | 23.2 | .618 |
| S1037. | b/a | X | 62b-95 | 33.4 | 96-117 | 22 | .603 |
| S1038. | a/b | X | 597-605 | 9 | 591-596 | 6 | .60 |
| S1039. | (a+c)/b | XI | 477-491 | 15 | 473-476<br>492-497 | 10 | .60 |
| S1040. | b/a | XII | 54-63 | 10 | 64-69 | 6 | .625 |
| S1041. | (a+d)/(b+c) | XII | 166-174<br>175-186 | 21 | 161-165<br>187-194 | 13 | .618 |
| S1042. | (b+d)/(a+c+e) | XII | 289-310<br>383-410<br>441-553 | 163 | 311-382<br>411-440 | 102 | .615 |
| S1043. | c/(a+b) | XII | 353-361<br>362-370 | 18 | 371-382 | 12 | .60 |
| S1044. | (a+c)/(b+d) | XII | 565-592<br>604-611 | 36 | 554-564<br>593-603 | 22 | .621 |

## TABLE VIII. THE *AENEID* AS A WHOLE

| No. | m/M: type | Major | Total lines | Minor | Total lines | Ratio: M/(M+m) |
|---|---|---|---|---|---|---|
| 1045. | c/(a+b) | I-II<br>III-IV | 2968.3 | V-VI | 1766.6 | .627 |
| 1046. | c/(a+b) | VII-VIII<br>IX-X | 3255.1 | XI-XII | 1862.2 | .636 |
| 1047. | (b+d)/(a+c) | I-IV<br>VII-X | 6223.4 | V-VI<br>XI-XII | 3628.8 | .632 |
| 1048. | c/(a+b) | I-IV<br>V-VIII | 6276.9 | IX-XII | 3575.3 | .637 |

**TABLE IX**
**CHART-INDEX OF THE** *AENEID*

## Aeneid I

I. 1a-222 Prologue. Juno and the storm
   (1) 1a-80    Prologue. Juno and Aeolus
   (2) 81-156    The storm and Neptune
   (3) 157-222    Aeneas and the Trojans in Africa

II. 223-417 The Venus episodes
   (1) 223-296    Venus and Jupiter
   (2) 297-417    Venus and Aeneas

III. 418-756 The Trojans at Carthage
   (1) 418-519    Carthage and Dido
   (2) 520-656    The Trojans welcomed
   (3) 657-747    Cupid and the banquet
        748-756    Epilogue to introduce *Aeneid* II-III

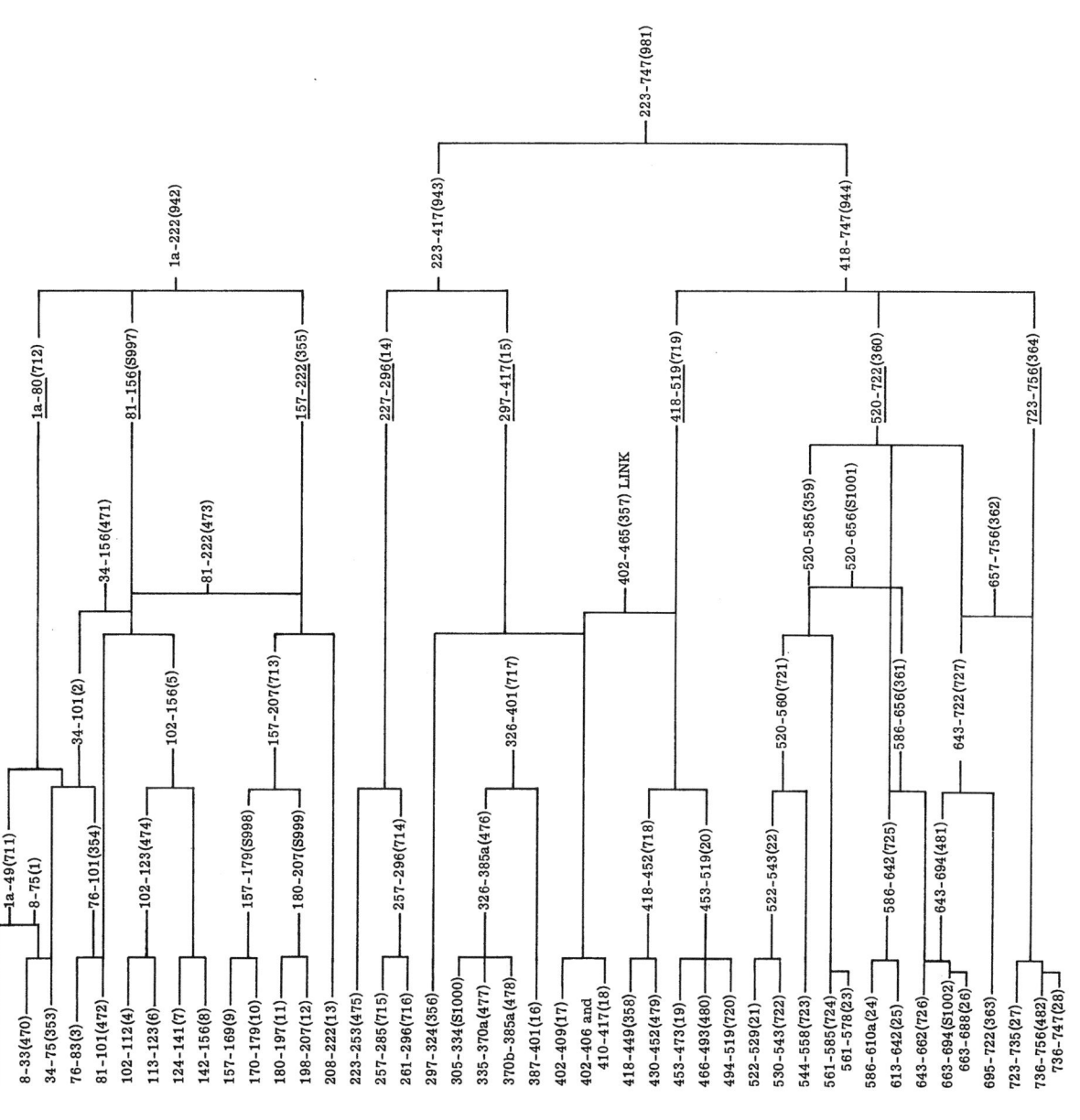

## Aeneid II

I. 1-249 Sinon, Laocoon, and the wooden horse
   (1) 1-39 Wooden horse on shore
   (2) 40-56 Laocoon
   (3) 57-194 Sinon
   (2) 195-227 Laocoon
   (1) 228-249 Wooden horse in city

II. 250-558 Fall of Troy
   (1) 250-369 Return of Greeks
   (2) 370-505 Capture of Troy
   (3) 506-558 Death of Priam

III. 559-804 Aeneas' departure
   (1) 559-633 Aeneas-Venus episode
   (2) 634-729 Aeneas at home
   (3) 730-804 Departure and loss of Creusa

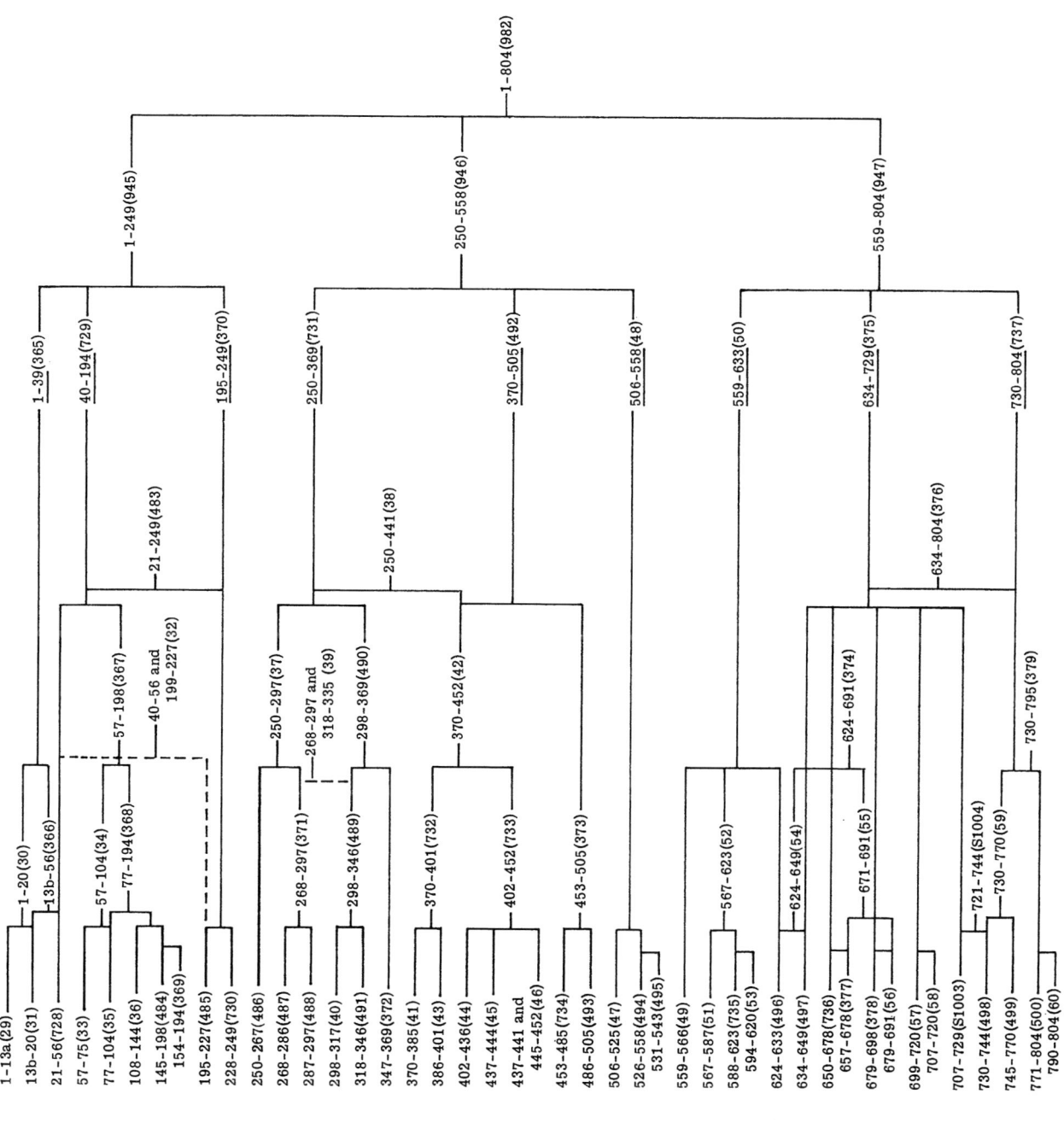

## Aeneid III

I.  1-191   Aegean area
    (1) 1-68      Departure from Troy. Thrace
    (2) 69-120    Delos
    (3) 121-191   Crete

II. 192-505  Western Greece
    (1) 192-273   Strophades
    (2) 274-293   Actium
    (3) 294-505   Buthrotum (Helenus and Andromache)

III. 506-718 Magna Graecia and Sicily
    (1) 506-547   Journey to Italy
    (2) 548-587   Scylla and Charybdis
    (3) 588-718   Rescue of Achaemenides. Conclusion

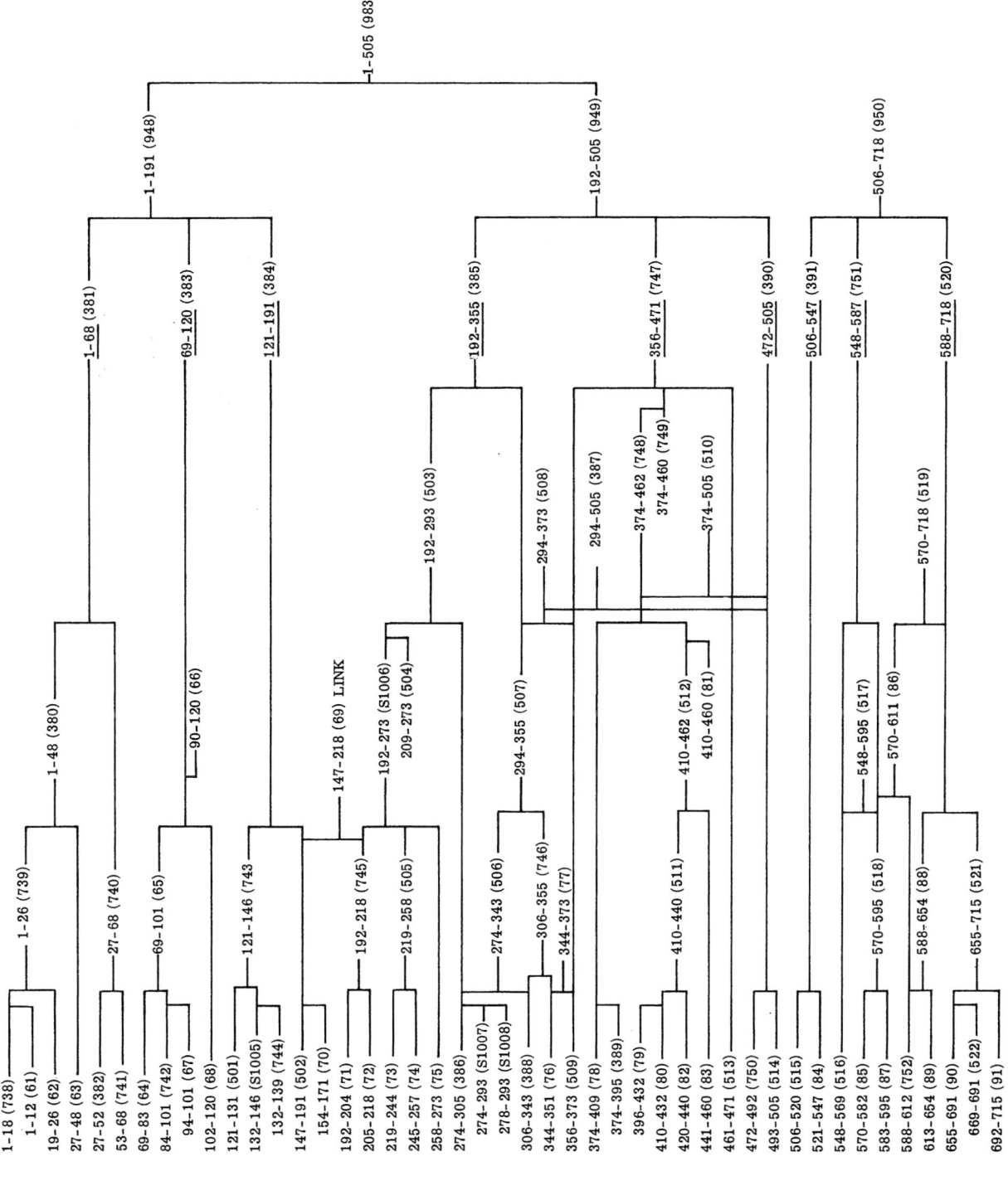

## Aeneid IV

I. 1-172   Dido's love and its consummation
   (1) 1-89       Growth of Dido's love
   (2) 90-128    Juno-Venus scene
   (3) 129-172   Hunting scene and "coniugium"

II. 173-449   Aeneas' determination to leave
   (1) 173-278   Fama—Iarbas—Jupiter—Mercury Narrative
   (2) 279-415   Speeches (Dido, Aeneas, Dido) Narrative
   (3) 416-449   Attempted reconciliation fails

III. 450-705   Aeneas' departure and Dido's suicide
   (1) 450-552   Magic rites and Dido's lament
   (2) 553-583   Aeneas' departure
   (3) 584-705   Dido's curses and suicide

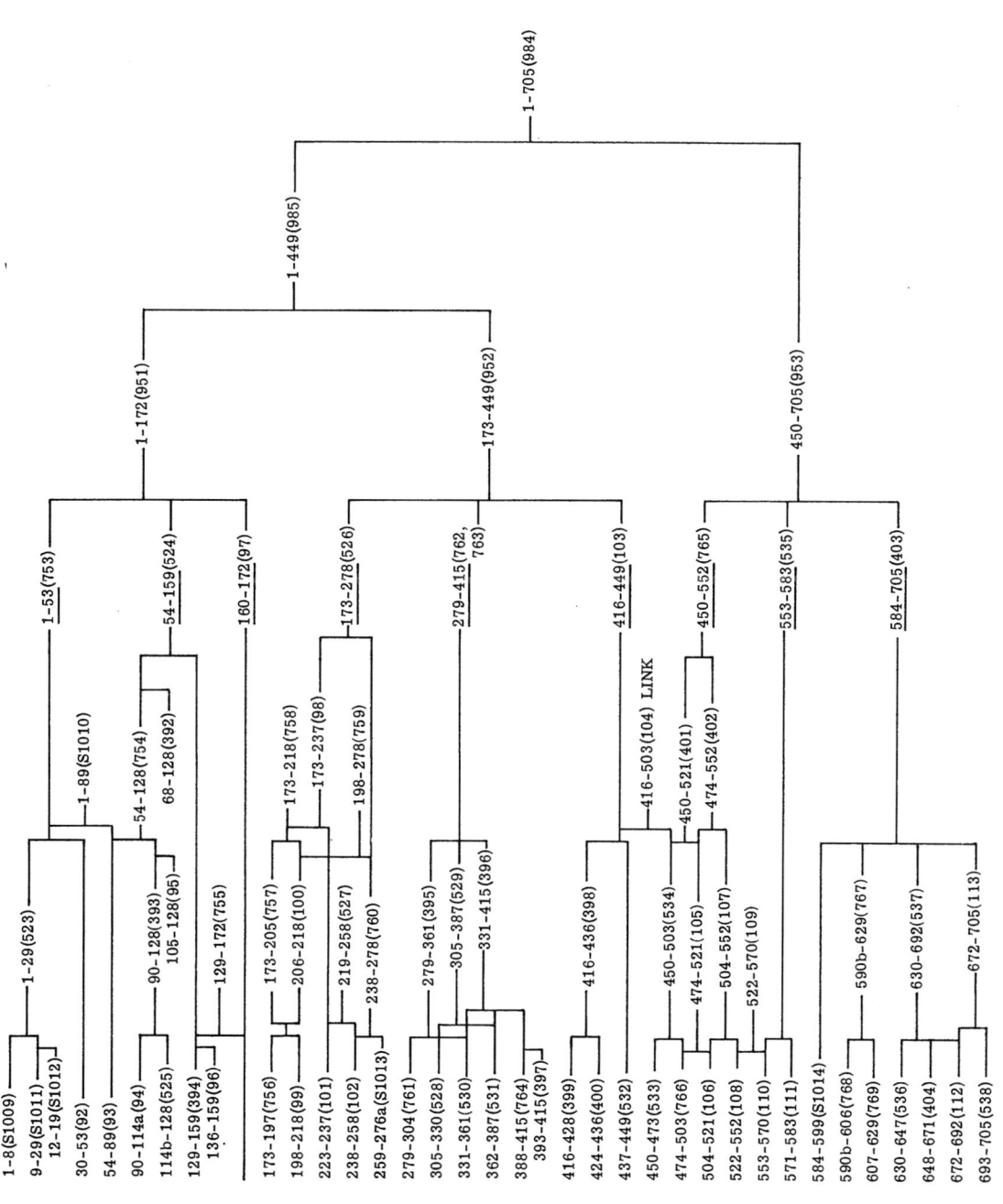

## Aeneid V

I. 1-544 Contests
    1-34    Arrival. Palinurus
    35-544  Contests
        35-113    Preparations
        (1)  114-285    Boat race
        (2)  286-361    Foot race
        (3)  362-484    Boxing match
        (4)  485-544    Archery contest

II. 545-603  *Ludus Troiae*

III. 604-871  Burning of the ships. Departure
     604-663  Iris and the Trojan women
     664-778  Burning of the ships
     779-826  Venus and Neptune
     827-871  Death of Palinurus

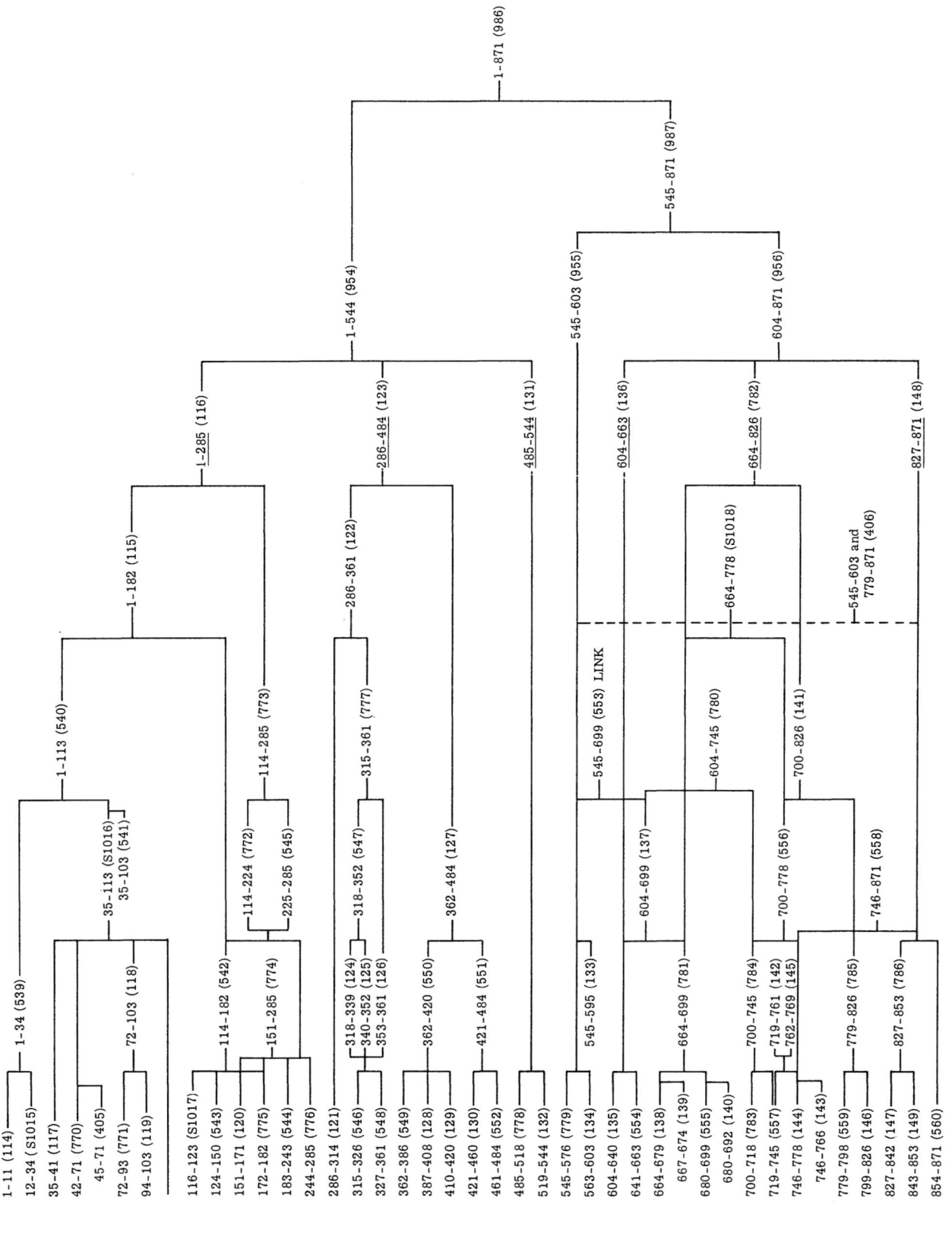

## Aeneid VI

I. 1-235 At Cumae
  (1) 1-123 Cumae and the Sibyl
  (2) and (3) 124-155 Instructions concerning Misenus and the Golden Bough
    (2) 156-182 Death of Misenus
    (3) 183-211 The Golden Bough
    (2) 212-235 Burial of Misenus

II. 236-547 Journey to Underworld. Three encounters
    236-267 Preliminary sacrifices (cf. 153-155)
    268-336 First stage of journey
  (1) 337-383 Palinurus
    384-449 Charon, Cerberus, untimely deaths
  (2) 450-476 Dido
    477-493 Warriors
  (3) 494-534 Deiphobus
    535-547 At the entrance to Elysium

III. 548-901 The Underworld proper
  (1) 548-636 Tartarus
  (2) 637-702 Elysium
  (3) 703-755 Lethe
  Anchises' forecast of Roman history
  (1) 756-807 Julian line
    (a) 756-776 Alban kings
    (b) 777-787 Romulus
    (c) 788-807 Augustus
  (2) 808-853 Roman line
    (a) 808-818 Roman kings
    (b) 819-846 Roman heroes
    (c) 847-853 Task of the Roman
  (3) 854-901 Marcellus. Conclusion

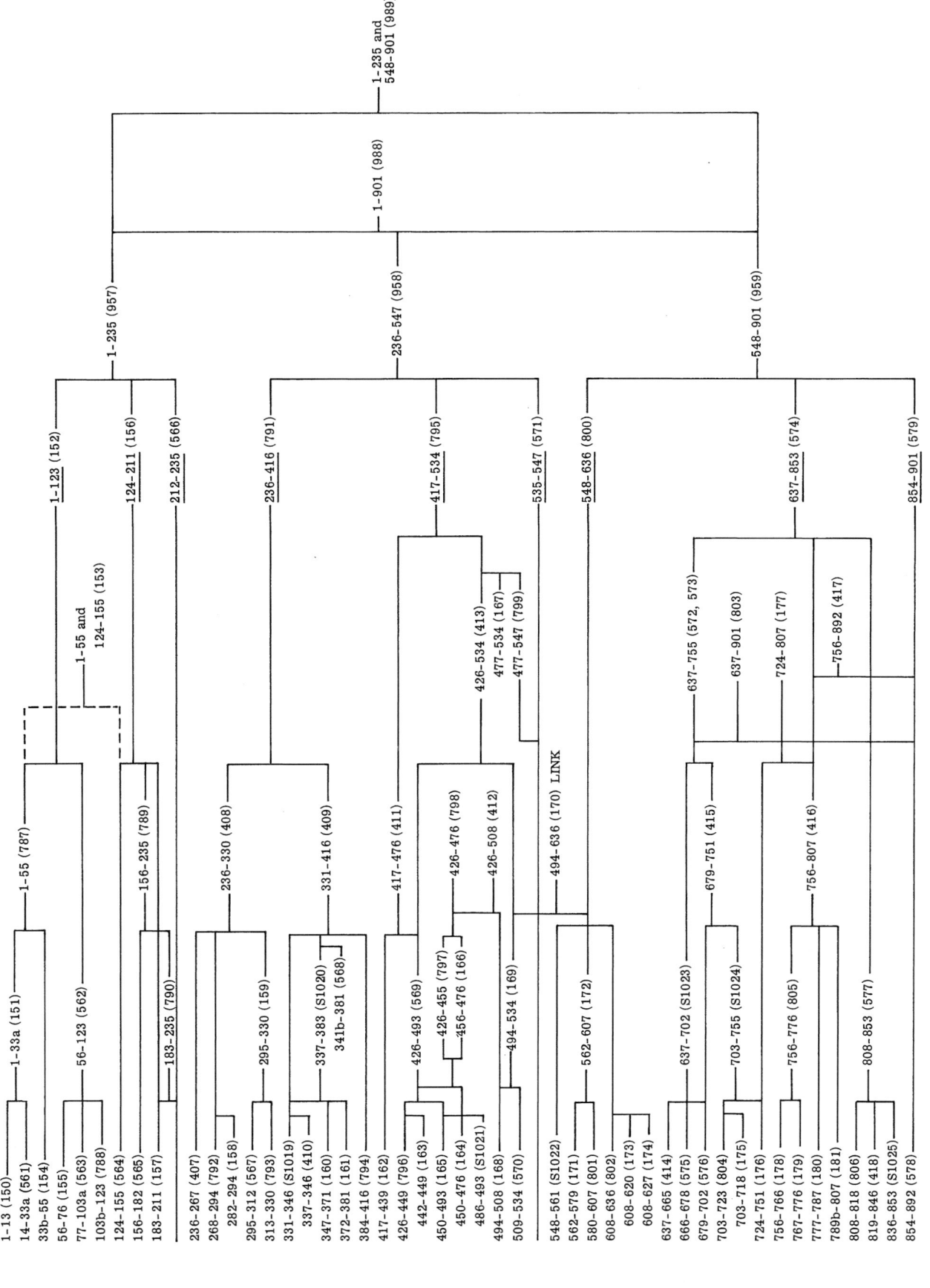

## Aeneid VII

I. 1-285 Arrival in Latium and welcome by Latinus
   (1) 1-106    Arrival. Latinus and omens
   (2) 107-169   Eating of tables. Embassy
   (3) 170-285   Reception of embassy. Three speeches
      (a) 192-211  Latinus
      (b) 212-248  Ilioneus
      (c) 249-285  Latinus

II. 286-640 Juno, Allecto, and outbreak of war
   (1) 286-322   Juno's lament
   (2) 323-539   The Allecto episode
      323-340  Juno summons Allecto
      341-539  Threefold activity of Allecto
         (a) 341-405  Maddens Amata
         (b) 406-474  Maddens Turnus
         (c) 475-539  Maddens hounds of Ascanius
   (3) 540-640   Preparations for war

III. 641-817 Catalogue of Latin warriors
Three groups of three less important warriors each, enclosed by important leaders:
   (a) 647-654  Mezentius and Lausus
   (b) 691-705  Messapus
   (c) 783-817  Turnus and Camilla

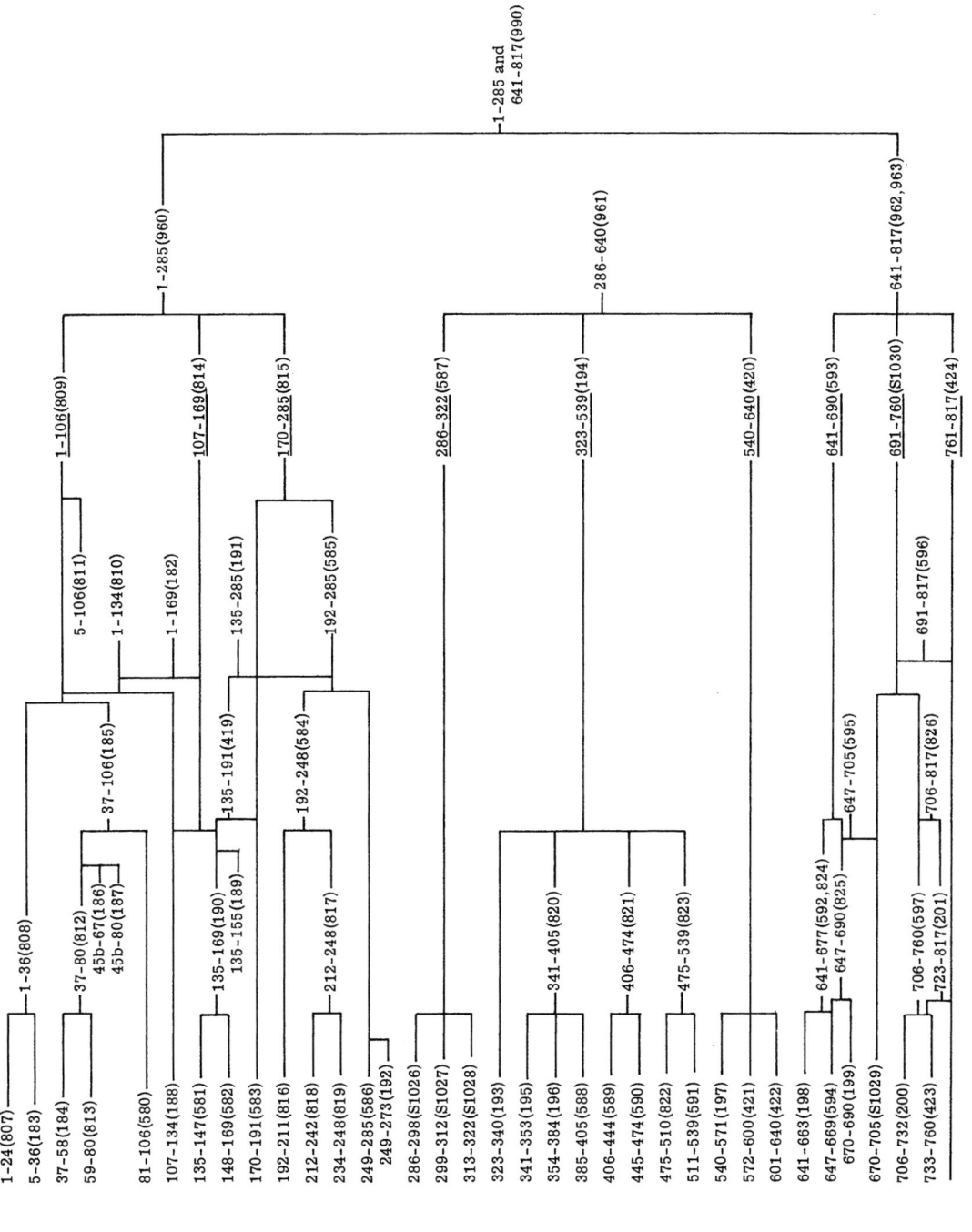

## Aeneid VIII

I. 1-369 Aeneas and Evander
    1-101 Aeneas at his camp
    102-369 DAY Aeneas at Rome
    (1) 102-183 Welcome by Evander
    (2) 184-305 Festival of Hercules
    (3) 306-369 The site of Rome

II. 370-453 NIGHT Venus and Vulcan. Making of armor

III. 454-731 DAY Journey and the shield
    (1) 454-596 Departure from Pallanteum
    (2) 597-625 Venus brings the armor
    (3) 626-728 Description of the shield
        729-731 Conclusion

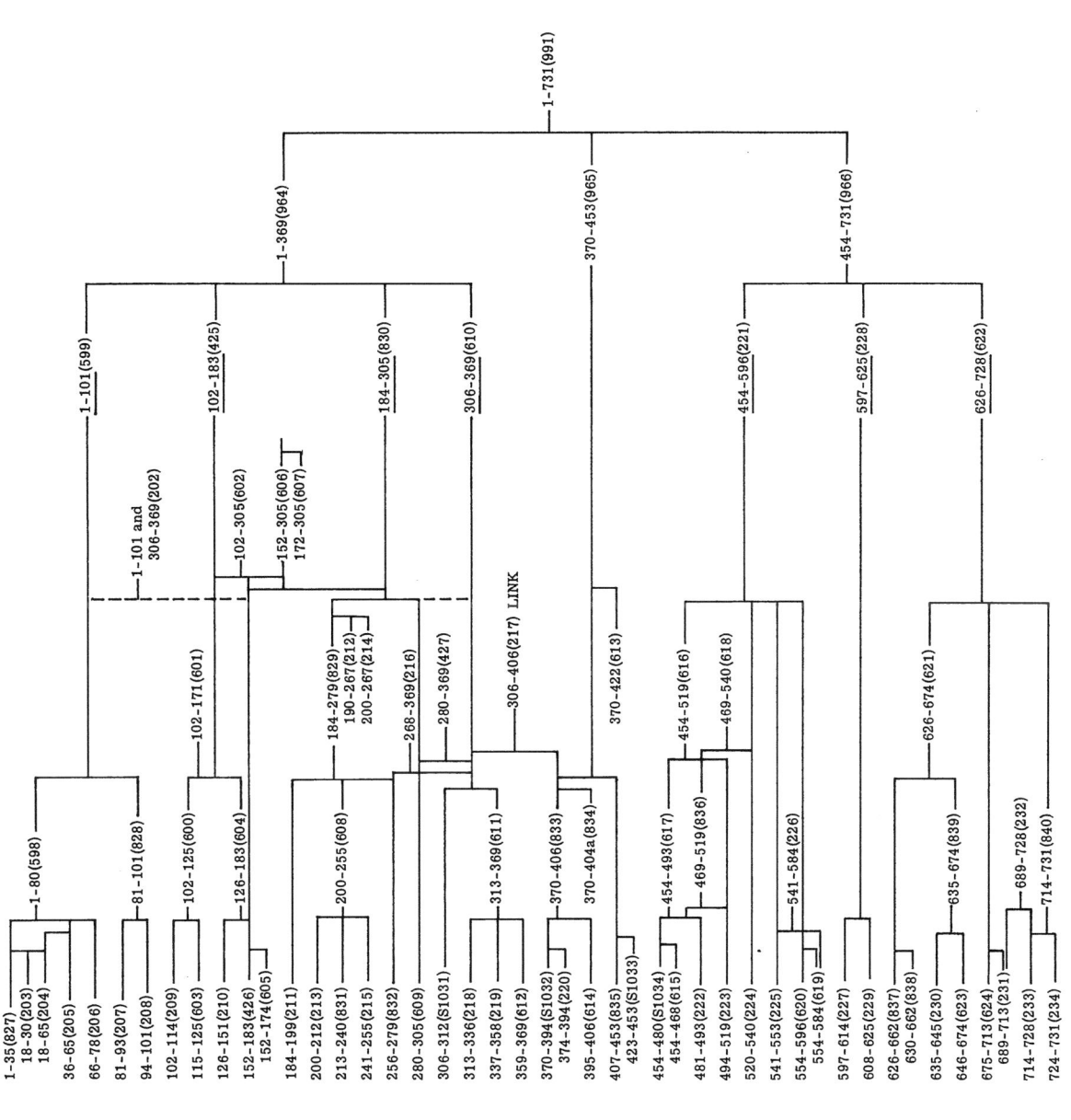

## Aeneid IX

| | | | |
|---|---|---|---|
| I. | 1-175 | DAY | Attack on Trojan camp |
| (1) | 1-76 | | Attack on camp |
| (2) | 77-122 | | Metamorphosis of the ships |
| (3) | 123-175 | | Turnus' speech. Preparations |
| II. | 176-449 | NIGHT | Nisus and Euryalus |
| (1) | 176-313 | | In the Trojan camp |
| (2) | 314-366 | | In the camp of the enemy |
| (3) | 367-449 | | Departure and death |
| III. | 450-818 | DAY | Battle at the Trojan camp |
| (1) | 450-589 | | Fighting |
| (2) | 590-671 | | Ascanius-Numanus episode |
| (3) | 672-818 | | Turnus inside the camp |

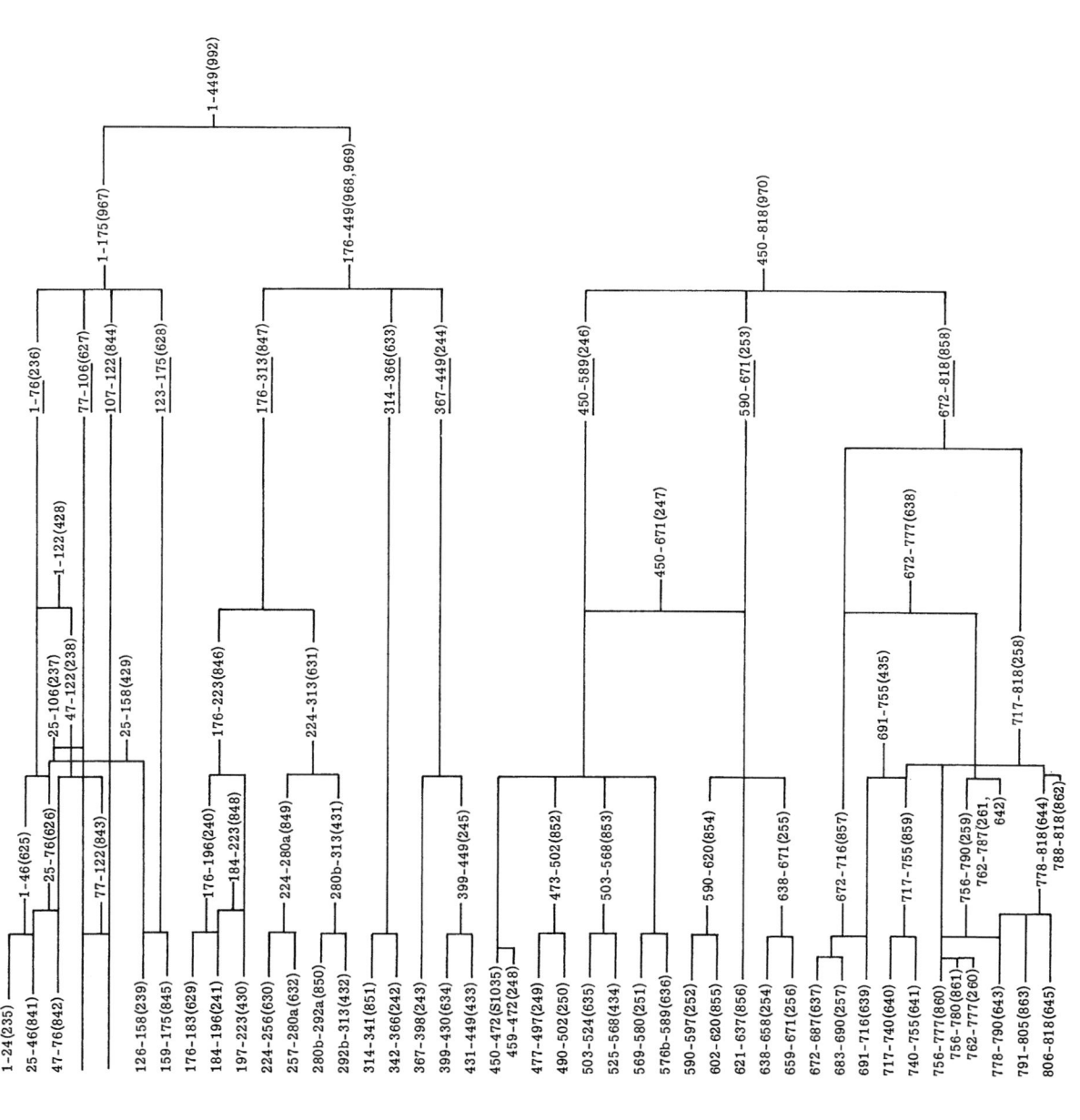

## Aeneid X

I. 1-361 Return of Aeneas
   (1) 1-117 Council of the gods
      (a) 1-17 Speech of Jupiter
      (b) 18-62a Speech of Venus
      (c) 62b-95 Speech of Juno
      (a) 96-117 Speech of Jupiter
   (2) 118-255 Return of Aeneas. Catalogue of ships
   (3) 256-361 Landing and battle

II. 362-688 Death of Pallas
   (1) 362-478 Aristeia of Pallas
   (2) 479-605 Death of Pallas and effect on Aeneas
   (3) 606-688 Removal of Turnus from battle

III. 689-908 Death of Lausus and Mezentius
   (1) 689-746 Aristeia of Lausus
   (2) 747-832 Death of Lausus
   (3) 833-908 Death of Mezentius

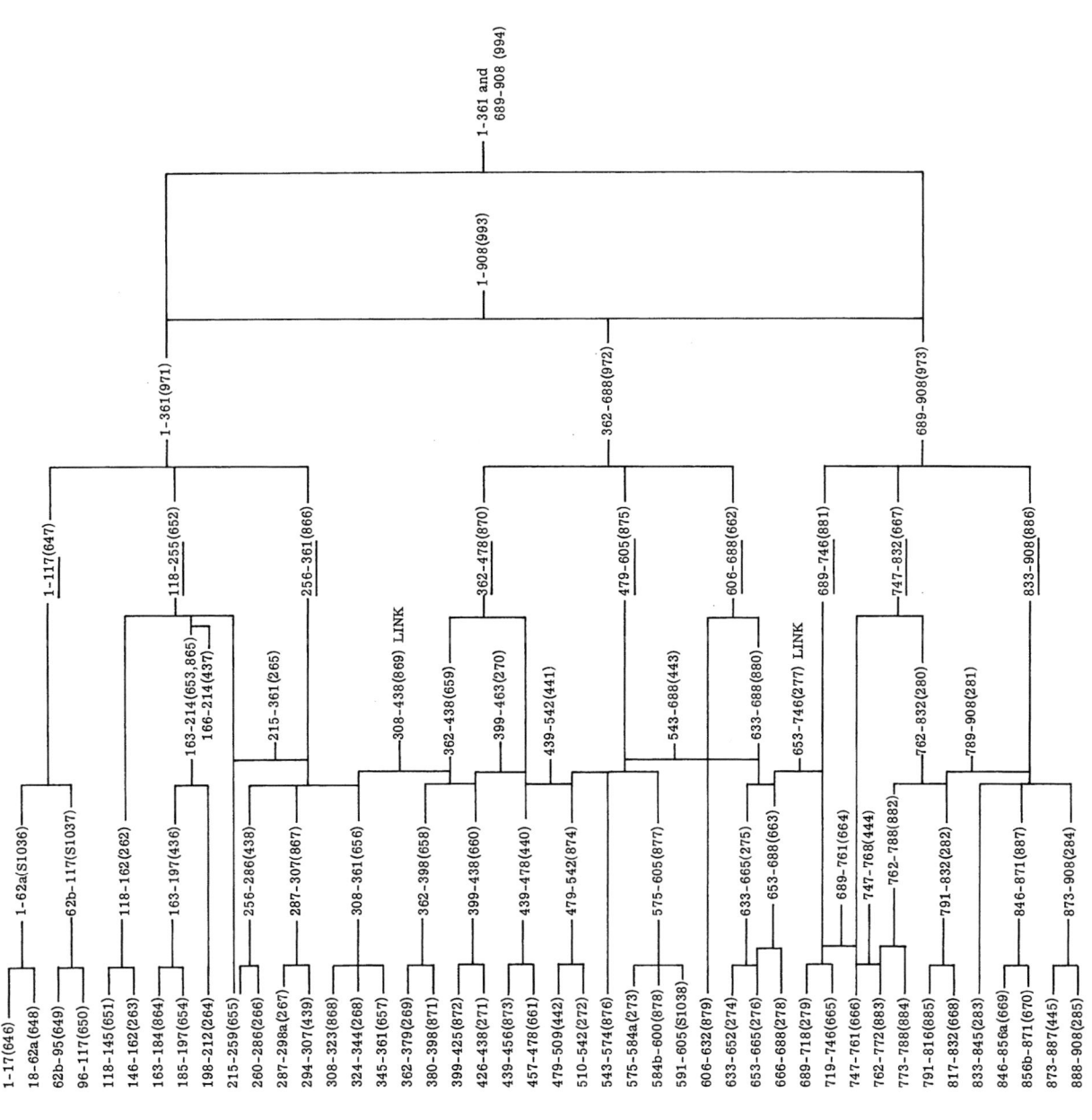

## Aeneid XI

I. 1-224 Truce and burial of the dead
   (1) 1-99 Mourning for Pallas
   (2) 100-138 Embassy of Latins
   (3) 139-224 Grief of Evander. Burial of dead

II. 225-467 Council and speeches
   (1) 225-295 Speech of Venulus
   (2) 296-375 Latinus' speech and Drances' reply
   (3) 376-467 Turnus' speech. Renewed attack

III. 468-867 Cavalry battle and death of Camilla
   (1) 468-497 Preparations for battle
   (2) 498-647 Camilla and the cavalry battle
      (a) 498-531 Camilla enters the conflict
      (b) 532-596 Speech about Camilla
      (c) 597-647 Cavalry battle
   (3) 648-867 Aristeia and death of Camilla

Epilogue 868-915 Transition to Book XII

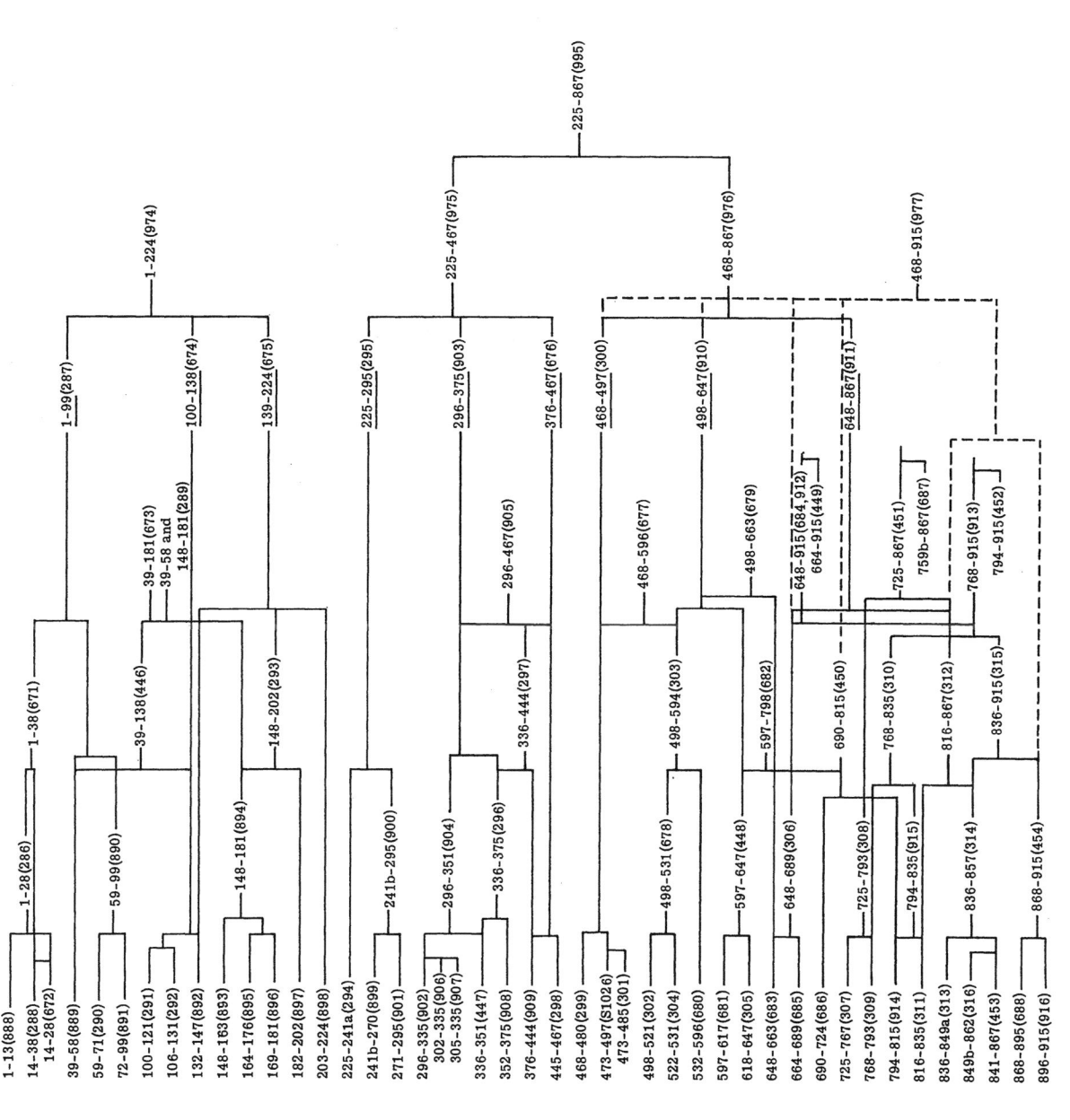

## Aeneid XII

I. 1-288 Breaking of the truce
   (1) 1-80   Turnus resolves to fight
      (a) 1-9     Description of Turnus
      (b) 10-53   Turnus and Latinus
      (c) 54-80   Turnus and Amata
   (2) 81-215   Preparations for the truce
      (a) 81-133   Preparations
      (b) 134-160   Juno and Juturna
      (c) 161-215   Vows of Aeneas and Latinus
   (3) 216-288   Breaking of the truce

II. 289-696 Turnus on the battlefield
   (1) 289-553   Turnus in battle. Parallel aristeias
   (2) 554-611   Assault on Laurentum. Death of Amata
   (3) 614-696   Turnus again resolves to face Aeneas

III. 697-952 Turnus and Aeneas
   (1) 697-790   First encounter
   (2) 791-842   Jupiter and Juno
   (3) 843-952   Death of Turnus

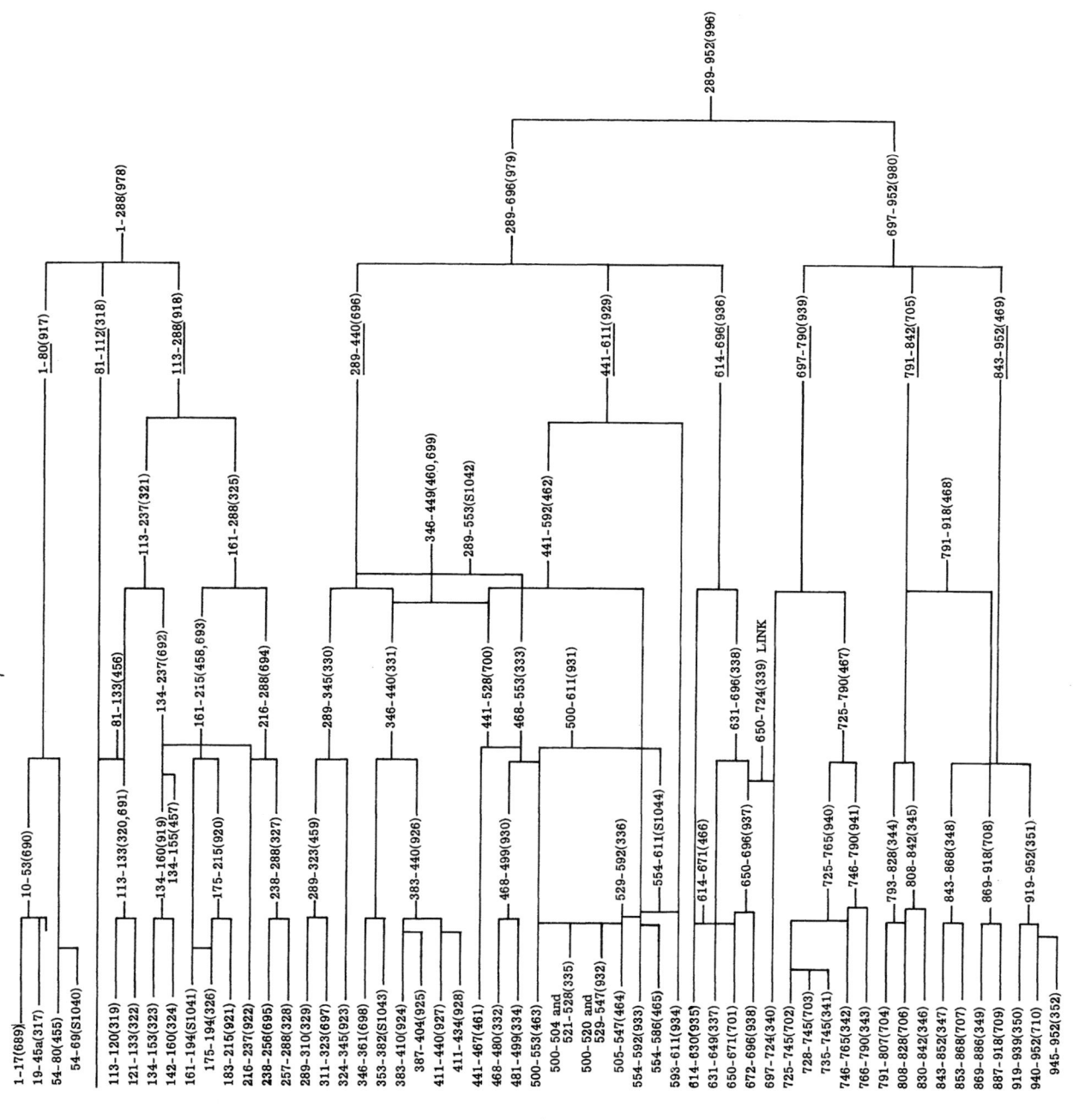

## TABLE X. RATIOS MORE ACCURATE WITH HALF-LINES AS FRACTIONS

| Half-line | Value of half-line | Proportions containing half-line | Ratio as fraction | Ratio as whole line |
|---|---|---|---|---|
| I, 534 | .3 | No. 722 | .624 | .643 |
|  |  | 22 | .624 | .636 |
| I, 636 | .6 | No. 25 | .608 | .60 |
| II, 66 | .2 | No. 33 | .615 | .632 |
|  |  | 34 | .606 | .596 |
| II, 233 | .4 | No. 730 | .617 | .60 |
|  |  | 370 | .607 | .60 |
| II, 346 | .2 | No. 491 | .603 | .586 |
|  |  | 489 | .627 | .633 |
|  |  | 490 | .604 | .597 |
| II, 468 | .7 | No. 734 | .633 | .636 |
|  |  | 373 | .620 | .623 |
| II, 614 | .4 | No. 53 | .621 | .630 |
| II, 720 | .2 | No. 58 | .621 | .643 |
|  |  | 57 | .623 | .636 |
| III, 218 | .2 | No. 504 | .617 | .609 |
| III, 316 and | .6 |  |  |  |
| III, 340 | .4 | No. 507 | .616 | .613 |
|  |  | 508 | .620 | .625 |
| III, 470 | .6 | No. 513 | .604 | .582 |
| III, 527 | .4 | No. 84 | .606 | .593 |
| IV, 44 | .4 | No. 92 | .615 | .625 |
|  |  | 753 | .618 | .623 |
| IV, 361 | .6 | No. 530 | .621 | .613 |
| IV, 400 | .6 | No. 397 | .619 | .609 |
| IV, 503 | .4 | No. 766 | .626 | .633 |
| V, 322 | .4 | No. 546 | .614 | .583 |
|  |  | 124 | .607 | .591 |
|  |  | 547 | .622 | .629 |
| V, 792 | .6 | No. 559 | .612 | .60 |
| VI, 94 | .6 | No. 563 | .634 | .639 |
|  |  | 562 | .612 | .609 |
| VII, 129 | .6 | No. 188 | .623 | .629 |
|  |  | 814 | .626 | .629 |

TABLE X. RATIOS MORE ACCURATE WITH HALF-LINES AS FRACTIONS

| Half-line | Value of half-line | Proportions containing half-line | Ratio as fraction | Ratio as whole line |
|---|---|---|---|---|
| VII, 248 | .6 | No. 819 | .616 | .60 |
|  |  | 817 | .601 | .595 |
| VII, 439 | .4 | No. 589 | .635 | .641 |
| VII, 455 | .6 | No. 590 | .628 | .633 |
| VII, 702 | .2 | No. 595 | .618 | .610 |
|  |  | S1029 | .631 | .639 |
| VII, 760 | .6 | No. 423 | .616 | .607 |
| VIII, 469 | .2 | No. S1034 | .618 | .630 |
|  |  | 617 | .617 | .625 |
|  |  | 836 | .618 | .608 |
| IX, 167 | .4 | No. 845 | .622 | .60 |
| IX, 295 | .4 | No. 432 | .619 | .630 |
|  |  | 431 | .636 | .643 |
| IX, 467 | .4 | No. 248 | .627 | .643 |
| IX, 520 | .4 | No. 635 | .607 | .591 |
| IX, 721 | .7 | No. 640 | .620 | .625 |
| IX, 761 | .4 | No. 860 | .607 | .591 |
|  |  | 861 | .631 | .640 |
|  |  | 259 | .622 | .629 |
| X, 17 | .2 | No. S1036 | .618 | .610 |
|  |  | 646 | .617 | .588 |
| X, 490 | .6 | No. 442 | .621 | .613 |
|  |  | 874 | .616 | .606 |
| X, 580 | .2 | No. 273 | .605 | .638 |
|  |  | 877 | .616 | .626 |
| X, 728 | .2 | No. 665 | .632 | .643 |
| XI, 375 | .2 | No. 296 | .612 | .60 |
| XI, 391 | .2 | No. 909 | .616 | .609 |
|  |  | 676 | .616 | .633 |
| XII, 218 | .6 | No. 922 | .629 | .636 |
| XII, 631 | .2 | No. 337 | .615 | .632 |

## TABLE XI. RATIOS LESS ACCURATE WITH HALF-LINES AS FRACTIONS

| Half-line | Value of half-line | Proportions containing half-line | Ratio as fraction | Ratio as whole line | Golden Mean series |
|---|---|---|---|---|---|
| II, 640 | .4 | No. 497 | .610 | .625 | (1,1,2,) 3, 5, 8 |
|  |  | 54 | .606 | .615 | (1,1,2,3,) 5, 8, 13 |
|  |  | 374 | .623 | .618 | (1,1,2,3,5,8,) 13, 21, 34 |
| II, 767 | .2 | No. 499 | .603 | .615 | (1,1,2,3,) 5, 8, 13 |
| II, 787 | .7 | No. 500 | .614 | .618 | (1,1,2,3,5,8,) 13, 21, 34 |
| III, 640 | .2 | No. 89 | .631 | .619 | (1,1,2,3,5,) 8, 13, 21 |
| III, 661 | .4 | No. 90 | .632 | .622 | (1,4,5,9,) 14, 23, 37 |
| IV, 516 | .6 | No. 107 | .630 | .625 | (1,1,2,) 3, 5, 8 |
| V, 574 | .2 | No. 779 | .615 | .625 | (1,1,2) 3, 5, 8 |
| V, 653 | .2 | No. 554 | .604 | .617 |  |
|  |  | 136 | .625 | .617 | (1,4,5,9,14,) 23, 37, 60 |
| V, 815 | .7 | No. 146 | .603 | .607 | (1,5,6,) 11, 17, 28 |
| VI, 835 | .7 | No. 418 | .603 | .607 | (1,5,6,) 11, 17, 28 |
|  |  | 577 | .606 | .609 | (1,4,5,) 9, 14, 23 |
| VIII, 41 | .4 | No. 205 | .613 | .621 | (1,3,4,7,) 11, 18, 29 |
|  |  | 204 | .612 | .617 | (1,3,4,7,11) 18, 29, 47 |
| VIII, 536 | .4 | No. 224 | .608 | .619 | (1,1,2,3,5,) 8, 13, 21 |
| X, 284 | .6 | No. 266 | .625 | .615 | (1,1,2,3,) 5, 8, 13 |
| X, 876 | .6 | No. 445 | .603 | .613 |  |
|  |  | 284 | .607 | .611 | (1,3,4,) 7, 11, 18 |

## TABLE XII. THE INTERPOLATIONS

| Interpolated verse | No. of proportion | Ratio with verse omitted | Ratio with verse included |
|---|---|---|---|
| II, 76 | No. 33 | .615 | .583 |
|  | 34 | .606 | .593 |
| III, 230 | No. 73 | .60 | .615 |
|  | 505 | .615 | .60 |
|  | 504 | .617 | .623 |
|  | 503 | .630 | .634 |
| IV, 273 | No. S1013 | .630 | .651 |
|  | 760 | .610 | .620 |
|  | 759 | .60 | .593 |
|  | 526 | .619 | .623 |
| IV, 528 | No. 108 | .633 | .613 |
|  | 107 | .630 | .638 |
|  | 109 | .625 | .633 |
|  | 402 | .618 | .623 |
|  | 765 | .614 | .608 |
| IV, 528, with IV, 516, as a whole line | No. 107 | .625 | .633 |
|  | 402 | .620 | .612 |
|  | 765 | .615 | .610 |
| VI, 242 | No. 407 | .60 | .581 |
|  | 408 | .617 | .621 |
| VIII, 46 | No. 205 | .613 | .626 |
|  | 204 | .612 | .620 |
|  | 598 | .617 | .622 |
|  | 599 | .618 | .622 |
| VIII, 46, with VIII, 41, as a whole line | No. 205 | .621 | .633 |
|  | 204 | .617 | .625 |
|  | 598 | .620 | .625 |
|  | 599 | .620 | .624 |
| IX, 29 | No. 841 | .619 | .636 |
|  | 626 | .608 | .615 |
|  | 625 | .60 | .587 |
|  | 237 | .630 | .638 |
| IX, 121 | No. 844 | .613 | .575 |
|  | 843 | .622 | .630 |
| IX, 151 | No. 239 | .625 | .606 |
|  | 628 | .623 | .630 |
| IX, 529 | No. 434 | .628 | .614 |
|  | 853 | .612 | .602 |
|  | 246 | .620 | .622 |

| Interpolated verse | No. of proportion | Ratio with verse omitted | Ratio with verse included |
|---|---|---|---|
| X, 278 | No. 266 | .625 | .602 |
|  | 438 | .628 | .641 |
| X, 278, with X, 284, as a whole line | No. 266 | .615 | .593 |
|  | 438 | .633 | .645 |
| X, 872 | No. 670 | .620 | .643 |
|  | 887 | .623 | .60 |
|  | 886 | .614 | .606 |
| but with X, 876, as a whole line | No. 886 | .616 | .608 |
| XII, 612-613 | No. 934 | .632 | .666 |
|  | S1044 | .621 | .633 |
|  | 931 | .607 | .596 |

# TABLE XIII. A GUIDE TO PARAGRAPHING

| Book | Paragraphing *not* after | Paragraphing *better* after | Supported by ratios | Additional comments |
|---|---|---|---|---|
| I | 418 (M) | 417 (HJS) | Nos. 18, 15, 943; 358, 718, 719, 944; cf. No. 981 | 418 begins subdivision and third main division |
|  | 463 (M), 459 (S) | 465 (J) | Nos. —; 480; cf. No. 19 |  |
|  | 696 (M) | 694 (HJS) | Nos. S1002, 481; 363; cf. Nos. 727, 362 |  |
| II | 233 (S) | 227 (J) | Nos. 485; 730; cf. No. 370 | cf. Chapter 5, note 3 |
|  | 313 (M) | 317 (HJS) | Nos. 40; 491; cf. Nos. 489, 490 |  |
|  | 335 (JM) | 346 (S) | Nos. 491, 489; 372; cf. Nos. 490, 731 |  |
|  | 430 (JS), 437 (J) | 436 (M) | Nos. 44; 45; cf. No. 733 | cf. punctuation: comma after 436 (JHS), period (M) |
|  | 620 (M) | 623 (HJS) | Nos. 735, 52; 496, 54, 374 |  |
|  | 672 (M) | 670 (HJS) | Nos. —; 55; cf. Nos. 736, 377, 374 |  |
|  | 794 (JMS) | 795 (H) | No. 379; —; cf. Nos. 60, 500, 737 |  |
| III | 46 (M) | 48 (HJS) | Nos. 63, 380; —; cf. No. 381 |  |
|  | 72 (M) | 68 (HJS) | Nos. 741, 740, 381; 64, 65, 383 | 68 ends first subdivision |
|  | 134 (HJ) | 131 (M) | Nos. 501; S1005, 744 |  |
|  | 246 (M) | 244 (JS) | Nos. 73; 74; cf. No. 505 |  |
|  | 269 (M) | 267 (S) | cf. No. 75 | Better after 273; see p. 89 |
|  | 320 (M) | 319 (JS) | cf. Nos. 746, 507, 388 |  |

| Book | Paragraphing *not* after | Paragraphing *better* after | Supported by ratios | Additional comments |
|------|--------------------------|------------------------------|---------------------|---------------------|
|      | 550 (M) | 547 (HJS) | Nos. 84, 391; 516, 517, 751 | 547 ends subdivision |
| IV   | 30 (HJS) | 29 (M) | Nos. S1011, 523; 92; cf. No. 753 | |
|      | 194 (M) | 197 (HJS) | Nos. 756; 99; cf. Nos. 757, 758, 98, 526 | |
|      | 396 (M) | 392 (HJS) | Nos. —; 397; cf. No. 764 | |
|      | 553 (JS) | 552 (HM) | Nos. 108, 107, 402, 765; 110, 535; cf. No. 953 | 552 ends subdivision |
|      | 631 (J) | 629 (HMS) | Nos. 769, 767; 536, 537; cf. No. 403 | |
|      | 641 (S) | 640 (M) | cf. No. 536 | |
|      | 662 (M) | 671 (JS) | Nos. 404; 112, 113; cf. No. 537 | |
| V    | 31 (M) | 34 (HJS) | Nos. S1015, 539; 117, 541, S1016; cf. No. 540 | 34 ends first subdivision |
|      | 44 (M) | 41 (HJS) | Nos. 117; 770 | |
|      | 209 (S) | 200 (M) | cf. No. 544 | |
|      | 226 (M) | 224 (HJ) | Nos. 772; 545; cf. Nos. 544, 773 | |
|      | 574 (S) | 576 (JM) | No. 779; —; cf. No. 955 | |
|      | 602 (M) | 603 (HJS) | Nos. 134, 955; 135, 137, 780, 136, 956; cf. Nos. 553, 987, 986 | 604 begins subdivision and final main division |
|      | 658 (M) | 663 (HJS) | Nos. 554; 138, 781, S1018, 782; cf. Nos. 137, 780, 956 | 663 ends subdivision |
|      | 684 (JM) | 679 (HS) | Nos. 138; 140, 555; cf. Nos. 780, 781 | |
|      | 739 (M) | 745 (HJS) | Nos. 557, 784, 780; 143, 144; cf. Nos. 142, 782 | |
|      | 834 (J) | 832 (M) | cf. Nos. 147, 786 | |

TABLE XIII. A GUIDE TO PARAGRAPHING

| Book | Paragraphing *not* after | Paragraphing *better* after | Supported by ratios | Additional comments |
|---|---|---|---|---|
| VI | 12 (M) | 13 (HJS) | Nos. 150; 561; cf. Nos. 151, 787 | |
| | 39 (M) | 41 (HJS) | cf. Nos. 154, 787 | |
| | 189 (M) | 182 (HJ) | Nos. 565; 157, 790; cf. No. 789 | 182 ends section on death of Misenus |
| VII | 194 (DM) | 191 (HS) | Nos. 583, 419; 816, 584, 585; cf. Nos. 191, 815 | |
| | 403 (M) | 405 (DHS) | Nos. 588, 820; 589, 821; cf. No. 194 | 405 ends episode of Allecto and Amata |
| VIII | 126 (M) | 125 (DHS) | Nos. 603, 600; 210, 604; cf. Nos. 601, 425 | |
| | 368 (D) | 369 (HMS) | Nos. 612, 611, 610; S1032, 834, 833, 613, 965; cf. Nos. 217, 991 | 369 ends subdivision and first main division |
| | 469 (M) | 468 (DS) | Nos. 615; 836, 618; cf. Nos. S1034, 617, 616 | |
| | 557 (M) | 553 (DHS) | Nos. 225; 619, 620 | cf. No. 226; minor ends at 557 |
| | 603 (M) | 607 (DHS) | Nos. —; 229; cf. Nos. 227, 228 | |
| IX | 125 (M) | 122 (DHS) | Nos. 844, 843, 238, 428; 628; cf. No. 967 | 122 ends subdivision |
| X | 259 (HS) | 255 (DM) | Nos. 652; 438, 866; cf. No. 971 | 255 ends subdivision; cf. Chapter 4, note 27; Chapter 5, note 3 |
| | 378 (MS) | 379 (DH) | Nos. 269; 871; cf. No. 658 | |
| | 396 (HM) | 398 (D) | Nos. 871, 658; 872, 660, 270; cf. No. 870 | |
| | 473 (DMS) | 478 (H) | Nos. 661, 440, 870; 442, 874, 875; cf. Nos. 441, 972 | 478 ends subdivision |
| | 667 (M) | 665 (DS) | Nos. 276, 275; 278; cf. Nos. 663, 880 | |

| Book | Paragraphing *not* after | Paragraphing *better* after | Supported by ratios | Additional comments |
|---|---|---|---|---|
|  | 768 (DH) | 761 (MS) | Nos. 666, 664; 883, 882, 280; cf. No. 444 | See above, Chapter 4, note 20 |
|  | 866 (M), 869 (D) | 871 (HS) | Nos. 670, 887; 445, 284; cf. No. 886 | cf. No. 670; minor ends at 866 |
| XI | 119 (M) | 121 (DH) | No. 291; —; cf. No. 292 |  |
|  | 519 (M) | 521 (DH) | Nos. 302; 304 |  |
|  | 628 (M) | 617 (DH) | Nos. 681; 305; cf. No. 448 |  |
|  | 798 (M) | 793 (DH) | Nos. 309, 308; 914, 915; cf. No. 310 |  |
| XII | 175 (M) | 174 (DHS) | Nos. —; 326, 920; cf. Nos. S1041, 458, 693 |  |
|  | 282 (S), 286 (DH) | 288 (M) | Nos. 328, 327, 694, 325, 918, 978; 329, 459, 330, 696, 979, 996 | 288 ends subdivision and first main division; see Chapter 5, note 3 |
|  | 664 (M), 680 (M) | 671 (DH) | Nos. 701; 938; cf. Nos. 937, 338, 936 |  |
|  | 727 (HM) | 724 (DS) | Nos. 340, 339; 702, 940, 467; cf. No. 939 | cf. Chapter 4, p. 52 and note 18 |

## TABLE XIV. PROPORTIONS IN THE *APPENDIX VERGILIANA*

| No. | m/M: type | Major | Total lines | Minor | Total lines | Ratio: M/(M+m) |
|---|---|---|---|---|---|---|
| *Culex* | | | | | | |
| A1. | (b+d+f+h)/(a+c+e+g) | 1-41<br>58-97<br>123-156<br>231b-371 | 255.4 | 42-57<br>98-122<br>157-231a<br>372-414 | 158.6 | .617 |
| A2. | b/(a+c) | 1-41<br>202-414 | 254 | 42-201 | 160 | .614 |
| A3. | (b+d)/(a+c) | 1-7<br>11-17 | 14 | 8-10<br>18-23 | 9 | .609 |
| A4. | (a+d)/(b+c) | 11-23<br>24-36 | 26 | 1-10<br>37-41 | 15 | .634 |
| A5. | a/(b+c) | 31-36<br>37-41 | 11 | 24-30 | 7 | .611 |
| A6. | a/b | 48-57 | 10 | 42-47 | 6 | .625 |
| A7. | b/(a+c) | 42-57<br>72-78 | 23 | 58-71 | 14 | .622 |
| A8. | b/(a+c) | 42-57<br>79-97 | 35 | 58-78 | 21 | .625 |
| A9. | (a+c)/(b+d) | 79-97<br>123-201 | 98 | 42-78<br>98-122 | 62 | .613 |
| A10. | (a+c)/(b+d) | 62-71<br>76-78 | 13 | 58-61<br>72-75 | 8 | .619 |
| A11. | b/a | 58-97 | 40 | 98-122 | 25 | .615 |
| A12. | a/b | 86-97 | 12 | 79-85 | 7 | .632 |
| A13. | (a+c)/(b+d) | 86-97<br>107-122 | 28 | 79-85<br>98-106 | 16 | .636 |
| A14. | (b+d)/(a+c) | 98-106<br>109-114 | 15 | 107-108<br>115-122 | 10 | .60 |
| A15. | (a+c)/(b+d) | 127-136<br>146-156 | 21 | 123-126<br>137-145 | 13 | .618 |
| A16. | (b+d)/(a+c+e) | 123-136<br>157-182a<br>193-201 | 48.6 | 137-156<br>182b-192 | 30.4 | .615 |
| A17. | (a+c)/(b+d) | 163-182a<br>193-201 | 28.6 | 157-162<br>182b-192 | 16.4 | .636 |
| A18. | a/(b+c) | 206-209<br>210-212 | 7 | 202-205 | 4 | .636 |

| No. | m/M: type | Major | Total lines | Minor | Total lines | Ratio: M/(M+m) |
|---|---|---|---|---|---|---|
| *Culex* | | | | | | |
| A19. | (a+b)/c | 213-231a | 18.6 | 202-209<br>210-212 | 11 | .628 |
| A20. | b/(a+c) | 202-303<br>372-384 | 115 | 304-371 | 68 | .628 |
| A21. | b/(a+c) | 202-303<br>385-414 | 132 | 304-384 | 81 | .620 |
| A22. | a/b | 213-217 | 5 | 210-212 | 3 | .625 |
| A23. | a/b | 223-231a | 8.6 | 218-222 | 5 | .632 |
| A24. | b/a | 231b-247 | 16.4 | 248-257 | 10 | .621 |
| A25. | a/b | 258-303 | 46 | 231b-257 | 26.4 | .635 |
| A26. | (a+c)/b | 268-295a | 27.6 | 258-267<br>295b-303 | 18.4 | .60 |
| A27. | (b+d)/(a+c) | 268-276<br>286-293 | 17 | 277-285<br>294-295a | 10.6 | .616 |
| A28. | b/a | 304-314 | 11 | 315-321 | 7 | .611 |
| A29. | b/a | 304-321 | 18 | 322-333 | 12 | .60 |
| A30. | (b+d)/(a+c) | 304-321<br>334-357 | 42 | 322-333<br>358-371 | 26 | .618 |
| A31. | c/(a+b) | 315-321<br>322-326 | 12 | 327-333 | 7 | .632 |
| A32. | (b+d)/(a+c) | 334-338<br>342b-352a | 15 | 339-342a<br>352b-357 | 9 | .625 |
| A33. | b/a | 334-357 | 24 | 358-371 | 14 | .632 |
| A34. | b/(a+c) | 372-375<br>381-384 | 8 | 376-380 | 5 | .615 |
| A35. | (b+d)/(a+c) | 385-391a<br>398b-410 | 19 | 391b-398a<br>411-414 | 11 | .633 |
| *Ciris* | | | | | | |
| A36. | (a+c)/(b+d) | 101-348<br>459-541 | 331 | 1-100<br>349-458 | 210 | .612 |
| A37. | (a+c)/(b+d) | 12-26<br>29-34 | 21 | 1-11<br>27-28 | 13 | .618 |
| A38. | b/(a+c) | 1-53<br>92-100 | 62 | 54-91 | 38 | .620 |
| A39. | b/a | 1-100 | 100 | 101-162 | 62 | .617 |
| A40. | a/b | 42-53 | 12 | 35-41 | 7 | .632 |
| A41. | b/a | 54-82 | 29 | 83-100 | 18 | .617 |
| A42. | b/a | 54-100 | 47 | 101-128 | 28 | .627 |

TABLE XIV. PROPORTIONS IN THE APPENDIX VERGILIANA

| No. | m/M: type | Major | Total lines | Minor | Total lines | Ratio: M/(M+m) |
|---|---|---|---|---|---|---|
| *Ciris* | | | | | | |
| A43. | (a+c)/(b+d) | 110-114a 116-128 | 17.2 | 101-109 114b-115 | 10.8 | .614 |
| A44. | b/(a+c) | 101-128 163-190 | 56 | 129-162 | 34 | .622 |
| A45. | (b+d)/(a+c) | 101-162 191-282 | 154 | 163-190 283-348 | 94 | .621 |
| A46. | b/(a+c) | 129-132 146-162 | 21 | 133-145 | 13 | .618 |
| A47. | b/a | 163-179 | 17 | 180-190 | 11 | .607 |
| A48. | b/(a+c) | 191-194 206-219 | 18 | 195-205 | 11 | .621 |
| A49. | (b+d)/(a+c) | 191-205 220-240 | 36 | 206-219 241-249 | 23 | .610 |
| A50. | a/(b+c) | 250-282 283-348 | 99 | 191-249 | 59 | .627 |
| A51. | (a+c)/(b+d) | 224-233 241-249 | 19 | 220-223 234-240 | 11 | .633 |
| A52. | (b+d)/(a+c) | 220-240 250-267 | 39 | 241-249 268-282 | 24 | .619 |
| A53. | (a+c)/(b+d) | 257-261 268-282 | 20 | 250-256 262-267 | 13 | .606 |
| A54. | (a+c+e)/(b+d) | 257-282 306-339 | 60 | 250-256 283-305 340-348 | 39 | .606 |
| A55. | c/(a+b) | 283-285 286-296 | 14 | 297-305 | 9 | .609 |
| A56. | b/a | 286-292 | 7 | 293-296 | 4 | .636 |
| A57. | b/(a+c) | 290-292 297-300 | 7 | 293-296 | 4 | .636 |
| A58. | b/a | 293-300 | 8 | 301-305 | 5 | .615 |
| A59. | (a+c)/(b+d) | 310-323 333-339 | 21 | 306-309 324-332 | 13 | .618 |
| A60. | (b+c)/a | 306-332 | 27 | 333-339 340-348 | 16 | .628 |
| A61. | (a+c)/b | 355-377 | 23 | 349-354 378-385 | 14 | .622 |
| A62. | (b+d)/(a+c) | 349-385 404-432 | 66 | 386-403 433-458 | 44 | .60 |
| A63. | b/(a+c) | 349-385 459-541 | 120 | 386-458 | 73 | .622 |
| A64. | (a+c)/b | 389-399 | 11 | 386-388 400-403 | 7 | .611 |

| No. | m/M: type | Major | Total lines | Minor | Total lines | Ratio: M/(M+m) |
|---|---|---|---|---|---|---|
| *Ciris* | | | | | | |
| A65. | b/(a+c) | 386-403<br>433-458 | 44 | 404-432 | 29 | .603 |
| A66. | (a+c)/b | 407-424 | 18 | 404-406<br>425-432 | 11 | .621 |
| A67. | a/b | 443-458 | 16 | 433-442 | 10 | .615 |
| A68. | b/(a+c) | 459-462<br>470-477 | 12 | 463-469 | 7 | .632 |
| A69. | (a+c)/(b+d) | 470-477<br>490-519 | 38 | 459-469<br>478-489 | 23 | .623 |
| A70. | a/(b+c) | 490-519<br>520-541 | 52 | 459-489 | 31 | .627 |
| A71. | (a+c)/b | 484-509 | 26 | 478-483<br>510-519 | 16 | .619 |
| A72. | b/(a+c) | 520-529<br>538-541 | 14 | 530-537 | 8 | .636 |
| *Aetna* | | | | | | |
| A73. | (b+d)/(a+c) | 1-218<br>386-568 | 399 | 219-385<br>569-646 | 244 | .621 |
| A74. | b/(a+c) | 1-8<br>24-40 | 25 | 9-23 | 15 | .625 |
| A75. | (b+d)/(a+c) | 1-8<br>24-73 | 58 | 9-23<br>74-93 | 35 | .624 |
| A76. | b/(a+c) | 1-93<br>175-218 | 135 | 94-174 | 81 | .625 |
| A77. | (a+c)/b | 41-84 | 44 | 24-40<br>85-93 | 26 | .629 |
| A78. | (a+c)/(b+d) | 102-122<br>146-174 | 50 | 94-101<br>123-145 | 31 | .617 |
| A79. | (b+d)/(a+c+e) | 94-101<br>137-174<br>189-218 | 76 | 102-136<br>175-188 | 47 | .618 |
| A80. | b/a | 102-122 | 21 | 123-136 | 14 | .60 |
| A81. | c/(a+b) | 102-145<br>146-174 | 73 | 175-218 | 42 | .635 |
| A82. | b/(a+c) | 123-136<br>153-164 | 26 | 137-152 | 16 | .619 |
| A83. | (a+c)/(b+d) | 165-174<br>189-218 | 40 | 153-164<br>175-188 | 24 | .625 |
| A84. | c/(a+b) | 175-188<br>189-202 | 26 | 203-218 | 16 | .619 |

TABLE XIV. PROPORTIONS IN THE APPENDIX VERGILIANA

| No. | m/M: type | Major | Total lines | Minor | Total lines | Ratio: M/(M+m) |
|---|---|---|---|---|---|---|
| *Aetna* | | | | | | |
| A85. | c/(a+b) | 219-251<br>252-273 | 54 | 274-306 | 33 | .621 |
| A86. | (a+c)/(b+d) | 252-273<br>307-385 | 101 | 219-251<br>274-306 | 65 | .608 |
| A87. | a/b | 274-306 | 33 | 252-273 | 22 | .60 |
| A88. | c/(a+b) | 307-318<br>319-321a | 14.6 | 321b-329 | 8.4 | .635 |
| A89. | b/(a+c) | 307-329<br>359-385 | 50 | 330-358 | 29 | .633 |
| A90. | b/(a+c) | 330-337<br>349-358 | 18 | 338-348 | 11 | .621 |
| A91. | b/(a+c) | 359-369<br>380-385 | 17 | 370-379 | 10 | .630 |
| A92. | b/(a+c) | 386-393<br>409-425 | 25 | 394-408 | 15 | .625 |
| A93. | b/(a+c) | 386-401<br>426-448 | 39 | 402-425 | 24 | .619 |
| A94. | (b+d)/(a+c) | 386-448<br>487-536 | 113 | 449-486<br>537-568 | 70 | .617 |
| A95. | (b+d)/(a+c) | 386-425<br>449-568 | 160 | 426-448<br>569-646 | 101 | .613 |
| A96. | b/(a+c) | 426-433<br>443-448 | 14 | 434-442 | 9 | .609 |
| A97. | b/a | 449-486 | 38 | 487-510 | 24 | .613 |
| A98. | (b+c)/(a+d) | 449-486<br>521-536 | 54 | 487-510<br>511-520 | 34 | .614 |
| A99. | (b+d)/(a+c) | 449-510<br>537-548 | 74 | 511-536<br>549-568 | 46 | .617 |
| A100. | a/b | 521-536 | 16 | 511-520 | 10 | .615 |
| A101. | a/b | 549-568 | 20 | 537-548 | 12 | .625 |
| A102. | b/a | 569-573 | 5 | 574-576 | 3 | .625 |
| A103. | b/a | 569-576 | 8 | 577-581 | 5 | .615 |
| A104. | a/b | 577-589 | 13 | 569-576 | 8 | .619 |
| A105. | (a+c)/(b+d) | 577-589<br>594-599 | 19 | 569-576<br>590-593 | 12 | .613 |
| A106. | a/b | 600-646 | 47 | 569-599 | 31 | .603 |
| A107. | a/b | 577-581 | 5 | 574-576 | 3 | .625 |
| A108. | a/b | 582-589 | 8 | 577-581 | 5 | .615 |
| A109. | a/b | 594-599 | 6 | 590-593 | 4 | .60 |

| No. | m/M: type | Major | Total lines | Minor | Total lines | Ratio: M/(M+m) |
|---|---|---|---|---|---|---|
| *Aetna* | | | | | | |
| A110. | c/(a+b) | 590-599<br>600-625a | 35.6 | 625b-646 | 21.4 | .625 |
| A111. | (b+d)/(a+c+e) | 600-603<br>615-629<br>637-646 | 29 | 604-614<br>630-636 | 18 | .617 |
| *Moretum* | | | | | | |
| A112. | b/(a+c) | 1-38<br>87-124 | 75 | 39-86 | 47 | .615 |
| A113. | a/b | 8-18 | 11 | 1-7 | 7 | .611 |
| A114. | (a+c)/(b+d) | 8-23<br>31b-38 | 22.4 | 1-7<br>24-31a | 14.6 | .605 |
| A115. | b/a | 1-38 | 37 | 39-60 | 22 | .627 |
| A116. | b/(a+c) | 19-23<br>31b-38 | 11.4 | 24-31a | 7.6 | .60 |
| A117. | b/a | 19-60 | 41 | 61-86 | 25 | .621 |
| A118. | a/b | 61-124 | 63 | 19-60 | 41 | .606 |
| A119. | (a+c)/b | 42-55 | 14 | 39-41<br>56-60 | 8 | .636 |
| A120. | (a+c)/(b+d) | 52-60<br>66-86 | 29 | 39-51<br>61-65 | 18 | .617 |
| A121. | (a+c)/(b+d) | 66-78<br>84-86 | 15 | 61-65<br>79-83 | 10 | .60 |
| A122. | a/b | 87-124 | 38 | 61-86 | 25 | .603 |
| A123. | b/a | 87-110 | 24 | 111-124 | 14 | .632 |
| *Dirae* | | | | | | |
| A124. | (a+c)/(b+d) | 48-81<br>104-183 | 114 | 1-47<br>82-103 | 69 | .623 |
| A125. | a/b | 104-183 | 80 | 1-47 | 47 | .630 |
| A126. | (a+c)/b | 54-74 | 21 | 48-53<br>75-81 | 13 | .618 |
| A127. | b/a | 48-81 | 34 | 82-103 | 22 | .607 |
| A128. | a/b | 104-141 | 38 | 82-103 | 22 | .633 |
| A129. | a/b | 111-122 | 12 | 104-110 | 7 | .632 |
| A130. | (a+c)/(b+d) | 111-122<br>131-141 | 23 | 104-110<br>123-130 | 15 | .605 |
| A131. | b/a | 104-141 | 38 | 142-165 | 24 | .613 |
| A132. | (b+d)/(a+c) | 104-141<br>166-176 | 49 | 142-165<br>177-183 | 31 | .613 |

## TABLE XIV. PROPORTIONS IN THE APPENDIX VERGILIANA

| No. | m/M: type | Major | Total lines | Minor | Total lines | Ratio: M/(M+m) |
|---|---|---|---|---|---|---|
| *Dirae* | | | | | | |
| A133. | b/(a+c) | 131-141<br>156-165 | 21 | 142-155 | 14 | .60 |
| A134. | b/a | 166-176 | 11 | 177-183 | 7 | .611 |
| *Copa* | | | | | | |
| A135. | b/a | 1-24 | 24 | 25-38 | 14 | .632 |
| *Catalepton* IX | | | | | | |
| A136. | c/(a+b) | 1-12<br>13-40 | 40 | 41-64 | 24 | .625 |

Table XV. Comparison of *Appendix* ratios with those in Vergil, Catullus LXIV, and Lucretius I

| | Total lines | No. of ratios | Average: one ratio in | .618 | | .615-.621 | | .610-.626 | | Fibonacci series | | | | Series 1, 3, 4, 7, ... | |
|---|---|---|---|---|---|---|---|---|---|---|---|---|---|---|---|
| | | | | No. | % | No. | % | No. | % | No. | Ratios used | % | | No. | % |
| *Aeneid* | 9852.2 | 1044 | 9.4 vv. | 45 | 4.3 | 301 | 28.8 | 622 | 59.6 | 318 | 671 | 47.4 | | 92 | 13.7 |
| *Culex* | 414 | 35 | 11.8 | 2 | 5.7 | 10 | 28.6 | 19 | 54.3 | 11 | 26 | 42.3 | | 2 | 5.7 |
| *Ciris* | 541 | 37 | 14.6 | 3 | 8.1 | 13 | 35.1 | 21 | 56.8 | 8 | 36 | 22.2 | | 7 | 20 |
| *Aetna* | 643 | 39 | 16.5 | 1 | 2.6 | 15 | 38.5 | 28 | 71.8 | 17 | 37 | 45.7 | | 4 | 11.1 |
| *Moretum* | 122 | 12 | 10.2 | 0 | | 3 | 25 | 4 | 33.3 | 1 | 10 | 10 | | 3 | 30 |
| *Dirae* | 183 | 11 | 16.6 | 1 | 9.1 | 1 | 9.1 | 5 | 45.4 | 2 | 11 | 18.2 | | 1 | 9.1 |
| Catullus LXIV | 408 | 31 | 13.2 | 1 | 3.2 | 9 | 29 | 16 | 51.6 | 15 | 31 | 48.4 | | 2 | 6.5 |
| Lucretius I | 1121 | 62 | 18.1 | 1 | 1.6 | 13 | 21 | 31 | 50 | 23 | 62 | 37 | | 6 | 9.7 |
| *Georg.* IV, 281-558 | 277 | 22 | 12.6 | 2 | 9.1 | 5 | 22.7 | 12 | 54.6 | 4 | 22 | 18.2 | | 3 | 13.6 |

**TABLE XVI**

**CHART-INDEX OF
THE LONGER POEMS
IN THE *APPENDIX***

## The *Culex*

1-41    Prooemium

42-201    The shepherd, the serpent, and the gnat
(58-97 Praise of country life)
(123-156 Catalogue of trees)

202-414    Lament of the gnat; tomb and inscription
(231b-371 Underworld catalogue)

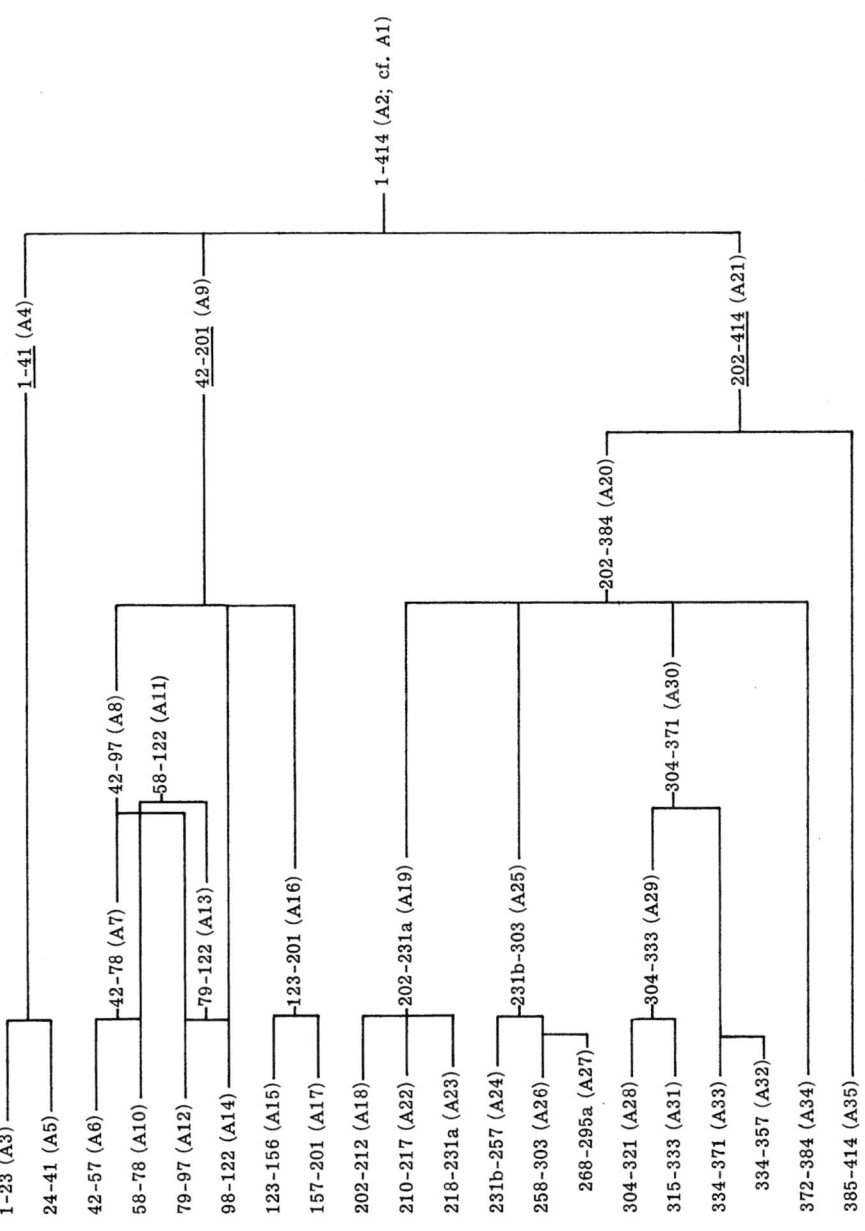

## The Ciris

| | |
|---|---|
| 1-100 | Prooemium, dedication, invocation |
| 101-348 | Preliminary situation: Scylla's passion for Minos; scene with Carme |
| 349-458 | Scylla cuts lock of Nisus; punishment and lament |
| 459-541 | Sea journey and metamorphoses |

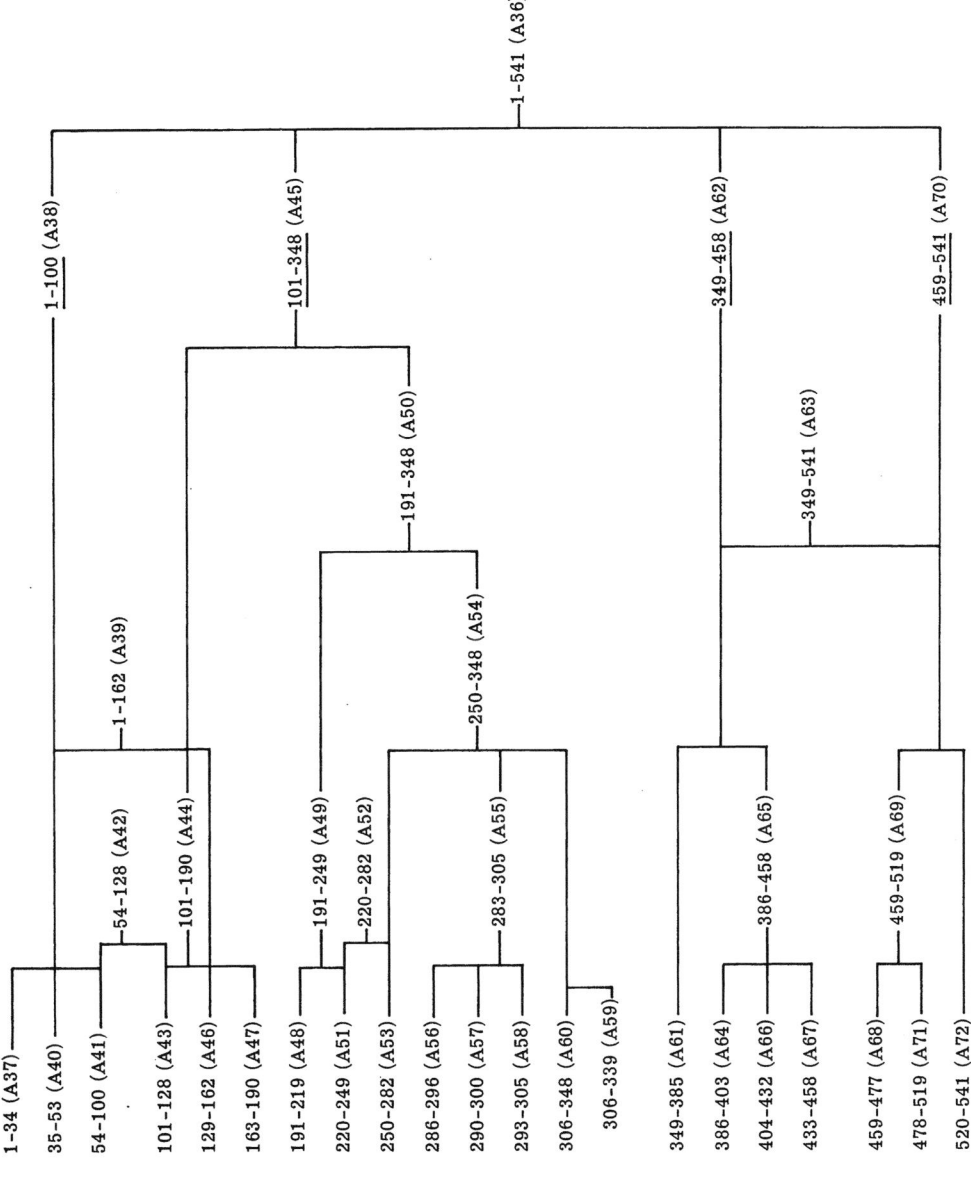

## The *Aetna*

| | | |
|---|---|---|
| 1-218 | 1-93 | Introduction |
| | 94-174 | Nature of the earth |
| | 175-218 | Activity of Aetna |
| 219-385 | | Winds and their causes |
| 386-568 | | The flames of Aetna and their fuels |
| 569-646 | | Achievements of man and nature; eruption of Aetna and its effect |

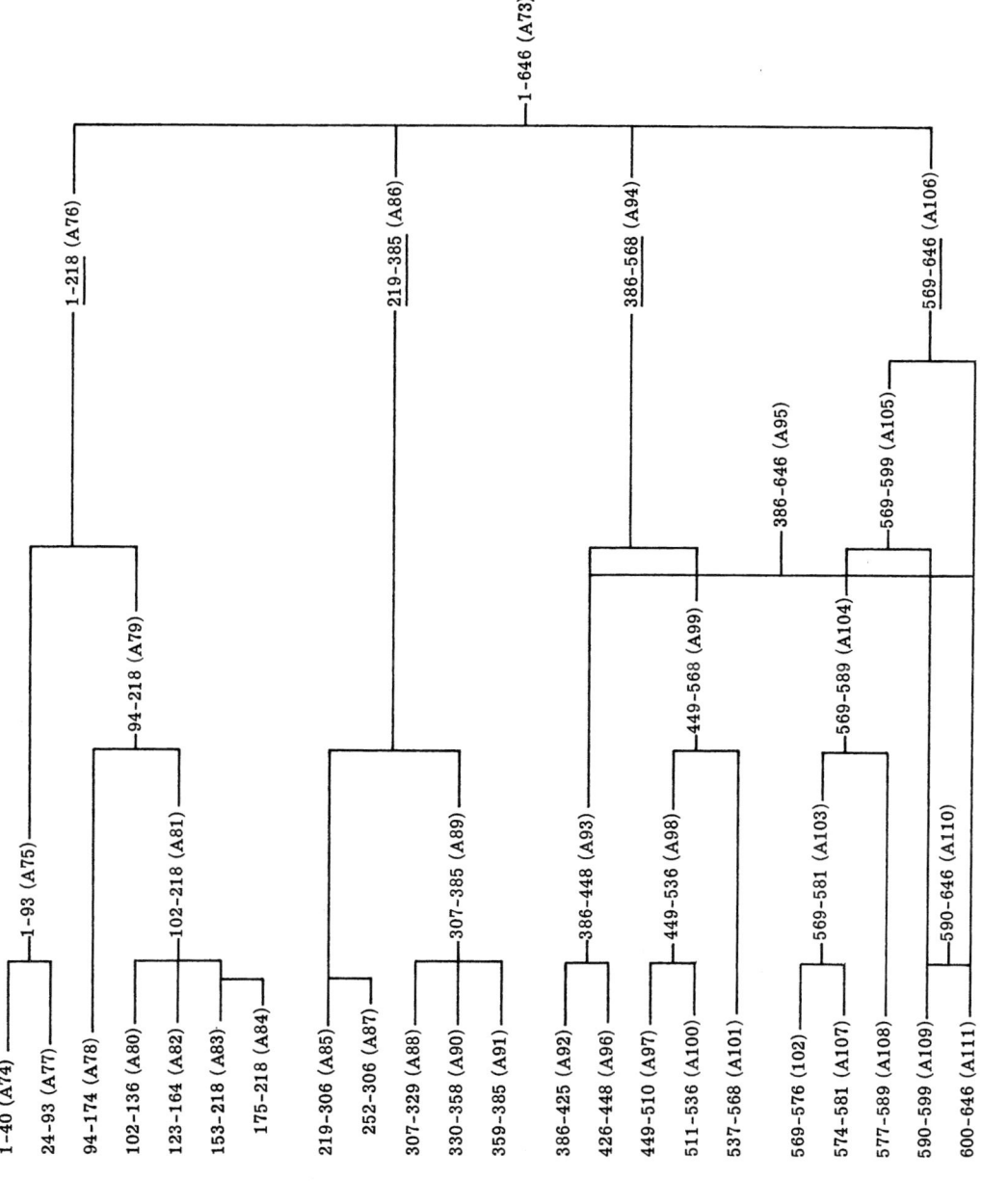

## The *Moretum*

1-38       Simylus grinds his corn and calls Scybale

39-86      The loaf prepared; description of garden and vegetables

87-124     Making of the salad; Simylus begins his day's work

## The *Dirae*

1-47       Imprecation by fire

48-81      Imprecation by water

82-103     Farewell to fields and Lydia

104-183    Memory of Lydia; cruel pangs of love

## The *Moretum*

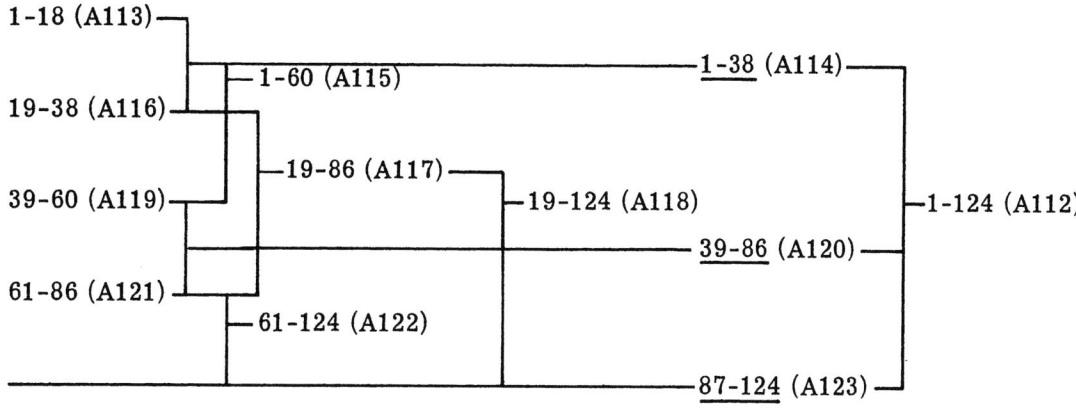

## The *Dirae*

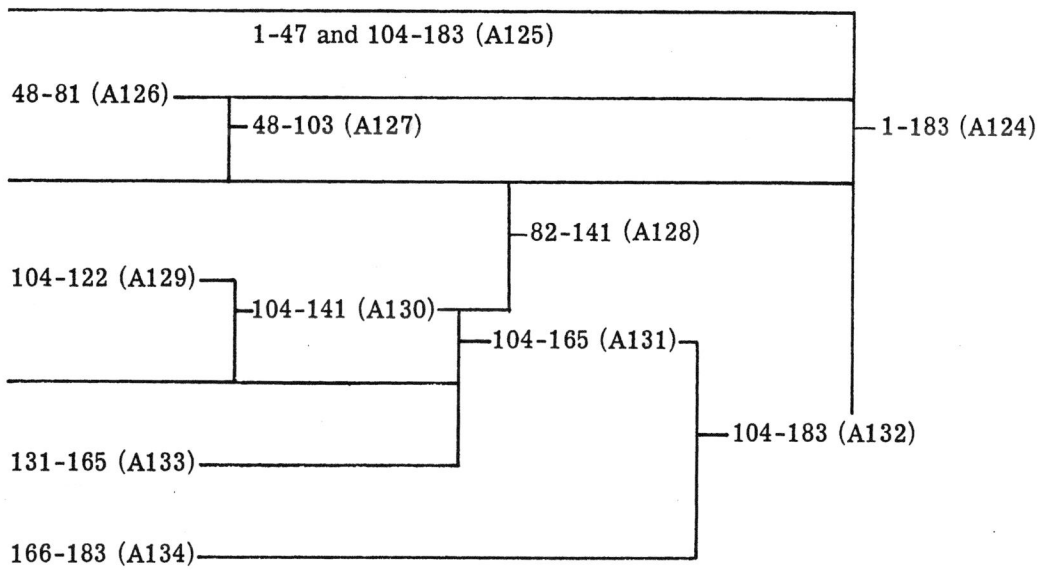

225

## TABLE XVII. PROPORTIONS IN CATULLUS LXIV

| No. | m/M: type | Major | Total lines | Minor | Total lines | Ratio: M/(M+m) |
|---|---|---|---|---|---|---|
| C1. | c/(a+b) | 1-51<br>52-250 | 250 | 251-408 | 158 | .613 |
| C2. | b/a | 1-7 | 7 | 8-11 | 4 | .636 |
| C3. | a/b | 12-30 | 19 | 1-11 | 11 | .633 |
| C4. | b/(a+c) | 1-11<br>31-51 | 32 | 12-30 | 19 | .627 |
| C5. | a/b | 19-30 | 12 | 12-18 | 7 | .632 |
| C6. | b/a | 12-30 | 19 | 31-42 | 12 | .613 |
| C7. | (a+c)/b | 34-46 | 13 | 31-33<br>47-51 | 8 | .619 |
| C8. | b/(a+c) | 34-37<br>43-46 | 8 | 38-42 | 5 | .615 |
| C9. | (a+c)/(b+d) | 58-67<br>71-75 | 15 | 52-57<br>68-70 | 9 | .625 |
| C10. | a/b | 76-115 | 40 | 52-75 | 24 | .625 |
| C11. | (a+c)/b | 116-237 | 122 | 52-115<br>238-250 | 77 | .613 |
| C12. | a/b | 80-85 | 6 | 76-79 | 4 | .60 |
| C13. | b/a | 86-104 | 19 | 105-115 | 11 | .633 |
| C14. | (a+c)/(b+d) | 124-131<br>143-163 | 29 | 116-123<br>132-142 | 19 | .604 |
| C15. | b/a | 116-163 | 48 | 164-191 | 28 | .632 |
| C16. | (b+d)/(a+c) | 132-142<br>149-157 | 20 | 143-148<br>158-163 | 12 | .625 |
| C17. | b/(a+c) | 132-163<br>192-201 | 42 | 164-191 | 28 | .60 |
| C18. | (a+c)/(b+d) | 171-183<br>192-201 | 23 | 164-170<br>184-191 | 15 | .605 |
| C19. | b/a | 171-183 | 13 | 184-191 | 8 | .619 |
| C20. | (b+d)/(a+c) | 171-183<br>192-197 | 19 | 184-191<br>198-201 | 12 | .613 |
| C21. | b/a | 192-197 | 6 | 198-201 | 4 | .60 |
| C22. | b/(a+c) | 202-206<br>212-214 | 8 | 207-211 | 5 | .615 |
| C23. | (a+c)/(b+d) | 207-214<br>228-250 | 31 | 202-206<br>215-227 | 18 | .633 |

TABLE XVII. PROPORTIONS IN CATULLUS LXIV

| No. | m/M: type | Major | Total lines | Minor | Total lines | Ratio: M/(M+m) |
|---|---|---|---|---|---|---|
| C24. | b/a | 212-227 | 16 | 228-237 | 10 | .615 |
| C25. | (a+c)/b | 241-248 | 8 | 238-240<br>249-250 | 5 | .615 |
| C26. | a/b | 278-322 | 45 | 251-277 | 28 | .616 |
| C27. | b/(a+c) | 251-322<br>384-408 | 98 | 323-383 | 60 | .620 |
| C28. | b/(a+c) | 323-327<br>334-337 | 9 | 328-333 | 6 | .60 |
| C29. | b/(a+c) | 338-352<br>366-371 | 21 | 353-365 | 13 | .618 |
| C30. | a/(b+c) | 376-381<br>382-383 | 7 | 372-375 | 4 | .636 |
| C31. | b/(a+c) | 384-396<br>407-408 | 15 | 397-406 | 10 | .60 |

**TABLE XVIII**

**CHART-INDEX OF CATULLUS LXIV**

1-51    Peleus and Thetis

    1-30    Meeting of Peleus and Thetis
    31-46    The wedding
    47-51    Drapery of the couch

        52-250    The story of Ariadne

            52-75      Ariadne deserted
            76-115     Theseus and the Minotaur
            116-201    Lament and curse of Ariadne
            202-250    Effect of curse; death of Aegeus

251-408    Peleus and Thetis

    251-266    Drapery of the couch
    267-322    The wedding and the Parcae
    323-383    The wedding song
    384-408    Gods and men

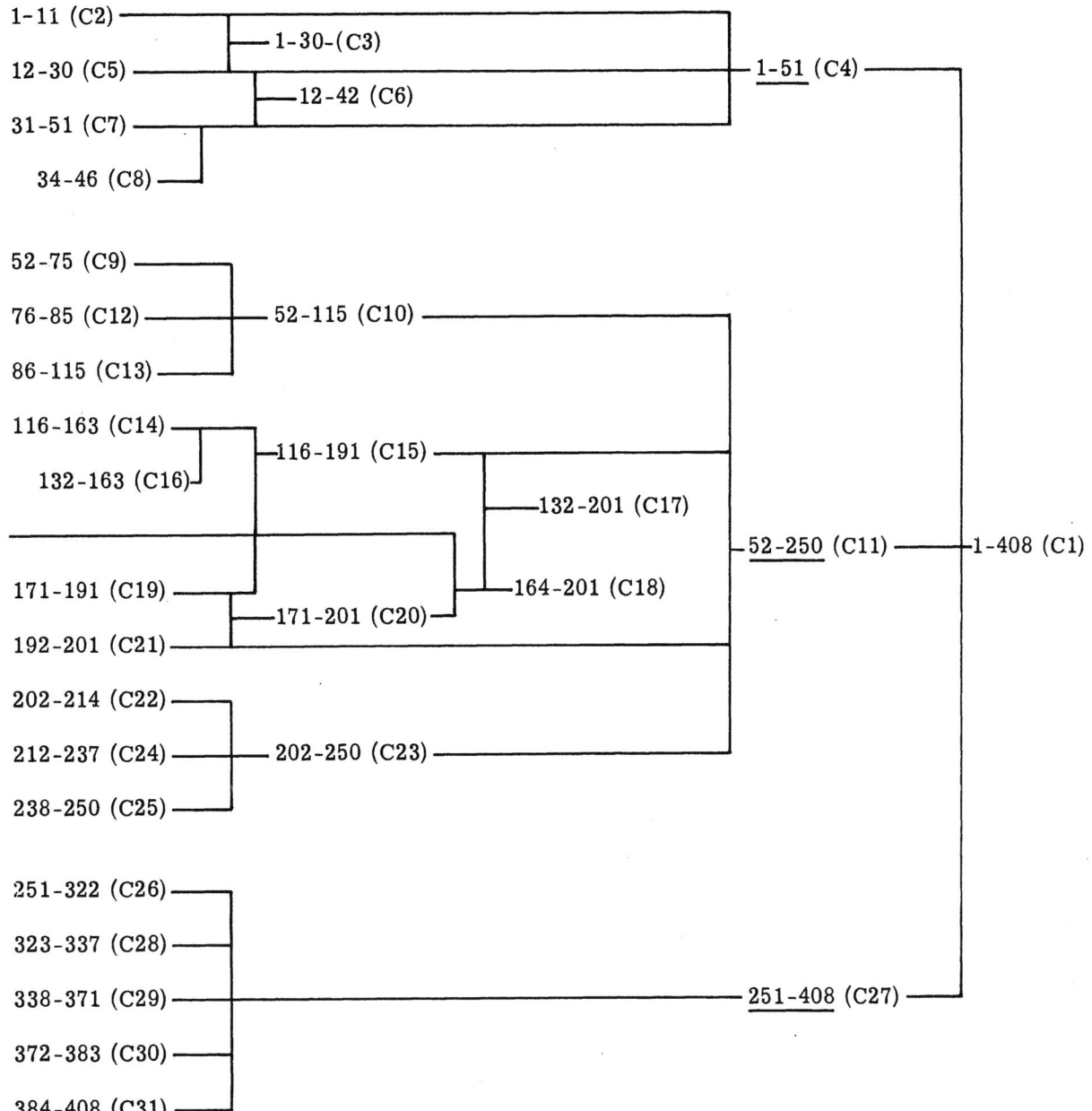

## TABLE XIX. THE MAIN DIVISIONS OF THE *DE RERUM NATURA* IN PROPORTION

| No. | m/M: type | Book | Major | Total lines | Minor | Total lines | Ratio: M/(M+m) |
|---|---|---|---|---|---|---|---|
| L1. | (a+c)/(b+d) | I | 146-634<br>921-1117 | 687 | 1-145<br>635-920 | 434 | .613 |
| L2. | c/b | | 146-634 | 489 | 635-920 | 288 | .629 |
| L3. | (c+d)/(a+b) | II | 1-332<br>333-729 | 732 | 730-1022<br>1023-1174 | 448 | .620 |
| L4. | b/(a+c) | | 1-332<br>730-1022 | 630 | 333-729 | 398 | .613 |
| L5. | c/(a+d) | | 1-332<br>1023-1174 | 486 | 730-1022 | 296 | .621 |
| L6. | (a+b)/(c+d) | III | 417-829<br>830-1094 | 677 | 1-93<br>94-416 | 417 | .619 |
| L7. | b/(a+c) | | 1-93<br>417-829 | 505 | 94-416 | 324 | .609 |
| L8. | d/c | | 417-829 | 412 | 830-1094 | 265 | .609 |
| L9. | (a+d+e)/(b+c) | IV | 26-215<br>216-822 | 790 | 1-25<br>823-1057<br>1058-1287 | 490 | .617 |
| L10. | (b+d)/(a+c) | | 1-25<br>216-822 | 634 | 26-215<br>823-1057 | 416 | .604 |
| L11. | (b+e)/(a+c) | | 1-25<br>216-822 | 634 | 26-215<br>1058-1287 | 411 | .607 |
| L12. | (a+d)/(b+e) | | 26-215<br>1058-1287 | 411 | 1-25<br>823-1057 | 260 | .612 |
| L13. | (a+e)/(b+d) | | 26-215<br>823-1057 | 416 | 1-25<br>1058-1287 | 255 | .620 |
| L14. | d/(b+e) | | 26-215<br>1058-1287 | 411 | 823-1057 | 235 | .636 |
| L15. | (b+d)/(a+c+e) | V | 1-90<br>509-770<br>925-1457 | 887 | 91-508<br>772-924 | 571 | .608 |
| L16. | (c+d)/(a+e) | | 1-90<br>925-1457 | 624 | 509-770<br>772-924 | 416 | .60 |
| L17. | a/d | | 772-924 | 153 | 1-90 | 91 | .627 |
| L18. | c/b | | 91-508 | 418 | 509-770 | 263 | .614 |

TABLE XIX. THE MAIN DIVISIONS OF THE DE RERUM NATURA IN PROPORTION

| No. | m/M: type | Book | Major | Total lines | Minor | Total lines | Ratio: M/(M+m) |
|---|---|---|---|---|---|---|---|
| L19. | d/c | | 509-770 | 263 | 772-924 | 153 | .632 |
| L20. | b/(a+c+d) | VI | 1-42<br>535-1089<br>1090-1286 | 798 | 43-534 | 491 | .619 |
| L21. | b/(c+d) | | 535-1089<br>1090-1286 | 756 | 43-534 | 491 | .606 |

## TABLE XX. PROPORTIONS IN THE *DE RERUM NATURA*, BOOK I

| No. | m/M: type | Major | Total lines | Minor | Total lines | Ratio: M/(M+m) |
|---|---|---|---|---|---|---|
| L1. | (a+c)/(b+d) | 146-634<br>921-1117 | 687 | 1-145<br>635-920 | 434 | .613 |
| L22. | a/b | 21-49 | 30 | 1-20 | 20 | .60 |
| L23. | (b+c)/a | 1-49 | 50 | 50-61<br>62-79 | 30 | .625 |
| L24. | (b+d)/(a+c) | 1-49<br>62-101 | 90 | 50-61<br>102-145 | 56 | .616 |
| L25. | a/b | 62-79 | 18 | 50-61 | 12 | .60 |
| L26. | (b+d)/(a+c) | 62-79<br>102-135 | 52 | 80-101<br>136-145 | 32 | .619 |
| L27. | b/(a+c) | 80-92<br>101 | 14 | 93-100 | 8 | .636 |
| L28. | a/b | 102-135 | 34 | 80-101 | 22 | .607 |
| L29. | (a+c)/(b+d) | 112-119<br>127-145 | 27 | 102-111<br>120-126 | 17 | .614 |
| L30. | a/b | 151-158 | 8 | 146-150 | 5 | .615 |
| L31. | a/(b+c) | 215-264<br>265-328 | 114 | 146-214 | 69 | .623 |
| L32. | a/(b+c) | 329-482<br>483-634 | 306 | 146-328 | 183 | .626 |
| L2. | b/a | 146-634 | 489 | 635-920 | 288 | .629 |
| L33. | (b+d)/(a+c) | 159-168<br>174-198 | 35 | 169-173<br>199-214 | 21 | .625 |
| L34. | b/(a+c) | 215-224<br>244-264 | 31 | 225-243 | 19 | .620 |
| L35. | (a+c)/b | 271-310 | 40 | 265-270<br>311-328 | 24 | .625 |
| L36. | b/a | 265-328 | 64 | 329-369 | 41 | .610 |
| L37. | b/a | 329-353 | 25 | 354-369 | 16 | .610 |
| L38. | (b+d)/(a+c+e) | 329-369<br>398-417<br>449-482 | 95 | 370-397<br>418-448 | 59 | .617 |
| L39. | b/(a+c) | 370-383<br>395-397 | 17 | 384-394 | 11 | .607 |
| L40. | b/a | 370-417 | 48 | 418-448 | 31 | .608 |
| L41. | b/a | 398-409 | 12 | 410-417 | 8 | .60 |

TABLE XX. PROPORTIONS IN THE DE RERUM NATURA, BOOK I

| No. | m/M: type | Major | Total lines | Minor | Total lines | Ratio: M/(M+m) |
|---|---|---|---|---|---|---|
| L42. | a/b | 418-448 | 31 | 398-417 | 20 | .608 |
| L43. | b/a | 398-448 | 51 | 449-482 | 34 | .60 |
| L44. | a/b | 430-448 | 19 | 418-429 | 12 | .613 |
| L45. | b/(a+c) | 449-450, 464-482 | 21 | 451-463 | 13 | .618 |
| L46. | b/a | 449-482 | 34 | 483-502 | 20 | .630 |
| L47. | (a+c)/b | 485-496 | 12 | 483-484, 497-502 | 8 | .60 |
| L48. | (a+c)/b | 503-598 | 96 | 483-502, 599-634 | 56 | .632 |
| L49. | (a+c)/b | 511-539 | 29 | 503-510, 540-550 | 19 | .604 |
| L50. | a/b | 551-634 | 84 | 503-550 | 48 | .636 |
| L51. | b/(a+c) | 551-564, 577-583 | 21 | 565-576 | 12 | .636 |
| L52. | a/(b+c) | 584-598, 599-634 | 51 | 551-583 | 33 | .607 |
| L53. | b/a | 584-598 | 15 | 599-608 | 10 | .60 |
| L54. | b/a | 584-614 | 31 | 615-634 | 20 | .608 |
| L55. | b/a | 599-608 | 10 | 609-614 | 6 | .625 |
| L56. | b/a | 615-622 | 8 | 623-627 | 5 | .615 |
| L57. | a/b | 623-634 | 12 | 615-622 | 8 | .60 |
| L58. | b/a | 635-640 | 6 | 641-644 | 4 | .60 |
| L59. | (b+d)/(a+c+e) | 635-644, 705-781, 830-920 | 180 | 645-704, 782-829 | 108 | .625 |
| L60. | b/(a+c+d) | 635-704, 830-920, 921-950 | 193 | 705-829 | 125 | .607 |
| L61. | (a+c)/b | 665-700 | 36 | 645-664, 701-704 | 24 | .60 |
| L62. | a/b | 716-733 | 18 | 705-715 | 11 | .621 |
| L63. | b/a | 705-781 | 77 | 782-829 | 48 | .616 |
| L64. | b/(a+c) | 734-741, 753-762 | 18 | 742-752 | 11 | .621 |
| L65. | b/a | 734-762 | 29 | 763-781 | 19 | .604 |
| L66. | a/b | 770-781 | 12 | 763-769 | 7 | .632 |
| L67. | a/b | 790-802 | 13 | 782-789 | 8 | .619 |
| L68. | (a+c)/(b+d) | 809-822, 827-829 | 17 | 803-808, 823-826 | 10 | .630 |

| No. | m/M: type | Major | Total lines | Minor | Total lines | Ratio: M/(M+m) |
|---|---|---|---|---|---|---|
| L69. | b/a | 830-858 | 29 | 859-874 | 18 | .617 |
| L70. | (a+c)/(b+d) | 843-874<br>897-920 | 58 | 830-842<br>875-896 | 35 | .624 |
| L71. | c/(a+b) | 859-874<br>875-896 | 40 | 897-920 | 24 | .625 |
| L72. | (b+d)/(a+c) | 897-903<br>907-914 | 15 | 904-906<br>915-920 | 9 | .625 |
| L73. | b/(a+c) | 921-930<br>943-950 | 18 | 931-942 | 12 | .60 |
| L74. | a/b | 951-1001 | 51 | 921-950 | 30 | .630 |
| L75. | (a+c)/(b+d) | 951-1007<br>1052-1117 | 123 | 921-950<br>1008-1051 | 75 | .621 |
| L76. | (a+c)/(b+d) | 958-983<br>998-1007 | 36 | 951-957<br>984-997 | 21 | .632 |
| L77. | c/(a+b) | 951-1001<br>1002-1051 | 102 | 1052-1113 | 62 | .622 |
| L78. | c/(a+b) | 951-1001<br>1002-1051 | 102 | 1052-1117 | 66 | .607 |
| L79. | (a+c)/b | 968-997 | 30 | 958-967<br>998-1007 | 20 | .60 |
| L80. | b/(a+c) | 1008-1020<br>1038-1051 | 28 | 1021-1037 | 17 | .622 |
| L81. | (b+d)/(a+c) | 1052-1060<br>1083-1113 | 40 | 1061-1082<br>1114-1117 | 26 | .606 |

**TABLE XXI**

**CHART-INDEX OF BOOK I OF**
***DE RERUM NATURA***

## De Rerum Natura, Book I

1-145 Introduction

- 1-49 Invocation to Venus
- 50-61 Address to Memmius
- 62-79 Praise of Epicurus
- 80-101 Wrongs committed in religion's name
- 102-135 Reasons for accepting the doctrine of Epicurus
- 136-145 Difficulty of treating Greek philosophy in Latin verse

146-634 The Atomic Theory

- 146-214 Nothing is created out of nothing
- 215-264 Nothing is destroyed into nothing
- 265-328 Matter exists in the form of small particles
- 329-417 Void exists
- 418-448 The universe consists of matter and void
- 449-482 Everything else is property or accident of these two
- 483-634 The first particles are solid, eternal, and indivisible

635-920 Criticism of rival theories

- 635-704 Heraclitus
- 705-829 Empedocles
- 830-920 Anaxagoras

921-1117 The Infinity of the Universe

- 921-950 Introduction: Lucretius' mission
- 951-1001 Infinity of the universe
- 1002-1007 Infinity of space
- 1008-1051 Infinity of matter
- 1052-1113 Refutation of a false theory
- 1114-1117 Conclusion

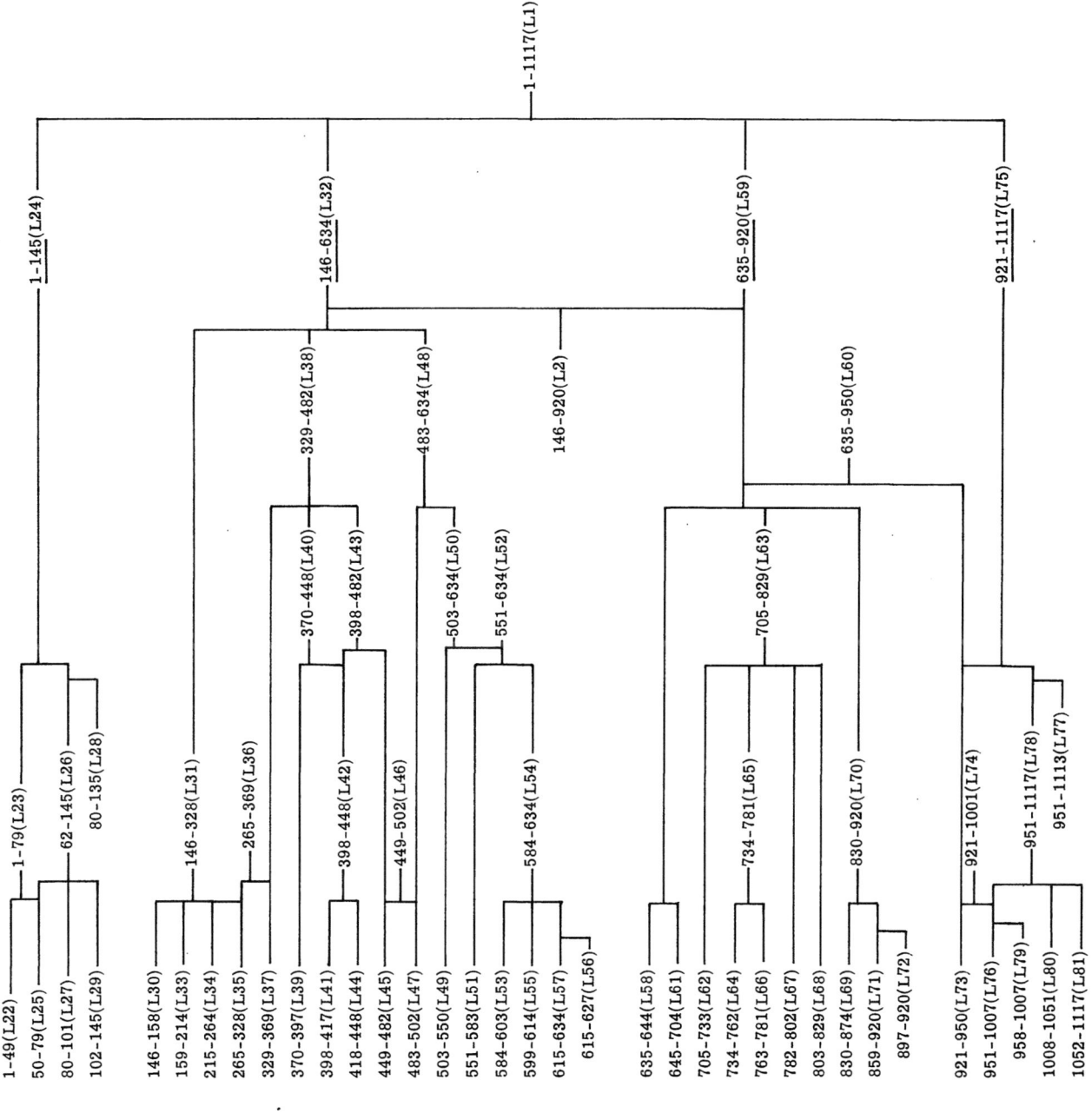

## TABLE XXII. PROPORTIONS IN *GEORG.* IV, 281-558

| No. | m/M: type | Major | Total lines | Minor | Total lines | Ratio: M/(M+m) |
|---|---|---|---|---|---|---|
| G1. | (b+c)/a | 281-452 | 171 | 453-529<br>530-558 | 106 | .617 |
| G2. | (a+c)/b | 287-307 | 21 | 281-286<br>308-314 | 13 | .618 |
| G3. | (a+c)/(b+d) | 287-307<br>321-332 | 33 | 281-286<br>308-320 | 19 | .635 |
| G4. | (b+c)/a | 287-307 | 21 | 308-314<br>315-320 | 13 | .618 |
| G5. | c/(a+b) | 281-332<br>333-386 | 105 | 387-452 | 66 | .614 |
| G6. | a/b | 303-314 | 12 | 295-302 | 8 | .60 |
| G7. | (b+d)/(a+c) | 295-314<br>317-320 | 24 | 315-316<br>321-332 | 14 | .632 |
| G8. | b/a | 333-347 | 14 | 348-356 | 9 | .609 |
| G9. | (a+c)/(b+d) | 348-356<br>363-386 | 33 | 333-347<br>357-362 | 20 | .623 |
| G10. | b/(a+c) | 357-362<br>374-386 | 19 | 363-373 | 11 | .633 |
| G11. | a/b | 398-414 | 17 | 387-397 | 11 | .607 |
| G12. | (b+c)/(a+d) | 387-397<br>423-452 | 41 | 398-414<br>415-422 | 25 | .621 |
| G13. | b/a | 415-431 | 17 | 432-442 | 11 | .607 |
| G14. | b/(a+c) | 415-422<br>437-452 | 24 | 423-436 | 14 | .632 |
| G15. | b/(a+c) | 453-466<br>485-498 | 28 | 467-484 | 18 | .609 |
| G16. | c/(a+b) | 453-466<br>467-498 | 46 | 499-527 | 29 | .613 |
| G17. | b/(a+c) | 453-498<br>528-529 | 48 | 499-527 | 29 | .623 |
| G18. | b/a | 499-503 | 5 | 504-506 | 3 | .625 |
| G19. | b/(a+c) | 499-515<br>528-529 | 19 | 516-527 | 12 | .613 |
| G20. | b/a | 530-540 | 11 | 541-547 | 7 | .611 |
| G21. | b/a | 530-547 | 18 | 548-558 | 11 | .621 |
| G22. | a/b | 552-558 | 7 | 548-551 | 4 | .636 |

**TABLE XXIII**

**CHART-INDEX OF**
***GEORG.*** **IV, 281-558**

**The Aristaeus story**
(*Georg.* IV, 281-558)

281-452 The story of Aristaeus
    281-294    Introduction
    295-314    Aristaeus' device to produce bees
    315-332    Aristaeus' lament to Cyrene
    333-356    The sister nymphs
    357-386    Aristaeus visits Cyrene
    387-414    Cyrene's instructions about Proteus
    415-452    Aristaeus and Proteus

        453-529    Proteus tells the story of Orpheus and Eurydice
            453-484    Orpheus in the Underworld
            485-498    Loss of Eurydice
            499-529    Grief and death of Orpheus

530-558 The birth of the bees
    530-547    Instructions of Cyrene
    548-558    Aristaeus follows the instructions

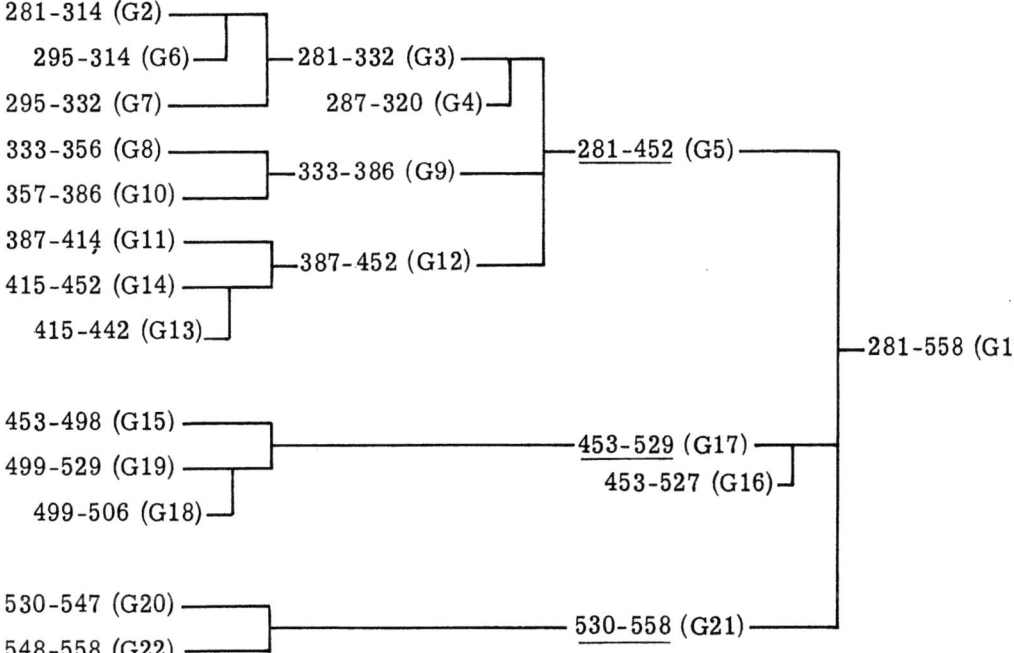

## TABLE XXIV. PROPORTIONS IN THE *SATIRES* AND *EPISTLES* OF HORACE

| Poem | m/M: type | Major | Total lines | Minor | Total lines | Ratio: M/(M+m) |
|---|---|---|---|---|---|---|
| *Satires* | | | | | | |
| I, 1 | b/(a+c) | 1-60<br>108-121 | 74 | 61-107 | 47 | .612 |
| I, 2 | (a+c)/b | 28-110 | 83 | 1-27<br>111-134 | 51 | .619 |
| I, 3 | a/b | 55-142 | 88 | 1-54 | 54 | .620 |
| I, 4 | (a+c)/b | 38b-126a | 87.8 | 1-38a<br>126b-143 | 55.2 | .614 |
| I, 5 | b/(a+c) | 1-46<br>86-104 | 65 | 47-85 | 39 | .625 |
| I, 7 | b/a | 1-21 | 21 | 22-35 | 14 | .60 |
| I, 8 | (b+d)/(a+c) | 1-13<br>23-39 | 30 | 14-22<br>40-50 | 18 | .625 |
| I, 10 | a/b | 36-92 | 57 | 1-35 | 35 | .620 |
| II, 2 | a/b | 53-136 | 84 | 1-52 | 52 | .618 |
| II, 3 | b/(a+c) | 1-157<br>281-326 | 203 | 158-280 | 123 | .623 |
| II, 6 | b/(a+c) | 1-15<br>59-117 | 74 | 16-58 | 43 | .632 |
| II, 7 | b/(a+c) | 1-37a<br>83-118 | 72.2 | 37b-82 | 45.8 | .612 |
| II, 8 | b/(a+c) | 1-17<br>54-95 | 59 | 18-53 | 36 | .621 |
| *Epistles* | | | | | | |
| I, 1 | b/(a+c) | 1-27<br>70-108 | 66 | 28-69 | 42 | .611 |
| I, 2 | b/a | 1-43 | 43 | 44-71 | 28 | .606 |
| I, 4 | b/(a+c) | 1-5<br>12-16 | 10 | 6-11 | 6 | .625 |
| I, 6 | a/b | 28-68 | 41 | 1-27 | 27 | .603 |
| I, 7 | a/b | 40-98 | 59 | 1-39 | 39 | .602 |
| I, 10 | (a+c)/b | 12-41 | 30 | 1-11<br>42-50 | 20 | .60 |
| I, 11 | b/(a+c) | 1-10<br>22-30 | 19 | 11-21 | 11 | .633 |

TABLE XXIV. PROPORTIONS IN THE SATIRES AND EPISTLES OF HORACE

| Poem | m/M: type | Major | Total lines | Minor | Total lines | Ratio: M/(M+m) |
|---|---|---|---|---|---|---|
| *Epistles* | | | | | | |
| I, 12 | a/b | 12-29 | 18 | 1-11 | 11 | .621 |
| I, 14 | b/(a+c) | 1-13<br>31-44 | 27 | 14-30 | 17 | .614 |
| I, 17 | (a+c)/b | 13-51 | 39 | 1-12<br>52-62 | 23 | .629 |
| I, 18 | (b+d)/(a+c) | 1-20<br>37-85 | 69 | 21-36<br>86-112 | 43 | .616 |
| I, 19 | a/b | 19-49 | 31 | 1-18 | 18 | .633 |
| II, 1 | a/(b+c) | 103-207<br>208-270 | 168 | 1-102 | 102 | .615 |
| II, 2 | (b+d)/(a+c) | 1-105<br>145-174 | 135 | 106-144<br>175-216 | 81 | .625 |
| II, 3 | b/a | 1-294 | 294 | 295-476 | 182 | .618 |

## TABLE XXV. FIBONACCI SERIES IN THE *ARS POETICA* OF HORACE

| m/M: type | Major | Total lines | Minor | Total lines | Ratio: M/(M+m) |
|---|---|---|---|---|---|
| a/b | 6-13 | 8 | 1-5 | 5 | .615 |
| b/a | 24-28 | 5 | 29-31 | 3 | .625 |
| b/a | 32-37 | 6 | 38-41 | 4 | .60 |
| (a+c)/b | 75-82 | 8 | 73-74<br>83-85 | 5 | .615 |
| b/(a+c) | 86-87<br>93-98 | 8 | 88-92 | 5 | .615 |
| a/b | 99-119 | 21 | 86-98 | 13 | .618 |
| b/a | 99-111 | 13 | 112-119 | 8 | .619 |
| b/a | 120-124 | 5 | 125-127 | 3 | .625 |
| b/a | 128-139 | 12 | 140-147 | 8 | .60 |
| b/a | 140-147 | 8 | 148-152 | 5 | .615 |
| (b+d)/(a+c) | 140-147<br>153-157 | 13 | 148-152<br>158-160 | 8 | .619 |
| b/a | 153-157 | 5 | 158-160 | 3 | .625 |
| (b+d)/(a+c) | 153-157<br>161-165 | 10 | 158-160<br>166-168 | 6 | .625 |
| b/a | 153-168 | 16 | 169-178 | 10 | .615 |
| b/a | 161-165 | 5 | 166-168 | 3 | .625 |
| b/a | 169-174 | 6 | 175-178 | 4 | .60 |
| b/a | 275-280 | 6 | 281-284 | 4 | .60 |
| (b+c)/(a+d) | 275-280<br>289-294 | 12 | 281-284<br>285-288 | 8 | .60 |
| a/b | 289-294 | 6 | 285-288 | 4 | .60 |
| a/b | 312-316 | 5 | 309-311 | 3 | .625 |
| (a+c)/(b+d) | 312-316<br>323-332 | 15 | 309-311<br>317-322 | 9 | .625 |
| a/b | 323-332 | 10 | 317-322 | 6 | .625 |
| b/a | 366-373 | 8 | 374-378 | 5 | .615 |
| b/(a+c) | 453-460<br>470-476 | 15 | 461-469 | 9 | .625 |

## TABLE XXVI. PROPORTIONS IN THE "THIRTEENTH BOOK" OF MAPHAEUS VEGIUS

| No. | m/M: type | Major | Total lines | Minor | Total lines | Ratio: M/(M+m) |
|---|---|---|---|---|---|---|
| M1. | b/(a+c) | 1-22<br>49-66 | 40 | 23-48 | 26 | .606 |
| M2. | a/b | 49-124 | 76 | 1-48 | 48 | .613 |
| M3. | b/a | 23-28 | 6 | 29-32 | 4 | .60 |
| M4. | a/b | 36-40 | 5 | 33-35 | 3 | .625 |
| M5. | b/a | 36-40 | 5 | 41-43 | 3 | .625 |
| M6. | a/b | 44-48 | 5 | 41-43 | 3 | .625 |
| M7. | b/a | 49-54 | 6 | 55-58 | 4 | .60 |
| M8. | (a+c)/(b+d) | 51-58<br>64-66 | 11 | 49-50<br>59-63 | 7 | .611 |
| M9. | b/a | 59-63 | 5 | 64-66 | 3 | .625 |
| M10. | c/(a+b) | 67-84<br>85-102 | 36 | 103-124 | 22 | .621 |
| M11. | b/a | 103-116 | 14 | 117-124 | 8 | .636 |
| M12. | (a+c)/(b+d) | 142-184<br>216-251 | 79 | 125-141<br>185-215 | 48 | .622 |
| M13. | b/(a+c) | 125-184<br>252-301 | 110 | 185-251 | 67 | .621 |
| M14. | a/b | 302-592 | 291 | 125-301 | 177 | .622 |
| M15. | a/(b+c) | 158-176<br>177-184 | 27 | 142-157 | 16 | .628 |
| M16. | b/a | 185-203 | 19 | 204-215 | 12 | .613 |
| M17. | b/(a+c) | 185-215<br>241-251 | 42 | 216-240 | 25 | .627 |
| M18. | (b+d)/(a+c) | 216-231<br>241-246 | 22 | 232-240<br>247-251 | 14 | .611 |
| M19. | (b+d)/(a+c) | 252-273<br>284-292 | 31 | 274-283<br>293-301 | 19 | .620 |
| M20. | (a+c)/b | 329-373 | 45 | 302-328<br>374-376 | 30 | .60 |
| M21. | b/a | 302-477 | 176 | 478-592 | 115 | .605 |
| M22. | a/b | 321-328 | 8 | 316-320 | 5 | .615 |
| M23. | b/(a+c) | 329-343<br>362-376 | 30 | 344-361 | 18 | .625 |
| M24. | b/a | 377-391 | 15 | 392-401 | 10 | .60 |

| No. | m/M: type | Major | Total lines | Minor | Total lines | Ratio: M/(M+m) |
|---|---|---|---|---|---|---|
| M25. | b/a | 424-431 | 8 | 432-436 | 5 | .615 |
| M26. | b/a | 432-436 | 5 | 437-439 | 3 | .625 |
| M27. | b/a | 447-465 | 19 | 466-477 | 12 | .613 |
| M28. | b/(a+c) | 447-465<br>490-508 | 38 | 466-489 | 24 | .613 |
| M29. | a/(b+c) | 490-500<br>501-508 | 19 | 478-489 | 12 | .613 |
| M30. | b/a | 478-508 | 31 | 509-528 | 20 | .608 |
| M31. | b/a | 529-592 | 64 | 593-630 | 38 | .627 |
| M32. | b/(a+c) | 593-605<br>620-630 | 24 | 606-619 | 14 | .632 |
| M33. | a/b | 606-622 | 17 | 595-605 | 11 | .607 |
| M34. | a/b | 626-630 | 5 | 623-625 | 3 | .625 |

**TABLE XXVII**

**CHART-INDEX OF THE "THIRTEENTH BOOK"**

## The "Thirteenth Book" of the *Aeneid* by Maphaeus Vegius

1-124     Aeneas after the death of Turnus

     1-22        Effect of Aeneas' victory
     23-48       Aeneas' speech over Turnus' body
     49-74       Return to the Trojan camp
     75-82       Aeneas' speech to Iulus
     83-102      Aeneas addresses his comrades
     103-124     Attitude of Aeneas

125-301    Grief at the death of Turnus

     125-141     Rutulians take Turnus' body to Latinus
     142-184     Lament of Latinus
     185-251     Turnus' body taken to Ardea
     252-301     Lament of Daunus

302-592    Aeneas at Laurentum

     302-328     Aeneas invited to Laurentum
     329-376     Speech of Drances
     377-391     Speech of Aeneas
     392-401     Burning of the dead
     402-446     Meeting of Aeneas and Latinus; speeches
     447-477     Lavinia and her beauty
     478-489     Wedding gifts
     490-535     The wedding feast
     536-592     Founding of the city; speeches of Aeneas and Venus

593-630    The death of Aeneas

     593-605     Venus requests immortality for Aeneas
     606-622     Jupiter's assent
     623-630     Venus and the immortality of Aeneas

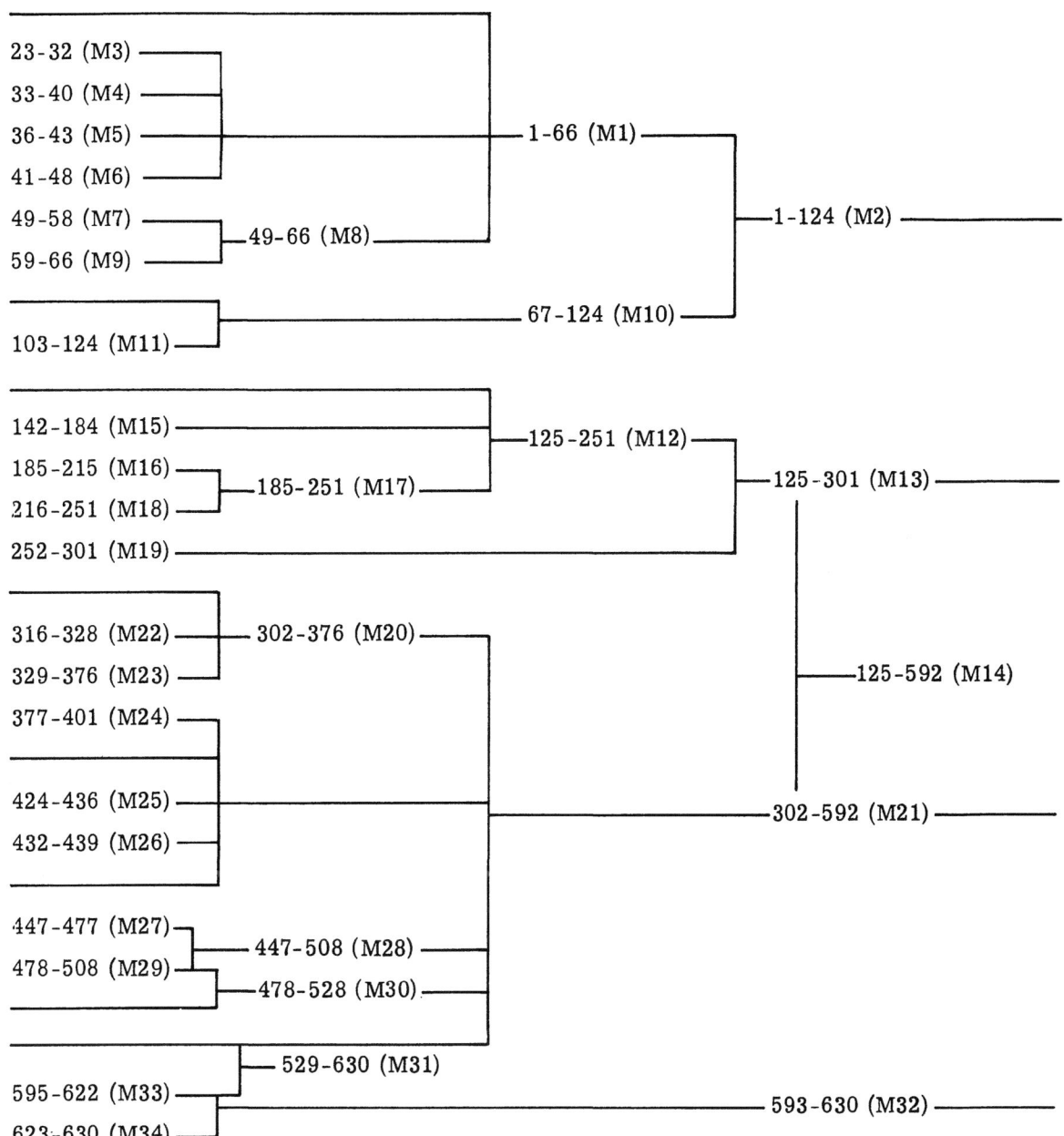

# ABBREVIATIONS

(The following abbreviations are used for references to periodicals and texts both in the Bibliography and in the notes to each chapter.)

| | |
|---|---|
| AJPh | American Journal of Philology |
| AMM | American Mathematical Monthly |
| CAH | Cambridge Ancient History |
| CJ | Classical Journal |
| CQ | Classical Quarterly |
| CPh | Classical Philology |
| CR | Classical Review |
| CW | Classical World (formerly Classical Weekly) |
| HSCPh | Harvard Studies in Classical Philology |
| IL | Information Littéraire |
| JEP | Journal of Experimental Psychology |
| JRS | Journal of Roman Studies |
| LEC | Les Études Classiques |
| LCL | Loeb Classical Library |
| OCT | Oxford Classical Texts |
| PBA | Proceedings of the British Academy |
| RA | Revue Archéologique |
| REA | Revue des Études Anciennes |
| RhM | Rheinisches Museum für Philologie |
| TAPhA | Transactions of the American Philological Association |
| YClS | Yale Classical Studies |
| ZAnt | Ziva Antika. Antiquité vivante |

# BIBLIOGRAPHY

## TEXTS, COMMENTARIES, AND TRANSLATIONS

Ennius

    I. Vahlen, *Ennianae Poesis Reliquiae* (2nd ed.; Leipzig, 1903).
    E. H. Warmington, *Remains of Old Latin. I. Ennius and Caecilius* (LCL, Cambridge, Mass., 1935).

Catullus

    R. Ellis, *Catulli Carmina* (OCT, Oxford, no date).
    F. W. Cornish, *The Poems of Gaius Valerius Catullus* (LCL, New York, 1924).
    R. A. B. Mynors, *C. Valerii Catulli Carmina* (OCT, Oxford, 1958).

Lucretius

    C. Bailey, *Lucreti De Rerum Natura Libri Sex* (OCT, 2nd ed.; Oxford, 1921).
    W. E. Leonard and S. B. Smith, *T. Lucreti Cari De Rerum Natura Libri Sex. Edited with Introduction and Commentary* (Madison, 1942).
    C. Bailey, *Titi Lucreti Cari De Rerum Natura Libri Sex. Edited with Prolegomena, Critical Apparatus, Translation, and Commentary* (Oxford, 1947). 3 vols.

Vergil

    C. G. Heyne, *Publius Virgilius Maro varietate lectionis et perpetua adnotatione illustratus.* 4th ed. by G. P. E. Wagner (Leipzig, 1830-41). 5 vols.
    A. Forbiger, *P. Vergili Maronis Opera* (Leipzig, 1872-75). 3 vols.
    J. Conington, *The Works of Virgil with a Commentary.* Revised by H. Nettleship. Vol. I, 5th ed. by F. Haverfield (London, 1898); Vol. II, 4th ed. (London, 1884); Vol. III, 3rd ed. (London, 1883).
    O. Ribbeck, *P. Vergili Maronis Opera* (2nd ed.; Leipzig, 1894-95). 4 vols. in 2.
    F. A. Hirtzel, *P. Vergili Maronis Opera* (OCT, Oxford, 1900).
    T. E. Page, *The Aeneid of Virgil* (London, 1900). 2 vols.
    Th. Ladewig, C. Schaper, and P. Deuticke, *Vergils Gedichte* (Berlin). 1. *Bukolika und Georgika*, 9th ed. by P. Jahn (1915); 2. *Buch I-VI der Äneis*, 13th ed. by P. Jahn (1912); 3. *Buch VII-XII der Äneis*, 9th ed. by P. Deuticke (1904).
    G. Janell, *P. Vergili Maronis Opera* (Leipzig, 1920).
    H. R. Fairclough, *Virgil* (LCL, New York, 1922). 2 vols.
    E. Norden, *P. Vergilius Maro. Aeneis, Buch VI* (3rd ed.; Leipzig, 1926).
    J. W. Mackail, *The Aeneid Edited with Introduction and Commentary* (Oxford, 1930).
    A. S. Pease, *Publi Vergili Maronis Aeneidos Liber Quartus* (Cambridge, Mass., 1935).
    R. Durand, *Virgile. Énéide, Livres VII-XII* (Paris, 1936).
    R. Sabbadini, *P. Vergili Maronis Opera* (2nd ed.; Rome, 1937). 2 vols.
    H. Goelzer, *Virgile. Énéide, Livres I-VI* (6th ed.; Paris, 1948).
    R. Humphries, *The Aeneid of Vergil. A Verse Translation* (New York, 1951).
    K. Guinagh, *The Aeneid of Vergil. Newly Translated with an Introduction* (New York, 1953).
    E. de Saint-Denis, *Virgile, Géorgiques* (Paris, 1956).
    W. F. J. Knight, *Virgil, The Aeneid. A New Translation* (Harmondsworth, 1956).
    W. Richter, *Vergil, Georgica* (München, 1957).

*Appendix Vergiliana*

    R. Ellis, *Appendix Vergiliana sive Carmina Minora Vergilio adtributa* (OCT, Oxford, 1907).
    J. W. Duff and A. M. Duff, *Minor Latin Poets* (LCL, Cambridge, Mass., 1934). For the *Aetna*, pp. 349-419.
    C. Van der Graaf, *The Dirae, with Translation, Commentary, and an Investigation of Its Authorship* (Leiden, 1945).
    R. Giomini, *Appendix Vergiliana. Testo, Introduzione e Traduzione* (Firenze, 1953).

Horace

    J. C. Rolfe, *Q. Horati Flacci Sermones et Epistulae* (New York, 1901).
    E. C. Wickham, *Q. Horati Flacci Opera* (OCT, 2nd ed. by H. W. Garrod; Oxford, 1912).
    A. Rostagni, *Arte Poetica di Orazio. Introduzione e Commento* (Torino, 1930).

Augustus

    J. Gagé, *Res Gestae Divi Augusti* (2nd ed.; Paris, 1950).

*Vitae Vergilianae*

    A. Rostagni, *Suetonio, De poetis e biografi minori* (Torino, 1944).
    C. Hardie, *Vitae Vergilianae Antiquae* (OCT, Oxford, 1954).

Maphaeus Vegius

    A. C. Brinton, *Maphaeus Vegius and His Thirteenth Book of the Aeneid* (Stanford, 1930).

## GENERAL

A. Alföldi, "Der neue Weltherrscher der vierten Ekloge Vergils," *Hermes* 65 (1930), pp. 369-384.
―――――, "Le basi spirituali del principato romano," *Corvina* Ser. III, 1 (1952), pp. 24-37.
―――――, "The Main Aspects of Political Propaganda on the Coinage of the Roman Republic," in R. A. G. Carson and C. H. V. Sutherland (eds.), *Essays in Roman Coinage Presented to Harold Mattingly* (Oxford, 1956), pp. 63-95.
R. C. Archibald, "Golden Section"; "A Fibonacci Series," *AMM* 25 (1918), pp. 232-238. Reprinted with corrections and additions in Hambidge, *Dynamic Symmetry. The Greek Vase* (New Haven, 1920), pp. 152-157.
C. Becker, "Virgils Eklogenbuch," *Hermes* 83 (1955), pp. 314-349.
H. Belling, *Studien über die Compositionskunst Vergils in der Aeneide* (Leipzig, 1899).
J. M. Benario, "Book 4 of Horace's *Odes*: Augustan Propaganda," *TAPhA* 91 (1960), pp. 339-352.
E. Bickel, "Syllabus Indiciorum Quibus Pseudovergiliana et Pseudoovidiana Carmina Definiantur: Symbolae ad Cirim, Culicem, Aetnam," *RhM* 93 (1949-50), pp. 289-324.
G. D. Birkhoff, *Aesthetic Measure* (Cambridge, Mass., 1933).
M. Borissavlievitch, *The Golden Number and the Scientific Aesthetics of Architecture* (New York, 1958).
B. Bosanquet, *A History of Aesthetic* (2nd ed.; London, 1904).
P. Boyancé, "Sur quelques vers de Virgile (*Géorgiques*, II, v. 490-492)," *RA* 25 (1927), pp. 361-379.
B. Brotherton, "Vergil's Catalogue of the Latin Forces," *TAPhA* 62 (1931), pp. 192-202.
E. L. Brown, *Studies in the Eclogues and Georgics of Vergil* (Ann Arbor, 1961). [Princeton University dissertation, microfilmed.]
K. Büchner, *P. Vergilius Maro, der Dichter der Römer* (Stuttgart, 1956).
E. Burck, "Der korykische Greis in Vergils Georgica (IV, 116-148)," *Navicula Chiloniensis. Studia Philologa Felici Jacoby . . . oblata* (Leiden, 1956), pp. 156-172.
G. Caiati, *Vita di Virgilio* (Padova, 1952).
W. A. Camps, "A Note on the Structure of the *Aeneid*," *CQ* N.S. 4 (1954), pp. 214-215.

_____, "A Second Note on the Structure of the *Aeneid*," *CQ* N.S. 9 (1959), pp. 53-56.
V. Capparelli, "Ludus Pythagoricus e Divina proporzione," *Sophia* 26 (1958), pp. 197-210.
J. Carcopino, *Virgile et le mystère de la IV$^e$ Églogue* (2nd ed.; Paris, 1943).
A. Cartault, *Étude sur les Bucoliques de Virgile* (Paris, 1897).
_____, *L'Art de Virgile dans l'Énéide* (Paris, 1926).
M. P. Charlesworth, "The Virtues of a Roman Emperor: Propaganda and the Creation of Belief," *PBA* 23 (1937), pp. 105-133.
B. H. Clark. See under E. O'Neill.
L.-A. Constans, *L'Énéide de Virgile. Étude et analyse* (Paris, 1938).
R. S. Conway, *Harvard Lectures on the Vergilian Age* (Cambridge, Mass., 1928).
_____, "Vergil's Creative Art," *PBA* 17 (1931), pp. 17-38.
A. M. Cook, "Virgil, *Aen.* VII. 7. 641 ff.," *CR* 33 (1919), pp. 103-104.
P. Courcelle, "Interprétations néo-platonisantes du livre VI de l'*Énéide*," in *Recherches sur la tradition platonicienne* [= *Entretiens sur l'antiquité classique*, Tome III, Fondation Hardt] (Genève, 1955), pp. 95-136.
M. M. Crump, *The Growth of the Aeneid* (Oxford, 1920).
A. Delatte, *Études sur la littérature pythagoricienne* (Paris, 1915).
M. Desport, *L'Incantation Virgilienne, Virgile et Orphée* (Paris, 1952).
N. W. DeWitt, *Virgil's Biographia Litteraria* (Toronto, 1923).
D. L. Drew, *The Allegory of the Aeneid* (Oxford, 1927).
_____, "The Structure of Vergil's Georgics, "*AJPh* 50 (1929), pp. 242-254.
G. E. Duckworth, *Foreshadowing and Suspense in the Epics of Homer, Apollonius, and Vergil* (Princeton, 1933).
_____, "Turnus as a Tragic Character," *Vergilius* 4 (1940), pp. 5-17.
_____, "The Architecture of the *Aeneid*," *AJPh* 75 (1954), pp. 1-15.
_____, "Fate and Free Will in Vergil's *Aeneid*," *CJ* 51 (1955-56), pp. 357-364.
_____, "*Animae Dimidium Meae:* Two Poets of Rome," *TAPhA* 87 (1956), pp. 281-316.
_____, "The *Aeneid* as a Trilogy," *TAPhA* 88 (1957), pp. 1-10.
_____, "Recent Work on Vergil (1940-1956)," *CW* 51 (1957-58), pp. 89-92, 116-117, 123-128, 151-159, 185-193, 228-235.
_____, "Vergil's *Georgics* and the *laudes Galli*," *AJPh* 80 (1959), pp. 225-237.
_____, "Mathematical Symmetry in Vergil's *Aeneid*," *TAPhA* 91 (1960), pp. 184-220.
H. R. Fairclough, "The Helen Episode in Vergil's *Aeneid* ii. 559-623," *CPh* 1 (1906), pp. 221-230.
G. T. Fechner, *Vorschule der Aesthetik* (3rd ed.; Leipzig, 1925). 2 vols.
B. Fenik, "Parallelism of Theme and Imagery in *Aeneid* II and IV," *AJPh* 80 (1959), pp. 1-24.
_____, *The Influence of Euripides on Vergil's Aeneid* (Ann Arbor, 1960). [Princeton University dissertation, microfilmed.]
W. W. Fowler, *Virgil's "Gathering of the Clans," Being Observations on Aeneid VII. 601-817* (2nd ed.; Oxford, 1918).
_____, *Aeneas at the Site of Rome* (Oxford, 1918).
E. Fraenkel, "Some Aspects of the Structure of Aeneid vii," *JRS* 35 (1945), pp. 1-14.
_____, "The Culex," *JRS* 42 (1952), pp. 1-9.
T. Frank, *Vergil, A Biography* (New York, 1922).
K. Gantar, "Struktura Horacove epistule Numiciju," *ZAnt* 3 (1953), pp. 79-81.
_____, "De compositione Horatii 'Epistulae ad Pisones,' " *ZAnt* 4 (1954), p. 277.
J. Gerloff, *Vindiciae Vergilianae. Quaestiones Criticae de Aeneidis libri II 567-588* (Jena, 1911).
R. J. Getty, "Neopythagoreanism and Mathematical Symmetry in Lucan, *De bello civili* 1," *TAPhA* 91 (1960), pp. 310-323.
M. C. Ghyka, *Le nombre d'Or. Rites et rhythmes pythagoriciens dans le développement de la civilisation occidentale* (Paris, 1931). 2 vols.
_____, *The Geometry of Art and Life* (New York, 1946).
_____, "The Pythagorean and Platonic Scientific Criterion of the Beautiful in Classical Western Art," in F. S. C. Northrup (ed.), *Ideological Differences and World Order. Studies in the Philosophy and Science of the World's Cultures* (New Haven, 1949), pp. 90-116.

H. Graf, *Bibliographie zum Problem der Proportionen: Literatur über Proportionen, Mass und Zahl in Architektur, bildender Kunst und Natur. Teil I: von 1800 bis zur Gegenwart* (Speyer, 1958).

P. Grimal, *Les jardins romains à la fin de la république et aux deux premiers siècles de l'empire* (Paris, 1943).

_____, "La promenade d'Évandre et d'Énée à la lumière des fouilles récentes," *REA* 50 (1948), pp. 348-351.

A. Gudeman (ed.), *Latin Literature of the Empire* (New York, 1899). 2 vols.

A. M. Guillemin, *Virgile, Poète, Artiste et Penseur* (Paris, 1951).

E. A. Hahn, "Vergil's Catalogue of the Latin Forces: A Reply to Professor Brotherton," *TAPhA* 63 (1932), pp. lxii-lxiii.

_____, "The Characters in the *Eclogues*," *TAPhA* 75 (1944), pp. 196-241.

J. Hambidge, *Dynamic Symmetry. The Greek Vase* (New Haven, 1920).

_____, *The Elements of Dynamic Symmetry* (New York, 1926).

N. L. Hatch, "The Time Element in Interpretation of *Aeneid* 2. 575-76 and 585-87," *CPh* 54 (1959), pp. 255-257.

R. Heinze, *Virgils epische Technik* (3rd ed.; Berlin, 1915).

G. Hirst, "An Attempt to Date the Composition of *Aeneid* VII," *CQ* 10 (1916), pp. 87-96.

A. Klotz, "Das Ordnungsprinzip in Vergils Bucolica," *RhM* 64 (1909), pp. 325-327.

W. F. J. Knight, *Vergil's Troy. Essays on the Second Book of the Aeneid* (Oxford, 1932).

_____, *Accentual Symmetry in Vergil* (Oxford, 1939).

_____, "Integration of Plot in the Aeneid," *Vergilius* 6 (1940), pp. 17-25.

_____, *Roman Vergil* (2nd ed.; London, 1944).

U. Knoche, *Die römische Satire* (2nd ed.; Göttingen, 1957).

E. Krause, *Quibus temporibus quoque ordine Vergilius eclogas scripserit* (Berlin, 1884).

G. Le Grelle, S. J., "Le premier livre des *Géorgiques*, poème pythagoricien," *LEC* 17 (1949), pp. 139-235.

F. Leo, *Plautinische Forschungen zur Kritik and Geschichte der Komödie* (2nd ed.; Berlin, 1912).

F. J. H. Letters, *Virgil* (New York, 1946).

H. Liebing, *Die Aeneasgestalt bei Vergil* (Kiel, 1953).

R. B. Lloyd, "*Aeneid* III: A New Approach," *AJPh* 78 (1957), pp. 133-151.

_____, "*Aeneid* III and the Aeneas Legend," *AJPh* 78 (1957), pp. 382-400.

J. W. Mackail, "The *Aeneid* as a Work of Art," *CJ* 26 (1930-31), pp. 12-18.

L. A. MacKay, "Three Levels of Meaning in *Aeneid* VI," *TAPhA* 86 (1955), pp. 180-189.

P. MacKendrick, "The Pleasures of Pedagogy," *CJ* 54 (1958-59), pp. 194-200.

H. Markowski, "De quattuor virtutibus Augusti in clupeo aureo ei dato inscriptis," *Eos* 37 (1936), pp. 109-128.

P. Maury, "Le secret de Virgile et l'architecture des Bucoliques," *Lettres d'Humanité* 3 (1944), pp. 71-147.

C. W. Mendell, "The Influence of the Epyllion on the *Aeneid*," *YClS* 12 (1951), pp. 203-226.

C. Murley, "The Structure and Proportion of Catullus LXIV," *TAPhA* 68 (1937), pp. 305-317.

A. D. Nock, "Chapter XV. Religious Developments from the Close of the Republic to the Death of Nero," in *CAH*, Vol. X: *The Augustan Empire 44 B.C.-A.D. 70* (Cambridge, 1934), pp. 465-511.

F. S. C. Northrup. See under M. C. Ghyka.

G. Norwood, "Vergil, *Georgics* IV, 453-527," *CJ* 36 (1940-41), pp. 354-355.

F. Norwood, "The Tripartite Eschatology of *Aeneid* 6," *CPh* 49 (1954), pp. 15-26.

R. M. Ogden, *The Psychology of Art* (New York, 1938).

E. O'Neill, "Working Notes and Extracts from a Fragmentary Work Diary," in B. H. Clark, *European Theories of the Drama* (rev. ed.; New York, 1947), pp. 530-536.

H. Osborne, *Theory of Beauty, An Introduction to Aesthetics* (London, 1952).

C. Ottaviano, "Nuove ricerche intorno all' essenza del bello," *Sophia* 22 (1954), pp. 3-46.

L. R. Palmer, "Aris invisa sedebat," *Mnemosyne* 3rd Ser. 6 (1938), pp. 368-379.

J. Perret, *Virgile, l'homme et l'oeuvre* (Paris, 1952).

V. Pöschl, *Die Dichtkunst Virgils. Bild und Symbol in der Äneis* (Innsbruck, 1950).

W. Port, "Die Anordnung in Gedichtbüchern augusteischer Zeit," *Philologus* 81 (1925-26), pp. 280-308, 427-468.
H. W. Prescott, *The Development of Virgil's Art* (Chicago, 1927).
_____, "The Present Status of the Virgilian *Appendix*," *CJ* 26 (1930-31), pp. 49-62.
E. K. Rand, "Young Virgil's Poetry," *HSCPh* 30 (1919), pp. 103-185.
_____, "Virgil the Magician," *CJ* 26 (1930-31), pp. 37-48.
_____, *The Magical Art of Virgil* (Cambridge, Mass., 1931).
L. Richardson, Jr., *Poetical Theory in Republican Rome* (New Haven, 1944).
A. Rostagni, *Virgilio Minore. Saggio sullo svolgimento della poesia Virgiliana* (Torino, 1933).
E. de Saint-Denis, "Douze années d'études virgiliennes: l'architecture des 'Bucoliques,'" *IL* 6 (1954), pp. 139-147, 184-188.
G. Sarton, *A History of Science. Ancient Science through the Golden Age of Greece* (Cambridge, Mass., 1952).
P. Scazzoso, "Reflessi misterici nelle 'Georgiche' di Virgilio," *Paideia* 11 (1956), pp. 5-28.
F. W. Shipley, "The Vergilian Authorship of the Helen Episode, Aeneid II, 567-588," *TAPhA* 56 (1925), pp. 172-184.
O. von Simson, *The Gothic Cathedral. Origins of Gothic Architecture and the Medieval Concept of Order* (New York, 1956).
O. Skutsch, "Zu Vergils Eklogen," *RhM* 99 (1956), pp. 193-201.
J. Sparrow, *Half-Lines and Repetitions in Virgil* (Oxford, 1931).
T. W. Stadler, *Vergils Aeneis. Eine poetische Betrachtung* (Einsiedeln, 1942).
G. Stégen, *Commentaire sur cinq Bucoliques de Virgile (3, 6, 8, 9, 10)* (Namur, 1957).
_____, *Les Épîtres Littéraires d'Horace* (Namur, 1958).
N. Terzaghi, *Virgilio ed Enea* (Palermo, 1928).
D. W. Thompson, "Excess and Defect: or The Little More and The Little Less," *Mind* 38 (1920), pp. 43-55. Reprinted in *Science and the Classics* (Oxford, 1940), pp. 188-213.
_____, *On Growth and Form* (2nd ed.; Cambridge, 1942). 2 vols.
G. G. Thompson, "The Effect of Chronological Age on Aesthetic Preferences for Rectangles of Different Proportions," *JEP* 36 (1946), pp. 50-58.
H. L. Tracy, "The Pattern of Vergil's *Aeneid* I-VI," *Phoenix* 4 (1950), pp. 1-8.
A. W. Van Buren, "The Ara Pacis Augustae," *JRS* 3 (1913), pp. 134-141.
O. Walter, *Die Entstehung der Halbverse in der Aeneis* (Giessen, 1933).
S. Weinstock, "Pax and the 'Ara Pacis,'" *JRS* 50 (1960), pp. 44-58.
H. Weyl, *Symmetry* (Princeton, 1952).
J. Whaler, *Counterpoint and Symbol. An Inquiry into the Rhythm of Milton's Epic Style* (Copenhagen, 1956). [= *Anglistica*, Vol. VI]
C. H. Whitman, *Homer and the Heroic Tradition* (Cambridge, Mass., 1958).
W. Wili, *Vergil* (München, no date).
_____, *Horaz und die augusteische Kultur* (Basel, 1948).
R. Wittkower, "The Changing Concept of Proportion," *Daedalus* (Winter, 1960), pp. 199-215.

# INDEX

# INDEX

Modern scholars are cited when the reference contains a description or criticism of their views. References to important passages are in italics.

accentual symmetry, *see* Knight, metrical patterns of

Achilles, anger of, 12, 18 (n. 49)

Aeneas: anger of, 18 (n. 49); as ideal Roman, 7; *clementia* of, 7, 14, 50; decision of, 24; desires to kill Helen, 85 f.; *iustitia* of, 7, 14, 50; role in *Aeneid,* 5; sheds tears, 17 (n. 37); symbolizes Augustus, 14

*Aeneid:* alternation of books, 1, 2, 10, 13, 45, 60, 63; architecture of, 1, 2, 6, 13, 16 (n. 3), 20, 63, 103; as a poem of Rome and Augustus, vii, 11; as a trilogy, 1, *11-13,* 45, 63 f.; character delineation in, 7, 12, *see also* Aeneas, Dido, Turnus; composed for recitation, 47, 57, 78; contrasts in even-numbered books, 10; Fibonacci series in, *see* Fibonacci series, in *Aeneid;* focal points of books, 24, 32 f., golden mean ratios in, *see* golden mean ratios in *Aeneid and* patterns of ratios; half-lines in, 47 f., 64, 65 (n. 5), *77-80,* 90, 200-202; in secondary schools, 3; interpolations in, *81-83,* 203-204; lessening of tension in, 7; main divisions analyzed, 71 f.; paragraphing of text, *87-90,* 203-204, *see also* paragraphing; parallelism of halves, 1, 3, *5-10,* 45, 63; patriotic and nationalistic themes in, vii, 12; percentage of homodyne in, 111; prooemium of, 2 f.; ratios in, *see* golden mean ratios in *Aeneid and* patterns of ratios; revision of, 1, 59, 65 (n. 3), 77, 79, *90-93,* 97 (n. 4), 101 (nn. 70, 72); similarities in odd-numbered books, 10; spurious passages in, *83-86;* sudden reversals in, 20; summary of proportions in, *60-63;* "Thirteenth Book" of, 104, 110 f.; transpositions in text of, *86 f.,* 90; tripartite divisions, in short passages, 25; in books, *25-32,* 70-72

individual books: I-VI, as "Odyssey" of wanderings, 3, 11, 13; I, bipartite division in, 35 (n. 29); tripartite structure of, 29; I and V, 5; I and VII, 8; II, bipartite division in, 32; tripartite structure of, 29 f.; II and IV, 12 f.; II and VIII, 9; III, alternating patterns in, 20; criticism of, 92; tripartite structure of, 27, 92; III and IX, 9; IV, as recessed panel, 24; tripartite structure of, 30; value of, 7; IV and II, 12 f.; IV and X, 9 f.; V, alternating patterns in, 20; length of, 2, 67 (n. 40), 76; tripartite structure of, 25 f., 66 (n. 32); V and XI, 10; VI, keystone of poem, 5, 7, 8, 12; tripartite structure of, 27 f.; VI and VIII, 13; VI and XII, 10; VII-XII, character delineation in, 7, 12; as "Iliad" of battles, 3, 13; as *maius opus,* 3, 7; VII, tripartite structure of, 30 f.; VII and I, 8; VIII, tripartite structure of, 31; VIII and II, 9; VIII and VI, 13; IX, tripartite structure of, 28; IX and III, 9; X, framework pattern in, 58; tripartite structure of, 31 f.; value of, 7; X and IV, 9 f.; X and XII, 13; XI, length of, 2, 67 (n. 40), 76; possible revision of, 92; structure of, 22 f., 58, 66 (n. 30); tripartite structure of, 26; XI and V, 10; XII, possible revision of, 59, 92; tripartite structure of, 28 f.; value of, 7; XII and VI, 10; XII and X, 13

passages in books: I 1a-1d, authenticity of, 58, 64, 66 (n. 28), 84; I 8-296, symbolic of *Aeneid,* 11; I 257-296, Jupiter's prophecy in, 11, 18 (n. 46); I 305-418, recessed panel in, 22; II 40-56 and 199-227, *see* Laocoon episode; II 76, paragraphing of, 117 f.; II 345-346, rejection of, 92; II 567-588, *see* Helen episode; III 204a-204c spurious, 84; III 273, paragraphing of, 89; III 340, problem of half-line, 77; IV 276-415, ratios in, 55 f.; recessed panel in, 24, 55; IV 333-361 as focal point, 24, 30, 53, 55; IV 441-447, simile of *quercus,* 17 (n. 37); IV 449, tears of Aeneas, 17 (n. 37); IV 450-705, recessed panel in, 23 f.; VI 56-123, numerical symmetry in, 23; recessed panel in, 23; VI 289a-289d spurious, 85; VI 851-853, Anchises' instructions in, 7, 14, 50; VII 641-817, alphabetical order in, 34 (n. 5); alternation in, 20 f., 54, 58 f., 66 (n. 31); VII 664-669, transposition of, 87, 100 (n. 48); VIII 596, paragraphing of, 89; VIII 626-728, shield of Aeneas, 18 (nn. 56, 60); IX 176-449, *see* Nisus and Euryalus; IX 363, rejection of, 82; X 439-542, ratios in, 52; X 689-761, ratios in, 66 (n. 20); X 689-908, ratios in, 45 f.; XI 468-867, structure of, 22 f., 58; XI 868-915 as epilogue, 23, 58, 92; XII 593-611, 91; XII 919-952 and VI 853, 7, 14, 50; XII 939, punctuation of, 50, 65 (n. 11)

*Aetna,* 93, 94, 101 (n. 73); chart index of, 222 f.; golden mean ratios in, 95, 212-214, 216

Alexandrian poetic forms, 93

altar, sacrilege at, 86

alternation: of homodyne and heterodyne, *112-117;* in books of *Aeneid,* 1, 2, 10, 13, 45, 60, 63; in short passages, 20 f., 54; in *Eclogues,* 1, 4; in *Georgics,* 1 f., 41, 76. *See also* patterns of ratios

Amata, suicide of to be discarded, 91

Anchises, gives duty to Roman, 7, 14, 50

Andromache, 77

*Annales, see* Ennius

Apollo, in *Eclogue* IV, 73; temple to, 14

Apollonius, 24, 34 (n. 15)

*Appendix Vergiliana, 93-96,* 103; considered post-Vergilian, 94; Fibonacci series in, 95; golden mean ratios in, 209-215; influenced by Catullus and Lucretius, 95; nature of collection, 93. *See also Aetna, Catalepton* IX, *Ciris, Copa, Culex, Dirae, Moretum.*

Ara Pacis Augustae, 15, 18 f. (nn. 63, 64)

Aratus, 104 ( n. 6)

architecture: of *Aeneid,* 1, 2, 6, 13, 16 ( n. 3), 20, 63, 103; Golden Section in, 37, 43 (n. 13)

*Argonautica,* 24, 34 (n. 15)

Ariadne, 75, 107

Aristaeus, 21, 41, 75, 95, 109; in *Georgics* I and IV, 5; symbolic of Octavian, 5

*Ars Poetica,* Fibonacci series in, 77, 109, 244; golden mean ratios in, 76 f., 109, 244. *See also* Horace

Astyanax, 77

Augustus, vii, 11, 13, 14; *Aeneid* read to, 43 (n. 11), 77; golden shield given to, 14, 18 (n. 60); *in medio* in *Aeneid,* 14; *in medio* in Horace's Roman Odes, 14 f.; *in medio* on Aeneas' shield, 14; *in medio* on Ara Pacis, 15; interested in athletics, 18 (n. 50). *See also* Octavian

*aurea sectio, see* Golden Section

Beast, number of the, 74

bees, regeneration of in *Georgics* IV, 2, 5, 41 f.

Belling, 32, 35 (nn. 23, 29), 77

Benario, 19 (n. 64)

Bickel, 94

bipartite division: in *Aeneid* I, according to Belling, 35 (n. 29); in *Aeneid* II, according to Belling, 32; in *Aeneid* V, 25; in *Georgics* III and IV, 34 (n. 18); of *Georgics,* 1 f., 4. *See also* patterns of ratios, bipartite

Borissavlievitch, 98 ( n. 24)

Brotherton, 34 (n. 4)

Brown, 41, 43 (n. 16), 44 (n. 18), 104 (n. 6)

Büchner, 1, 34 (n. 19); analysis of *Aeneid,* 11; on *Appendix Vergiliana,* 94; on *Georgics* I, 43 (n. 1); on half-lines, 77, 80

Caesar, 2; in *Eclogue* V, 16 (nn. 3, 15), 74. *See also* Augustus, Octavian

Caiati, 17 f. (n. 44)

Camilla, role in *Aeneid* XI, 22 f.

Camps, 11, 17 ( n. 42)

Carcopino, 73 f.

Cartault, 52, 66 (n. 27)

Carthage, 11, 12, 13

*Catalepton* IX, golden mean ratio in, 95, 215

Catullus, influence of, 22, 74, 75, 93, 95; Poem LXIV, 21, 22, 75; chart-index of LXIV, 228 f.; compared to story of Aristaeus, 75; golden mean ratios in, 75, 95, 104, 107, 109, 226 f.; use of homodyne in, 111

chance, 41; ruled out, 46, 61, 63, 73, 103

characters, in *Aeneid* VII-XII, 7, 12; not fully described at first, 100 (n. 50). *See also* Aeneas, Dido, Turnus

chart-index: of *Aeneid,* 65 (n. 1), 70 f., *175-199;* of *Appendix Vergiliana,* 96, *217-225;* of Catullus LXIV, 228 f.; of *Georg.* IV 281-558, 239-241; of Lucretius I, 235-237; of Maphaeus Vegius, 247-249

chrysodes, 37

*Ciris:* authenticity of, 94, 95; chart-index of, 220 f.; golden mean ratios in, 95, 210-212; relative frequency of ratios, 216

*clementia,* 7, 14, 50

*clupeus aureus, see* golden shield of Augustus

concentric pattern, *see* patterns of structure, framework or recessed panel

contrast, *see Aeneid,* contrasts in even-numbered books *and* parallelism of halves; *Georgics,* parallels and contrasts in

Conway, 2, 5, 12, 16 (nn. 9, 10)

Cook, 34 (n. 5)

*Copa,* 93; golden mean ratio in, 95, 215

Cornelius Gallus, 3, 16 (nn. 15, 18), 74

correlation, *see* correspondence

correspondence, of ratios and metrical patterns, *112-117;* of ratios and narrative divisions: in *Aeneid,* 45, *48-56,* 64, *70-72,* 97 (n. 4), 103; in *Aeneid* III, 92; in *Eclogues,* 3, 39 f.; in *Georgics,* 36 f., 41 f.; in Lucretius I, 108; suggested revision of, 92 f., 101 (nn. 70, 72). *See also* chart-index

cosmology: in *Eclogue* VI, 39, 74; Vergil's interest in, 73

Courcelle, 97 (n. 6)

Crump, 16 (n. 4), 91, 100 (n. 50), 101 (n. 65)

*Culex:* alternating pattern in, 95, 96; authenticity

# INDEX

of, 94 f., 103; chart-index of, 218 f.; descriptive passages in, 95; external evidence for authenticity, 94; golden mean ratios in, 95, 96, 209 f.; percentage of homodyne Vergilian, 111; ratios favor authenticity, 96, 103; relative frequency of ratios, 216; resembles *Georgics,* 95, 96; Vergilian themes in, 94

*culpa:* of Dido, 9, 18 (n. 48); of Turnus, 7, 9

Daphnis, 3, 16 (nn. 3, 15, 18), 40, 74

decimals: exact ratios with, 61, 66 (n. 33); in approximate ratios, 66 f. (n. 36); passages with decimals ignored, 62

*De Rerum Natura:* chart-index of Book I, 235-237; compared to *Georgics,* 75; golden mean ratios in, 76, 95, 107 f., 230-234; homodyne in Book I, 111; interpolations and lacunae in, 107 f.; main divisions in, 108; ratios support MS readings, 108; relative frequency of ratios, 108, 216; structure of, 75 f. *See also* Lucretius

description, in ratio with speech, 49, 51. *See also* narrative

Deuticke, paragraphing in text of, 88-90, 205-208

DeWitt, 101 (n. 76)

Dido: a danger, 24; *culpa* of, 9, 18 (n. 48); death of, 11 f., 13, 23; death symbolizes fall of Carthage, 12; tragedy of, 11 f., 13

*Dirae:* chart-index of, 224 f.; golden mean ratios in, 95, 96, 214 f., 216; *Lydia* an integral part of, 96

Divine Proportion, *see* golden mean ratios; Golden Section

Donatus, *see* Donatus-Suetonius Life

Donatus-Suetonius Life: manuscripts of, 101 (n. 73); on *Aen.* I 1a-1d, 84; on *Appendix Vergiliana,* 93; on proposed revision, 77, 90, 93; on reading of Vergil's poetry, 43 (n. 11), 59, 77; on Vergil's critics, 93; on Vergil's interest in mathematics, 42, 63, 73, 75, 103; on Vergil's method of composition, 1, 68, 103

Drew, 16 (n. 19), 17 (n. 44), 18 (n. 60)

Durand, 91, 100 (n. 44)

*Eclogues:* alternation in, 1, 4; as a temple, 16 (nn. 3, 15), 74; Cornelius Gallus in, 3, 16 (nn. 15, 18), 74; correspondence in, 3, 40; golden mean ratios in, *39 f.,* 44 (n. 19); numerical symmetry in, 21 f., 39; percentage of homodyne in, 111; perfection of, 93; triadic structure in, 4, 11, 16 (n. 18), 40, 64
    individual poems: I, analysis of, 21, 39; golden mean ratio in, 39; II, analysis of, 39; golden mean ratio in, 39; IV, analysis of, 21 f.; background of, 73 f.; V, as central poem, 3, 6, 16 (nn. 3, 15), 40; VI, analysis of, 39 f.; bal-anced passages in, 43 (n. 16); cosmology in, 40, 41; golden mean ratios in, 39 f., 62; prooemium not included in ratio, 39, 43 (n. 16); X, as later addition, 3, 16 (n. 15), 40; golden mean ratio in, 39

*Elegiae in Maecenatem,* 93

Ennius, 77, 97 f. (n. 21), 104 (and n. 6)

Epicureanism, 73, 74

Eratosthenes, 104 (and n. 6)

Euclid, 38

Euryalus, *see* Nisus and Euryalus

Eurydice, 41 f.

even-numbered books of *Aeneid, see Aeneid,* alternation of books *and* contrasts in even-numbered books

expanded alternations, 112-117

Fechner, 77

Fenik, 18 (n. 51), 34 (n. 15)

Fibonacci series: description of, 37-39; in *Aeneid,* 51, 57, 61-63, 66 (n. 33), 79, 103; in *Appendix Vergiliana,* 95; in Ennius, 97 f. (n. 21); in fractions, 61, 66 (n. 33); in *Georgics* I, 39; in *Georgics* II, 44 (n. 33); in *Georgics* III, 41, 66 (n. 35); in Horace, *Ars Poetica,* 76 f., 244; in Lucan, 104 (n. 3); relative frequency of, 62 f., 216. *See also* golden mean series

focal point: of books of *Aeneid,* 32 f.; in *Aeneid* V, 25; in tripartite framework pattern, 53 f.; nature of, 21, 23. *See also* patterns of structure, framework or recessed panel

Fowler, 18 (n. 56), 78; on *Aen.* VII 664-669, 87

"foyer astronomique," in *Georgics* I, 36, 37, 41, 74

fractions: in Fibonacci series, 61, 66 (n. 33); in other golden mean series, 66 (n. 33); in ratio totals, 47 f., 65 (n. 5). *See also* half-lines

Fraenkel, 101 (n. 79)

framework patterns, *see* metrical framework patterns; patterns of ratios, tripartite framework; patterns of structure, framework or recessed panel

*furor,* 11

Gallus, *see* Cornelius Gallus

games, in *Aeneid* V, 12, 18 (n. 50), 25

Gantar, 109

geometric structure in *Iliad,* 18 (n. 58), 22

*Georgics:* alternation in, 1 f., 5, 41, 76; astronomy in, 4; compared to *De Rerum Natura,* 79 f.; correspondence of books, 4 f.; descriptive passages in, 44 (n. 21); division into halves, 1 f., 4 f.; double ratios in, 41; exact ratios in, 36 f., 41, 42; Fibonacci series in, 39, 41, 44 (n. 33), 66 (n. 35); indebtedness to Hesiod, 36; *labor* in, 5; Maecenas addressed in, 5, 75; Octavian in, 2, 4 f., 36; parallels and contrasts in, 4 f.; percentage of homodyne in, 111; praise of country life in, 2; praise of Italy in, 4 f.; prologues of, 4; ratio in alternating pattern, 41, 64, 93, 95, 96; read to Octavian, 43 (n. 11); structure of, 1 f., 4 f., 75 f.; symbolism in, 5, 16 (n. 21); universality of, 2, 76

individual books and passages: I, "foyer astronomique" in, 36, 37, 41; golden mean ratios in, 36 f., 39, 41; structure of, 36; II, golden age in, 2, 16 (n. 7), 74; structure of, 42; II 1-34, ratios in, 44 (n. 33); II 458-540, ratio in, 41; II 475-494 Epicurean or Pythagorean, 74; III, bipartite division of, 34 (n. 18), 36; III 1-48, double ratio in, 41; III 13 ff., temple to Octavian, 1, 14; III 295-383, double ratio in, 41; III 478-566, three stages of plague, 41, 44 (n. 24); ratio in, 41; IV, bipartite division of, 34 (n. 18), 36; IV 281-558, chart-index of, 239-241; compared to Catullus LXIV, 75; golden mean ratios in, 41 f., 44 (n. 26), 109, 216, 238

Getty, 104 (nn. 3, 6)

Goelzer, 85

golden age, 2, 16 (n. 7), 41, 74

golden mean ratios, vii, 33, *37-39*, 103 f.; achieved by golden mean series, 37-39, 62 f.; double ratios, 41, 50, 58, 66 (n. 22); exact ratio (.618) defined, 39; exact ratios, 36 f., 41, 42, 46, 50, 58, *60-62*, 66 (n. 34), 216; nature of, 36, 37-39; perfect .618 defined, 39; relative frequency of, 60 f., 95 f., 216

golden mean ratios in *Aeneid: Aeneid* as a whole, 63 f., 173; correspondence with metrical patterns, *112-117*; correspondence with narrative divisions, *see* correspondence of ratios and narrative divisions; double ratios in, 50, 58, 66 (n. 22); exact ratios (.618) in, 46, 50, 58, *60-62*, 66 (n. 34), 216; in range .615-.621, 46, 60 f., 95, 216; in range .610-.626, 46, 60 f., 216; in range .608-.628, 46, 60 f.; in range .600-.636, 47, 73; in main divisions, 45 f., 58 f., 72, 166-168; in six parts, 56, 59, 60; in X 689-908, 45 f.; link ratios in, 50; main divisions in proportion, 33, *59*, 169; noncontiguous ratios, 50, 54, 56 f., 65 (nn. 2, 12), 66 (n. 22); ratios and half-lines, *77-80*, 90, 200-202; ratios and interpolations, *81-83*, 203-204; ratios and paragraphing, *87-90*, 205-208; ratios and proposed revisions, *90-93*, 101 (nn. 70, 72); ratios and spurious passages, *83-86*; ratios and transpositions, *86 f.*; relative frequency of ratios, 60 f., 95 f., 216; small ratios combine into large, 68-71, 73; summary of ratios,*60-63;* supplementary list of, 48, 54, 59 f., 170-173; support MS readings, 82 f., 84, 87, 103

golden mean ratios in *Appendix Vergiliana*, *94-96*, 209-215; in Catullus LXIV, 75, 107, 226 f.; in *Eclogues*, as a whole, *40; see Eclogues*, individual poems; in Ennius, 97 f. (n. 21); in *Georgics*, 36 f., 39, 41 f., 66 (n. 35); *see Georgics, especially* individual books and passages; in Horace, 76 f., *109 f.*, 242-244; in Lucan, 104 (n. 3); in Lucretius, books of, 76, 107 f., 230 f.; in Lucretius, Book I, 108, 232-234; in Maphaeus Vegius, 104, 110 f., 245 f.

*See also* patterns of ratios; Vergil, use of Golden Section deliberate

golden mean series, 37-39, 62 f., 66 (n. 33), 216. *See also* Fibonacci series

golden rectangle, 77

Golden Section, vii, important in art and architecture, 37, 77, 104; in Roman poetic theory, 77, 104; nature of, *37-39;* value of, 46, 73, 77, 104. *See also* golden mean ratios

golden shield of Augustus, 14, 18 (n. 60)

grand tetractys, 74

Grimal, 16 (n. 21)

Gudeman, 94

Hahn, 16 (n. 18), 34 (n. 4), 40

half-lines, *77-80;* as fractions in ratio totals, 47 f., 64, 65 (n. 5), 78; less accurate as whole lines, 47 f., 79, 80, 90, 200 f.; more accurate as whole lines, 79 f., 90, 202; only in Vergil, 77; part of detachable passages, 91 f.; Sparrow considers effective, 78; Sparrow favors more half-lines, 99 (n. 42)

Heinze, 85

Helen episode (*Aen.* II, 567-588): authenticity of, 85 f., 103; golden mean ratios in, 86; Mackail on, 85

hemistichs, *see* half-lines

heptads, 22, 24

Hesiod, influence on *Eclogue* IV, 74; influence on *Georgics*, 36

heterodyne, 111-117; defined, 44 (n. 17), 111

Hirtzel, viii, 57, 84; lines bracketed by, 81 f.; paragraphing of, 52, 66 (nn. 20, 27), 87 f., 89, 205-208

Homer, symmetry in, 17 (n. 38), 37; Vergil's indebtedness to, 12, 13, 18 (nn. 49, 51), 22. *See also Iliad, Odyssey*

homodyne, 111-117; defined, 44 (n. 17), 111

Horace: *Ars Poetica*, Fibonacci series in, 77, 109, 244; golden mean ratios in, 76 f., 109, 244; *Carmen Saeculare*, golden mean ratio in, 110; hexameter poems: Fibonacci series in, 104, 110; golden mean ratios in, 104, 109 f., 242 f.;

*Odes* I 2, 37; *Odes* III 1-6 (Roman Odes): golden mean ratios in, 110; structure of, 14 f., 110; *violentia* vs. *consilium* in III 4, 13; Book IV, 19 (n. 64); *Satires:* divided into halves, 109; golden mean ratio in, 109; I 10, 109; II 3, 109

*Iliad: Aeneid* VII-XII an "Iliad," 3, 13; architecture of, 22; ring composition in, 18 (n. 58), 22; tripartite structure of, 18 (n. 58), 25. *See also* Homer

*impietas,* 11, 13

interpolations, *81-83;* lines not in best MSS omitted, 48, 81; rejection supported by ratios, 81 f.

intuition ruled out, 46, 63, 73, 103

*iustitia,* 7, 14, 50

Jahn, paragraphing in text of, 88-90, 205-208

Jupiter, prophecy of, 11

Knight, 16 (n. 10), 44 (n. 17), 100 (n. 46), 103; metrical patterns of, *111-117*

*labor,* in *Georgics,* 5

lacunae, 57, 76, 107 f.

Laocoon episode (*Aen.* II 40-56, 199-227): Mackail on, 90; ratios in, 90 f.

law of averages, 46

Le Grelle, vii, 36 f., 45; believes in number symbolism, 43 (n. 8), 74; first to discover Golden Section in Vergil, 36, 37; procedure of, 43 (nn. 6, 7)

Leo, 83, 85

link ratios, 50

Lloyd, 34 (n. 19), 92

Lucan, golden mean ratios in, 104 (n. 3)

Lucretius, 73, 74, 95; chart-index of Book I, 235-237; golden mean ratios in, 76, 95, 107 f., 230-234; influence on Vergil, 75. *See also De Rerum Natura*

*ludus Troiae,* 18 (n. 50), 20, 25, 66 (n. 32)

*Lydia,* 96. *See also Dirae*

Mackail: interpolations suggested by, 82 f.; on *Aen.* II 256-258, 99 (n. 40); on *Aeneid* III, 92; on *Aen.* III 204a-204c, 84; on *Aeneid* IV, 30; on *Aeneid* VI, 101 (n. 67); on *Aen.* VI 289a-289d, 85; on *Aen.* VI 601, 57; on *Aen.* VII 664-665, 100 (n. 48); on *Aen.* VIII 41, 100 (n. 44), on *Aeneid* XI, 23; on *Aeneid* XII, 7; on architecture of *Aeneid,* 16 (n. 3); on Helen episode, 85; on Laocoon episode, 90; on *ludus Troiae,* 18 (n. 50); on tripartite division of *Aeneid,* 17 (n. 40); on twofold division of *Aeneid,* 3; paragraphing in text of, 52, 66 (n. 27), 87-90, 205-208; revisions suggested by, 91 f., 100 (n. 57)

MacKendrick 30, 35 (n. 27)

Madvig, 57

Maecenas: addressed in *Georgics,* 5, 44 (n. 33), 75; elegies on, 93; *Georgics* read by, 43 (n. 11)

major, 36 f. *See also* golden mean ratios; Golden Section

major and minor in patterns: in *Aeneid,* 46 f.; in *Eclogues,* 39 f.; in *Georgics,* 37, 41 f. *See also* metrical patterns and golden mean ratios; patterns of ratios

manuscripts of *Aeneid,* supported by ratios, 82 f., 84, 87, 103

Maphaeus Vegius, 104, 110 f.; chart-index of, 247-249; golden mean ratios in, 111, 245 f.; relative frequency of ratios, 111; "Thirteenth Book" of *Aeneid,* 110 f.

Martial, 238

Maury, 16 (nn. 3, 15), 74 f.

meat, eating of, 74

Mendell: on framework or recessed panel patterns, 22-24, 35 (n. 25); on Nisus and Euryalus, 34 (n. 16); on structure of *Aeneid,* 17 (n. 27)

*Metamorphoses,* 74, 104

metrical framework patterns, 113-117

metrical patterns and golden mean ratios, 103, 111-117

Mezentius, symbolizes *impietas,* 11, 13

Milton, 104 (n. 7)

minor, 36 f.; minor divided by major omitted, 43 (n. 6), 65 (n. 7). *See also* golden mean ratios; Golden Section

minor poems of Vergil, *see Appendix Vergiliana*

*Mourning Becomes Electra,* 12

*Moretum:* chart-index of, 224 f.; golden mean ratios in, 95, 214, 216

Murley, 22, 107

mysticism, *see* number symbolism

narrative: in ratio with speech, 49, 51, 52 f., 55; in ratio with narrative, 49 f., 52, 53, 56, 65 (n. 14)

Neo-Pythagoreanism, 73 f., 75, 104 (n. 6)

neoteric poetry, 22. *See also novi poetae*

Nigidius Figulus, 73, 104 (n. 3)

Nisus and Euryalus: as a five-act tragedy, 34 (n.

16); as a three-act tragedy, 25; Mendell on, 34 (n. 16); tragic nature of episode, 12

Nonius, 101 (n. 74)

Norden, 100 (n. 45)

*novi poetae,* 104. *See also* neoteric poetry

number of the Beast, 74

number symbolism, 73, 74 f.

numerical symmetry, 21 f. 23, 24, 55, 75

*obtrectatores,* 93

Octavia, 43 (n. 11), 77

Octavian: Aristaeus symbolic of, 5: divinity of, 4 f., 17 (n. 22), 36; *Georgics* read to, 43 (n. 11); praised in *Georgics,* 2, 4 f., 36; temple to, 1, 14, 41. *See also* Augustus

odd-numbered books of *Aeneid, see Aeneid,* alternation of books

*Odyssey: Aeneid* I-VI an "Odyssey," 3, 11, 13; tripartite divisions in, 25, 34 (n. 19)

old man of Tarentum, 5; symbolizes Pythagoras, 17 (n. 21)

O'Neill, 12

Orpheus, 21, 41 f., 73

Ovid, 74, 104

Page, 34 (n. 15)

Pallas, death of, 7, 9, 11, 12, 52; swordbelt of, 7, 14, 50

Palmer, 86

paragraphing, 52, 66 (nn. 20, 37), *87-90,* 97 (n. 3), 100 (nn. 53, 54); suggestions for paragraphing, 89 f., *205-208*

parallelism, of books of *Aeneid,* vii, 1, 3, *5-10,* 45, 63. *See also* patterns of structure, framework or recessed panel

Parthenius, 104 (and n. 6)

patterns of ratios: alternating: in *Aeneid* (four or five parts), 47, *54-57,* 60, 64, 68, *152-165;* in *Culex,* 95, 96; in *Georgics,* 41, 42, 44 (n. 33), 64, 93, 95, 96

    bipartite: in *Aeneid,* 46, *48-51,* 60, 68, *121-131;* in *Eclogues,* 39, 40, 43 (n. 19); in *Georgics,* 36 f., 39, 42

    tripartite framework: in *Aeneid,* 47, *52-54,* 58, 60, 68, *139-151;* in *Eclogues,* 39, 40; in *Georgics,* 37, 41, 42, 44 (n. 26)

    tripartite (nonframework): in *Aeneid,* 46 f., *51 f.,* 60, 64, 68, 75, *132-138;* in Catullus LXIV, 75, 107; in *Eclogues,* 40; in *Georgics,* 41 f.; in Lucretius, 76, 107 f.

patterns of structure: alternating, *see* alternation; framework or recessed panel, 3 f., *21-24,* 188; in *Aen.* I 305-418, 22; in *Aen.* IV 276-415, 24, 55; in *Aen.* IV 450-705, 23 f.; in *Aen.* VI 56-123, 23; in *Eclogue* I, 21, 39; in *Eclogue* II, 39; in *Eclogue* IV, 21 f.; in *Georg.* IV 281-558, 21, 34 (n. 7); tripartite, *see* tripartite structure

Pease, 99 (n. 40)

Peleus and Thetis, 75, 107

perfect number, 22, 24, 75

Perret: on architecture of *Aeneid,* 5 f.; on *Georgics* I, 36; on *Georgics* II, 42

*Philebus,* 103

phyllotaxis, 43 (n. 15)

*pietas,* 14 f.

Plato, 17 (n. 38), 38, 73, 103

Plautus, text of, 83

Pöschl, 11, 17 (nn. 34, 40)

poetry as architecture, 1; *see Aeneid,* architecture of

praise of country life, in *Georgics,* 2

praise of Italy, in *Georgics,* 4 f.

Prescott, 94

Priam, 13

proportion, definition of, 36 f.. *See also* golden mean ratios; Golden Section

prose outline of *Aeneid,* 1, 68, 103

punctuation, 50, 65 (n. 11), 113, 205

Pythagoras, 17 (n. 21), 74

Pythagoreanism, 74, 104 (and n. 6). *See also* Neo-Pythagoreanism

Quintilian, 94

Rand, 11, 36, 44 (n. 29)

ratios, *see* golden mean ratios

recessed panel pattern, *see* patterns of structure, framework or recessed panel

reincarnation, 73

released movement, 112-117; defined, 112

*retractatio,* 83

*Revelation,* number of the Beast in, 74

revision of *Aeneid,* 1, 59, 65 (n. 3), 77, 79, *90-93,* 97 (n. 4), 101 (nn. 70, 72)

*Rhesus,* 18 (n. 51)

Ribbeck: brackets lines as spurious, 83 f.; transpositions in *Aeneid,* 87; transpositions in *Georgics* II, 44 (n. 33)

# INDEX

Richardson, 3, 16 (n. 6), 42

Richter, 43 (n. 9), 44 (n. 24)

ring composition in *Iliad*, 18 (n. 58), 22

Roman coinage, 73

Rome, vii, 11, 13, 15; site of, 18 (n. 55)

Sabbadini, 52, 66 (nn. 20, 27), 85; paragraphing in text of, 88-90, 205-208

Saint-Denis, 37

Sarton, 77

Scazzoso, 97 (n. 9)

*sectio,* see Golden Section

Servius, 84, 85, 93

shield, in *Aeneid* VIII, 14, 18 (nn. 56, 60). *See also* golden shield of Augustus

short passages, definition of, 48; supplementary ratios in, 59 f., 170-172. *See also* patterns of ratios

Sibyl, 23, 73

Silenus, songs of, 40

Silver Latin epic, 104

similarity, *see Aeneid,* parallelism of halves *and* similarities in odd-numbered books

simile: of burning city, 12; of *quercus,* 17 (n. 37)

six-part ratios, 56, 59, 60

Skutsch, 16 (n. 14)

Socrates, on measure and symmetry, 103

Sparrow, 78, 79, 92, 99 (n. 42)

speeches: as narrative, 48; problem of, 48; ratios within speeches, 48 f., 51, 52, 55; speeches and narrative, 49, 51, 53, 55, 65 (n. 14); speeches in proportion, 49, 51, 52 f., 55

spurious passages, *83-86; Aen.* I 1a-1d authentic, 84; *Aen.* III 204a-204c spurious, 84; *Aen.* VI 289a-289d spurious, 85; Helen episode (*Aen.* II 567-588) authentic, 85 f. *See also* interpolations

square root of 2 series, 63

square root of 5 series, 37, 39, 63. *See also* golden mean series; Golden Section

Stadler, 2, 11, 16 (n. 11)

Statius, 94

Stégen, 97 (n. 14)

Stoicism, 73

Suetonius, 18 (n. 50), 94; biography of Vergil, *see* Donatus-Suetonius Life

summary of book, vii f., 103

*superbia,* 7

supplementary list of ratios, 48, 54, 59 f., 170-172

symbolism, 11, 15. *See also* Aeneas, Aristaeus, Dido, old man of Tarentum, Mezentius, Turnus

symmetry, vii, 103. *See also* golden mean ratios; Knight, metrical patterns of; numerical symmetry

Tables I-VII summarized, 48

Tarentum, 5, 17 (n. 21)

temple to Octavian, 1, 14

tension, lessening of, 2, 7

Theocritus, 93

Thetis, *see* Peleus and Thetis

"Thirteenth Book" of *Aeneid, see* Maphaeus Vegius

Thompson, 37 f., 63

*Timaeus,* 17 (n. 38)

*tibicines,* 1, 77

tragedy in epic, 11 f. *See also* Dido, Nisus and Euryalus, Turnus

transpositions, in Aeneid, 86 f., 90; in *Georgics* II, 44 (n. 33)

triads, in *Eclogues,* 4, 11, 16 (n. 18), 40, 64

tripartite structure, in *Aeneid,* 20, *25-33,* 59, 70-72. *See also* patterns of ratios, tripartite framework; patterns of structure, framework or recessed panel

Trojan horse, 20

Tucca, 85, 86, 91

Turnus: *culpa* of, 7, 9; death of, 7, 11 f., 14, 50; in Perret's analysis, 6; kills Pallas, 7, 11, 52; symbolizes *furor* and *violentia,* 11, 13; tragedy of, 11 f.

Van Buren, 18 (n. 63)

Van der Graaf, 96

Varius, 85, 86, 91

Varus, dedication to, 39

Vegius, *see* Maphaeus Vegius

Vergil: achievement in epic, vii; *Aeneid* read to Augustus, 43 (n. 11), 77; as a Neo-Pythagorean, 73-75, 77; death of, 90; *Georgics* read to Octavian, 43 (n. 11); historical epic planned for Octavian, 1, 14; indebted to Hesiod, 36, 74; indebted to Homer, 12, 13, 18 (nn. 49, 51), 22; influenced by Catullus, 22, 74, 75, 93; influenced by Ennius, 97 f. (n. 21); influenced by Lucretius, 75 f.; interest in mathematics, 42, 63, 73, 75, 103; many ratios accidental, 47; method of composition, *68-73;* problem of early poems, 93 f.; proposed revision of epic, *see* revision of *Aeneid;* use of Golden Section deliberate, 37, 46, 47, 61, 63, 73, 103, 117; use of half-lines, 77, 79; use of homodyne and heterodyne, 111. *See also Aeneid, Appendix Vergiliana, Eclogues, Georgics*

*violentia,* 11, 13

*virtus,* 14 f.

*vis consili expers,* 13

*vis temperata,* 13

Vita, of Donatus, *see* Donatus-Suetonius Life; of Servius, 84, 85, 93

Walter, 78, 80

Weinstock, 19 (n. 64)

Weyl, 37

Whaler, 104 (n. 7)

Whitman, 17 (n. 38), 18 (n. 58), 22, 34 (n. 10), 37

Wili, 41